Crucial Issues in

Climate Change and the Kyoto Protocol

Asia and the World

Crucial Issues in Climate Change and the Kyoto Protocol

Asia and the World

Editors

KOH Kheng-Lian
LYE Lin-Heng
Jolene LIN

World Scientific

NEW JERSEY • LONDON • SINGAPORE • BEIJING • SHANGHAI • HONG KONG • TAIPEI • CHENNAI

Published by

World Scientific Publishing Co. Pte. Ltd.

5 Toh Tuck Link, Singapore 596224

USA office: 27 Warren Street, Suite 401-402, Hackensack, NJ 07601

UK office: 57 Shelton Street, Covent Garden, London WC2H 9HE

British Library Cataloguing-in-Publication Data
A catalogue record for this book is available from the British Library.

CRUCIAL ISSUES IN CLIMATE CHANGE AND THE KYOTO PROTOCOL
Asia and the World

ISBN-13 978-981-4277-52-5
ISBN-10 981-4277-52-5

Desk Editor: Tjan Kwang Wei

Typeset by Stallion Press
Email: enquiries@stallionpress.com

Printed in Singapore by Mainland Press Pte Ltd.

FOREWORD

The new US Assistant Secretary for East Asia and the Pacific, Dr Kurt Campbell, is a distinguished security analyst. I was, therefore, surprised and impressed to hear him declare, at a conference last year, that the world's biggest security threat is global warming and climate change. This realisation is slowly dawning on security analysts, economists, politicians, public intellectuals and journalists. It is, however, not making any significant impact on public opinion in most parts of Asia.

The attitudes I have encountered among my friends are, on the whole, discouraging. The majority are ignorant and do not understand the problem. The threat of global warming and climate change is like an invisible enemy. It is not like a meteorite headed towards earth. The earth is warming slowly but progressively. We are like the proverbial frog in a pot of boiling water. We say to ourselves, the water is getting warmer but we are okay.

The second attitude is one of scepticism. Some very intelligent and well-informed friends of mine are not convinced by the science and the empirical evidence. To them, the reports of the IPCC represent only one point of view and have been challenged by some dissenting scientists. They regard Al Gore as a polemicist and Nicholas Stern as an alarmist. They are not convinced that, if the earth is warming, it is due to man's activities. They prefer to believe that it is cyclical as the earth goes through historic periods of cooling and warming. The anti-climate change literature spread by the dissenting scientists and public intellectuals have provided ammunition to the sceptics.

The third attitude is one I have encountered among some public intellectuals in Asia. They say that global warming and climate change is caused by the West and it is the responsibility of the West to substantially reduce their carbon emissions and cool the earth. Some of my Indian and Chinese friends have argued, not without some merit, that on a per capita basis, their countries' emissions of CO_2 are substantially below those of the OECD countries and that it is both unjust and premature to ask them to reduce their carbon emissions when they are faced with so much poverty and backwardness. The Chinese and Indian intellectuals feel that this is their historic opportunity to catch up with the West and to become First World countries. They are not going to let the threat of global warming and climate change derail them from their historic quest. I can understand and, to some extent, even sympathise with this view.

When the world meets in Copenhagen in December this year, it is faced with a truly monumental task of negotiating a new consensus to replace the Kyoto Protocol which expires in 2012. Will we succeed or fail? At this point, I do not know. It will depend, in part, on President Obama's leadership and his skilful diplomacy. It will depend, in part, on whether the OECD countries can arrive at a compromise with China, India, Russia and Brazil. It will depend, in part, on whether the two biggest polluters, China and the US, can agree to cooperate with each other.

This volume makes an important contribution to the debate among Asian policy makers and intellectuals and between Asia and the West. There is no doubt in my mind that the West bears the historic burden. Asia, however, has the potential to prosper the world and to pollute the world. My hope is that Asia will catch up with the West, but will succeed in transiting to a low carbon economy. If Asia can succeed in embracing sustainable development and avoid the unsustainable model which the West has followed, then we can all heave a sigh of relief and say that there is still hope for planet earth. If Asia follows the path trodden by the West, then there is no hope for the future of human civilisation on our planet.

Professor Tommy Koh
Ambassador-at-Large at the Ministry of Foreign Affairs;
Chairman of the Institute of Policy Studies
and the National Heritage Board
Singapore

PREFACE

There is voluminous literature on climate change today. As the global community struggles to address one of the most challenging environmental issues of our day, the amount of research dedicated to the subject of climate change is both staggering and inspiring. However, there exists a lacuna in this large body of literature as few scholars have sought to address climate change issues in an Asian context. This is extraordinary given that Asia is the world's most populous region which is already experiencing climate change and is expected to bear the brunt of climate change impacts in time to come.

Crucial Issues in Climate Change and the Kyoto Protocol: Asia and the World is an attempt to fill this lacuna. This book focuses on Asia, on the impacts of climate change on the region, on the legal, socio-economic and governance issues raised by countries' attempts to mitigate and adapt to climate change, and on the involvement of Asian countries in the international climate change regime. Each chapter addresses themes and issues which are of significant contemporary interest to policy-makers, lawyers, and academics alike. The book begins with a broad examination of climate change within the constellation of multilateral environmental agreements (MEAs) and the science of climate change. The Clean Development Mechanism (CDM), arguably the most successful Kyoto Protocol Flexible Mechanism, is a thematic area in itself. Contributors share their knowledge of various countries' experience with implementing the CDM, as well as the emerging problems with the international CDM regulatory regime. Some authors discuss how Asian economies could effectively mitigate greenhouse gas emissions, thereby contributing to the global effort to tackle climate change. Improving energy efficiency and adopting co-benefit approaches that seek to achieve sustainable development and climate change mitigation at the same time are fundamental in this regard. Finally, while there are clear signs that there will be a future international climate change treaty, the details of the post-2012 treaty remain unclear. The concluding chapters therefore discuss alternatives to the Kyoto Protocol as well as significant post-2012 issues such as protecting forests to mitigate climate change.

Crucial Issues in Climate Change and the Kyoto Protocol: Asia and the World provides an insight into how Asia perceives the challenges of climate change, and how the various countries of this region, highly disparate in their cultures, socio-economic conditions and political systems, are responding to climate change. Our hope is that sharing these ideas and information will increase our understanding of the issues and challenges that climate change presents to the world, and to Asia, in particular. We also hope to inspire further intellectual dialogue, and more importantly, to action to address climate change.

Finally, we would like to express our heartfelt thanks to Ambassador Tommy Koh for his *Foreword*. Ambassador Koh served as Chairman of the Preparatory Main Committee of the United Nations Conference on Environment and Development from 1990 to 1992. One of the key outputs of the Conference was the United Nations Framework Convention on Climate Change (UNFCCC). Combining his deep understanding of climate change issues and personal dedication to the environmental cause, this book has benefited immensely from Ambassador Koh's support and encouragement.

KOH Kheng-Lian, *Emeritus Professor, Faculty of Law, National University of Singapore; Director, Asia-Pacific Centre for Environmental Law, NUS*

LYE Lin-Heng, *Associate Professor, Faculty of Law, National University of Singapore; Deputy Director, Asia-Pacific Centre for Environmental Law, NUS*

Jolene LIN, *Assistant Professor, Faculty of Law, University of Hong Kong; Associate Member, Asia-Pacific Centre for Environmental Law, NUS*

CONTRIBUTORS

Andrew BEATTY

Mr Andrew Beatty is a partner in the Environment and Environmental Markets Group of Baker & McKenzie's office in Sydney, Australia. His areas of expertise encompass environmental, planning and valuation law. He is a Director of Earthwatch Institute, an Organisational Stakeholder of the Global Reporting Initiative and a member of the IUCN's Commission on Environmental Law.

William I.Y. BYUN

William Byun is Vice President, AES Corporation, Climate Change & Technology Development Group and Managing Director, AES Asia & Middle East. He had practiced foreign investment and finance law in New York, Tokyo, Seoul and Jakarta and prior to joining AES, acted as the managing director of a boutique private equity and carbon finance firm based out of Singapore. A member of the New York State Bar, Mr. Byun was also a U.S. Fulbright Scholar to Korea and the first foreign advisor resident at the Ministry of Finance, Office for Financial Systems Reform of the Republic of Korea, and served with the Office of General Counsel of US OPIC.

Felix H C CHAN

Mr Felix Chan is Regional Manager of AES Corporation's Climate Change & Technology Development Group. A graduate of the University of Michigan (Mechanical Engineering, B.Eng.), Mr. Chan is leading the Group's Asian initiatives for technology review and for leading projects related to China, particularly in landfill gas and energy efficiency. He is also pursuing a M.Sc. in Financial Economics from the University of London.

Kurt DEKETELAERE

Prof Kurt Deketelaere is Chief of Staff at the Private Office of Mrs. Hilde Crevits, Minister for Public Works, Energy, Environment and Nature, Flemish Regional Government (Belgium); and Full Professor of Law at the University of Leuven (Belgium). He holds many positions, including Honorary Professor of Climate Change Law at the University of Dundee (Scotland); Director of the Institute for Environmental and Energy Law, Faculty of Law, University of Leuven; Visiting Professor, Law Faculty, NUS. He is Co-Founder and Co-Director of the LLM Program "Advanced Master of Energy and Environmental Law", and Chair of the Department of Public Law, Faculty of Law, University of Leuven.

Asanga GUNAWANSA

Dr Asanga Gunawansa is Assistant Professor in the Department of Building, School of Design and Environment, National University of Singapore (NUS). He holds an LLM from the University of Warwick and a PhD from NUS. He has worked with the UN and the government of Sri Lanka. His areas of interest include Development Projects and the Protection of Indigenous Communities; Legal Implications of Project Financing; and International Investment Projects and their Impact on the Environment

Hiroji ISOZAKI

Prof Hiroji ISOZAKI is a Professor of International Law at the Meiji Gakuin University, Japan; a Visiting Professor at the Institute for Advanced Studies of the United Nations University, and an Emeritus Professor at the Iwate University. He has been a Research Associate at the Faculty of Law of the Tokyo Metropolitan University, an Associate Professor and a Professor at the Iwate University. He has researched on North-South Problems, Law of the Sea Issues and Environmental Issues from the international law perspectives. He published the first textbook in Japanese on international environmental law.

Kishan KHODAY

Kishan Khoday has served with the United Nations since 1995, focusing on ways to address issues of climate change and environment in emerging economies. Kishan currently serves as Deputy Resident Representative of the United Nations Development Programme (UNDP) in Saudi Arabia, after having served as Assistant Resident Representative and Team Leader for Energy & Environment with UNDP in China (2005–2009) and Deputy Programme Coordinator for Environment with UNDP in Indonesia (2001–2005). Since 2005, he has also served as Research Fellow with the Center for International Sustainable Development Law (CISDL) in Canada, supporting research and publication activities on various topics including climate change and sustainable development.

KOH Kheng Lian

Prof Koh Kheng Lian is Emeritus Professor in the Faculty of Law, NUS. She is founder member and the Director of the Asia–Pacific Centre for Environmental Law (APCEL), established by the Faculty in 1996 in partnership with IUCN CEL (the World Conservation Union — Commission on Environmental Law) and in collaboration with UNEP (United Nations Environment Program). She was the former IUCN Commission on Environmental Law's Regional Vice Chair for South and East Asia, and a member of its Steering Committee from 1996–2004. After her retirement she continues to co-teach some environmental law courses at the Faculty and also in the MSc (Env Mgt) programme.

Tommy KOH

Prof Tommy KOH is currently Ambassador-At-Large at the Ministry of Foreign Affairs; Chairman of the Institute of Policy Studies and National Heritage Board. He is also a Professor of Law at the National University of Singapore, where he was Dean from 1971–1974. He was Singapore's Permanent Representative to the United Nations in New York from 1968 to 1971 (concurrently accredited as High

Commissioner to Canada) and from 1974 to 1984 (concurrently accredited as High Commissioner to Canada and Ambassador to Mexico). He was Ambassador to the United States of America from 1984 to 1990. He was President of the Third UN Conference on the Law of the Sea from 1980 to 1982. He was Chairman of the Preparatory Committee and the Main Committee of the UN Conference on Environment and Development from 1990 to 1992.

Jolene LIN Shuwen

Jolene Lin is Assistant Professor at the Faculty of Law, University of Hong Kong where she teaches International Environmental Law and Administrative Law. In 2003, she was a member of the World Bank delegation to a workshop, organized by the European Commission and the World Bank, on the legal aspects of implementing the Kyoto Protocol mechanisms. She has done research on the feasibility of introducing emissions trading in Singapore. She is the International Environmental Law editor of the Singapore Year Book of International Law as well as an associate member of the Asia-Pacific Centre for Environmental Law (APCEL).

LYE Lin-Heng

Assoc Prof Lye Lin-Heng is Deputy Director of the Asia-Pacific Centre for Environmental Law (APCEL), Law Faculty, National University of Singapore. She chairs the University's multi-disciplinary graduate program, the MSc (Environmental Management) [MEM]. She is a member of the Governing Board, IUCN Academy of Environmental Law, and co-chairs its Teaching and Capacity Building Committee. She is Visiting Associate Professor at Yale University's School of Forestry & Environmental Studies and at the School of Law, University of Sydney.

Maria Socorro MANGUIAT

Maria Socorro Manguiat is a graduate of Ateneo de Manila University, Philippines (JD) and Harvard Law School (LLM). She was Legal Officer at the Environmental Law Centre, IUCN, in Bonn Germany, from September 2001 to January 2006. She has worked with the government of the Philippines as legal officer, legal counsel, and attorney, in the Department of Environment and Natural Resources, and has taught at the College of Law, Ateneo de Manila University, Philippines.

Jeff OBBARD

Assoc Prof Jeff Obbard is Deputy Head (Research) in the Division of Environmental Science & Engineering at the National University of Singapore (NUS). He is also an Associate Director at the Tropical Marine Science Institute at NUS. Jeff has a first Class B.Sc. (Hon) degree in Ecology, and his Ph.D. was fully sponsored by the Water Research Centre in the UK. He is also Director (Research & Development) in the NUS spin-off company 'Advanced Clean Energy Solutions' that addresses climate change and carbon mitigation issues in Asia.

Michael I. Jeffery QC

Professor Michael Jeffery holds a Chair in Law in the Division of Law at Macquarie University, Sydney, Australia and is Director of the University's Centre for Environmental Law, having served briefly as Dean of Law during the Law School's recent restructuring. A former Chairman of the Environmental Assessment Board of Ontario, Canada in the 1980s, (a position analogous to that of Chief Judge of the NSW Land and Environment Court), Michael has published and lectured extensively on a wide range of environmental and administrative law issues. He is a former Deputy Chair of the IUCN's Commission on Environmental Law and a member of the IUCN's Global Energy Law Working Group.

Nicholas ROBINSON

Professor Nicholas A. Robinson is the Gilbert and Sarah Kerlin Distinguished Professor of Environmental Law at Pace University School of Law. He was among the founders of Pace Law School in 1978, and is the father of its environmental law programs. The World Conservation Congress elected him as Chair of the IUCN Commission on Environmental Law in 1996, and IUCN's Council appointed him as IUCN's Legal Advisor (1996–2004). He served on the IUCN team that drafted the UN World Charter for Nature. In concert with law professors in all regions, he founded the IUCN Academy of Environmental Law and serves as its first chairman. A graduate of Brown University (AB, 1967) and Columbia University School of Law (JD, 1970), his career has been devoted to establishing environmental law as a new field of law. The Université Libre du Bruxelles conferred upon him the Elizabeth Haub Prize in Environmental Law in 1992.

Marijke SCHURMANS

Marijke Schurmans is a Lead Lawyer with DLA Piper UK LLP (Brussels, Belgium) where she heads the "Energy, Environment & Economic Government Relations" Unit. She advises private investors and actors in relation to emission trading, JI and CDM-projects; assists the Flemish government in the redrafting of the Flemish legislation on the transfer of emission credits and its the impact on public, private and corporate law, follow up of carbon funding, and daily follow up of Point Carbon. She also assisted the Flemish Government in relation to closures, mergers and acquisitions and the impact on the greenhouse gas emission permit and trading system and in relation to emission rights and tax issues (VAT).

Charlotte STRECK

Dr. Charlotte Streck is Director of Climate Focus, a consultancy company specialized on climate change law and policy and the global carbon market based in Rotterdam, The Netherlands. She is an international legal expert in developing and structuring projects mitigating greenhouse gas reductions (carbon finance transactions, clean development mechanism and joint implementation). Until February 2005, Charlotte Streck was Senior Counsel with the World Bank in Washington, DC. She is a board member of the Global Public Policy Institute, an adjunct lecturer at the University of Potsdam, Germany and fellow of the Center of International Sustainable Development Law at McGill University, Montreal, Canada.

K SURESH

Since May 2005, Mr. K Suresh holds the post of Senior Engineer, Climate Change Unit, Resource Conservation Department, National Environment Agency (NEA), Singapore. Broadly, the Climate Change Unit is responsible for implementing programs and activities that support Singapore's National Climate Change Strategy as well as the work of the National Climate Change Committee. Mr Suresh has been with NEA, and its predecessor the Ministry of the Environment (ENV), now renamed the Ministry of the Environment and Water Resources (MEWR), since 1994. He has held posts in the Pollution Control Department and the Planning & Development Department before his present appointment.

Scott VALENTINE

Scott Valentine has over 20 years of diverse international business experience – specializing in business development and organizational development. In academia, he has been an Assistant Professor in two MBA programs and has worked as a Research Fellow in Energy at the National University of Singapore. He has a DBA (SCUPS), MBA (Adelaide), MSc in Environmental Management (NUS), and MA in Advanced Japanese Studies (Sheffield). He is currently a research scholar at the Lee Kuan Yew School of Public Policy and a founding Director of a new clean energy consultancy in Singapore.

WANG Xi

Dr. Wang Xi is Professor and Associate Dean of the Law School, and Director of the Environmental and Resources Law Institute, of Shanghai Jiaotong University. He is a leading environmental law scholar in China. He is a member of the Academic Committee of the Key Institute of Education Ministry of P.R.C for Environmental Law. He is Vice Chairman of the Chinese Society of Environmental and Resources Law. In addition to teaching, he has provided legal consultation on environmental law for Chinese national and local legislatures and international organizations. His major publications include a university textbook on International Environmental Law, a book on Environmental Law of USA and the Volume on Environmental Law of China for the International Encyclopedia of Laws published by Kluwer Law International (2006).

YANG Xing

Dr. Yang Xing is an Associate Professor of the Law School of Hunan Normal University, China. He has a Ph.D. in Law from the Law School of Wuhan University in 2005, supervised by Prof Wang Xi. He practiced law with the Hunan Wanhe Law Firm from 2003–2006 with which he continues to be affiliated. He was a legal expert at the Hunan CDM Project Service Center from 2005–2006, and advised many enterprises.

CONTENTS

THEME I

SETTING THE STAGE

CHAPTER 1

CLIMATE DISRUPTION: REMAKING THE AGENDA OF MEAs IN ASIA AND THE WORLD

Nicholas A. Robinson

Gilbert & Sarah Kerlin Distinguished Professor of Environmental Law
and
Chair, IUCN Academy of Environmental Law

Climate Change, as the successive reports of the Intergovernmental Panel on Climate Change ably documents, is a present and real phenomenon. There is a growing realization among policy-makers, and indeed all segments of society, that the effects of climate change will make significant demands on all aspects of life. The US Supreme Court ruling in *Massachusetts* v. *EPA* (April 2007) illustrates this facet of human thinking. For the implications of the worst case scenario, we need look no further than the sage of New Orleans following Hurricanes Katrina and Rita.

As the field of environmental law has matured, it has pioneered and refined most of the legal and administrative methodologies that governments will need to cope with the effects of climate change. Environmental law's holistic approach — as exemplified in the scope and inter-sectoral emphasis of Agenda 21, mirrors the omnipresent challenge posed by the effects of new climatic conditions, for example: "Technology forcing", environmental impact assessment, economic instruments for inducing sustainable practices, ambient environmental monitoring and reporting with feed-back loops to promote measures for constant improvement, public participation in environmental decision-making, and ecosystem management.

However, most nations have not established the pre-requisites for applying, much less observing, environmental laws. Lacking a sustainable legal framework, these societies and their economies are as vulnerable as New Orleans, or the coastal communities in Asia that suffered the effects of the tsunami, or the alpine communities that lack water and snow as glaciers melt.

Scientific assessment demonstrates that the growing volume of greenhouse gases in the atmosphere cannot be contained by cap and trade commercial systems. The melt of permafrost releases gases from the once frozen tundra, and forest fires or slash-and-burn agriculture fires release more gases than do industrial sources. After 2012, there will be some sort of global cap and trade system, with major funds flowing into sequestration projects, but all other sectors in the interdependent global environment will also need to be addressed. Article 4 of the UN Framework Convention on Climate Change sets forth the agenda of topics for negotiating the addition "sister" agreements to the Kyoto Protocol.

To leverage new agreements, and to build capacity to employ the beneficial tools of environmental law, it will be necessary to understand the dynamics of how public decision-making evolves. Analysis of the "stages of environmental awareness" provides useful insights in this respect. Ultimately, Earth and her people and her nature will evolve through the era of climate change. The resulting mosaic will record winners and losers. Those that master environmental law can be among the winners; those that pretend, but persist in business as usual, will be among the losers.

Human society evolved its legal systems, including its laws for environmental quality and sustainable natural resource use, during a time of relative stability in the functioning of geological and physical systems of the Earth. Stable climatic conditions accommodated a growth in the human population from less than 1 billion people two centuries ago, to some 6 billion today. Climate change is one of many stresses that the Earth experiences as the human "foot-print" grows with is numbers.

Humans have sustained this extraordinary growth in numbers through human ingenuity. We invented new seeds to feed ever increasing numbers of mouths, and designed electronic means of instant communications world-wide, such as the Internet. Revolutionizing agricultural production led us to — for the moment — tame famine. The revolution in information technology in turn led to a "flat Earth"[1] in which commercial and social interaction collapsed time-zones into a virtual reality of the moment. We moved beyond the cosmonaut and astronaut's competing visions of Earth from space, and produced uses of orbital space once unimaginable, yielding terrestrial systems such as universal global positioning systems (GPSs) or Google Search Systems through which "Everyman" has a window on everyone else backyard. Indeed, the capacity of satellites and the recent development of powerful software for ever more powerful computers, has enabled scientists to assemble data and model the changes in Earth's natural systems and cycles that scientists and lawyers and diplomats call "climate change."

There are many who hope that the ingenuity of the human mind will continue to invent new ways to cope with the effects that climate change produced. Hydrogen fuel cells might replace other carbon based fuels. Solar panels might reduce sunlight to energy directly, everywhere, bypassing the need to convert solar radiation into other forms of energy. New housing models might produce buildings that in fact reuse their own water and consume and recycle the waste of these using the buildings. Bio-engineering produces new plants and animals that promise food stuffs that are resistant to disease and blight, and increase productivity to accommodate the growing global population.

It is fair to say that when the League of Nations collapsed, or the United Nations was created, none could believe that a new technological era like ours would exist, save perhaps in the imagination of Jules Verne and later science fiction writers. If we could muddle through to a world of 8 billion people, surely our past suggests that we can muddle through the challenges posed by climate change. This view

[1] Thomas Freidman, The Earth is Flat (2005), explaining how new communications capabilities have remade both service and manufacturing industries through out-sourcing and team-work around the world.

is shared by Exxon/Mobil[2] and by the administration of President George W. Bush.[3] Many industrial or political leaders agree with them.

While securing production of oil and natural gas and coal in order to meet present economic and security needs, the Bush Administration aims to work with industry to develop, deploy and trade new "clean-energy" technologies, including bio-fuels, coal-methane electricity generating facilities.[4] The federal government under President Bush acknowledges its commitment "to the UN Framework Convention on Climate Change and its objective of stabilizing the concentration of greenhouse gases in the atmosphere at levels that would prevent dangerous anthropogenic interference with the climate system" and expects to reduce greenhouse gas emissions by 18% from 2002 to 2012.[5]

Unfortunately, unlike the United Kingdom and several other industrialized nations, the USA is far from reducing its emission of greenhouse gases to the level it experienced in 1990, as contemplated when the United Nations Framework Convention on Climate Change (UNFCCC) was signed in 1992.[6] Moreover, although there is a great disparity in levels of development and their history of emitting greenhouse gases, most nations are in the comparable positions to that for which the USA is criticized. Emissions of greenhouse gases have grown in tandem with population growth, and new technologies that might avert new emissions and reduce them to the 1990 base year have come on line neither widely nor timely. It matters little to the atmosphere whether the State is developed or developing from the perspective of the increase in emissions alone.

---

[2] See, e.g. "Hydrogen on Board," installment 3 in EXXONMOBIL's Clean Technologies Op-Ed Series, NY TIMES, p. A21, col. 3 (23 August 2007), also available at www.exxonmobil.com/opeds

[3] See Statement by Ambassador Zalmay Khalilzad, U.S. Permanent Representative, on the UN General Assembly Thematic Debate: Climate Change as a Global Challenge, August 1, 2007, UNUS Press Release # 186(07), available at www.un.int/usa/press_releases/2—70901_186.html

[4] Ibid.

[5] Ibid.

[6] http://unfccc.int/resource/docs/convkp/conveng.pdf

1. The Gathering "Storm"

The practical impact of the tepid pace by which nations act to implement Rio de Janiero's 1992 Framework Convention on Climate Change cannot be hidden. The several reports of the Intergovernmental Panel on Climate Change have documented how the increase in greenhouse gases in the atmosphere is irreparably altering Earth's natural systems as it traps solar radiation as heat in the atmosphere. Sea levels are rising, and will continue to rise for some decades as the ice melts from the North Pole and Greenland and Antarctica, and the glaciers of all high mountains. Rainfall in tropical and subtropical areas is declining and is increasing in temperate zones,[7] and in the first half of 2007 alone devastating increases in flood waters were experienced from England in the United Kingdom, to North Korea, to the Bihar Province in India, to Ohio or Wisconsin in the USA. At the same time, desertification patterns are widening and temperatures rising in many locations; China has reported that the high Qinghai-Tibetan plateau and mountain areas are warming faster than any other region in the world, rising an average of 0.7 degrees Fahrenheit every 10 years.[8] In South Asia, the Ganges River is loosing its source of water as the glaciers retreat in the Himalayan Mountains, while the Delta of the Ganges is becoming submerged as sea levels rise, displacing many communities. Pacific small island states already suffer the loss of land and the coral reefs which are essential to the life cycle of fish, their main food sources.

Actual reports of the effects of climate change confirm the impacts predicted in the models prepared by the Intergovernmental Panel on Climate Change.

From a scientific perspective, human impacts on the climate are like the genie, who, once out of the bottle, cannot easily be returned

[7] NATURE, 26 July 2007. Francis Zwiers, a co-author of the NATURE report and director of the climate research division of Environment Canada, is quoted as saying "It's the first time that we've detected in precipitation data a clear imprint of human influence on the climate system." Carline Alphonso, *Toronto Globe and Mail* (24 July 2007).

[8] "China Considering Trading Market, Quota System," Vol. 10, no 9 Air, Water & Climate (August 1, 2007).

to it. Some policy makers, who believe in the power of technology and have a vested interest in continuing the use of fossil fuels, believe we can get the genie back in the bottle. They will strive mightily to do so. However, what these persons fail to realize is that the bottle is now broken. The better analogy might be to Humpty Dumpty; all the King's Horses and all the King's men could not put Humpty together again.

Although there will be massive efforts at technological innovation in the coming years, and there will be a global market in emission trading to reduce greenhouse gas emissions from the industrial and transportation sectors, it may be questioned whether such measures to control industrial emissions can in fact stabilize Earth's climate conditions. There are many other sources of greenhouse gases, those which are not being addressed by technological innovation or emission trading, and these releases of greenhouse gases are very likely to continue to become even more pronounced than they are now. For instance, vast amounts of methane are encapsulated in the frozen tundra around the Artic Circle. The tundra is melting, crumbling the buildings of the far north, and releasing their store of methane gas; tundra melt seems irreversible, and we cannot stop the incremental releases of methane.

Moreover, in tropical and temperate areas alike, forest fires burn uncontrolled. We see this from Asia, as in parts of Indonesia, to Siberia into Eastern Europe, to the Amazon and the Andes. Even States with advanced forest fire fighting capacity, such as Australia, have not been able to quell their recent vast forest fires. Although collectively these fires release as much carbon dioxide annually as do industrial processes, and forests are needed for sequestration of carbon through sustaining the photosynthesis of their flora, as yet there is *no* global campaign to help prevent and extinguish forest fires. The opposite phenomena can be seen, as economic forces clear forests in South America or Southeast Asia. For instance, the mania to find new biofuels,[9] such as produced from palm oil, has been a cause of the

[9] See C. Ford Runge and Benjamin Senauer, "How Biofuels Could Starve the Poor," Vol. 86, No. 2 FOREIGN AFFAIRS, p. 41 (May/June 2007).

forest fires in Indonesia, with sad side effects. The Center for Orangutan Protection reports that workers on Borneo palm oil plantations have deliberately killed some Orangutans because they were eating palm seedlings; the Center reports that such killings, plus the loss of habitat for this endangered primate, has caused 1600 to die.[10]

2. Responding to Climate Disruption

Atmospheric warming causing climatic disruption will not stop soon, and the tundra will continue to melt just as the glaciers or polar ice caps continue to melt. Nations *could* assemble a global campaign to save the forests from forest fires, but since timber is a prime commodity and nations are reluctant to limit exploiting forests, there has been no significant movement to advancing global management of forests since the adoption of the 1992 Rio Conference's "Non-Legally Binding Authoritative Statement of Principles for a Global Consensus on the Management, Conservation and Sustainable Development of All Kinds of Forests."[11] Even this weak document of recommended policies managed to ignore the historic phenomenon of forest fires, evidently regarding them to be essentially a domestic matter not requiring priority international attention. As tropical and sub-tropical regions become drier, the likelihood of dangerous fires in their forests increases.

Releases of greenhouse gases from non-human phenomena and the loss of forest complicate achieving the legal objective of achieving climate stabilization through the UNFCCC's Kyoto Protocol. The focus on technological innovation to find new fuels, or use fuel more efficiently, as a higher priority than addressing forest fires, is motivated by the traditional approach of meeting the demands for profitable economic growth by finding ways to increase the supply of fuels. It is characteristically a "supply-side" management effort. There is money to be made in developing and selling new supplies of energy, and the markets invest in such efforts, and the laws promote such ends.

[10] Greenwire, 27 July 2007, Environment and Energy Publishing LLC.

[11] A/CONF.151/26 (vol. III), 14 Aug. 2992, reprinted in 11 I.L.M. 881 (1992).

However, if we are to reduce greenhouse gas emissions, we must work much more effectively on "demand-side" management. Society has too little experience with demand-side management to understand it. Some fear that by reducing demand they will lose money, or suffer a declining quality of life. These fears, while not grounded in rational assessment, are nonetheless real, can drive public policy, and thus need to be addressed if legal techniques to advance demand-side management are to be realized. Without curbing demand, mitigation of greenhouse gases from anthropogenic sources has little chance to succeed.

Even if the methane in the tundra could be kept in a frozen encapsulation and the forest fires world-wide contained, the magnitude of the remaining controllable human activities that release greenhouse gases is formidable. In their classic essay in 2004 defining the scale of the problem, Stephen Pacala and Robert Socolow demonstrate the deep and pervasive changes that the industrial and agricultural sectors would need to make to reduce the greenhouse gases to levels that will not cause further climate disruption.[12] They project that fundamental changes would be required (1) in energy efficiency and conservation, and (2) in "decarbonizing" electrical generation and fuels, and (3) in enhancing use of natural sinks in forests and farm lands. However, while all the technology needed presently exists to implement their plan, when one considers the political, economic and social obstacles to deploying this technology, one may question whether such change will come, either slowly or at all. Some of the obstacles to phasing out our now obsolete technologies and practices and introducing replacements illustrate the challenges:

(1) To realize the vision articulated by Pacala and Socolow, all cities would need rapidly to put on line the mass transit systems, and the curbs on the use of automobiles, that now exist in Singapore. Vast investment in infrastructure is needed, plus life-style changes as we remove older model cars from use and reduce use of cars

[12] S. Pacala & R. Socolow, "Stabilization Wedges: Solving the Climate Problem for the Next 50 Years with Current Technologies," SCIENCE, Vol. 305, pp. 968–972 (August 13, 2004). See also its supporting material on line.

generally. Although public health benefits would come as a major source of urban smog is eliminated, somehow smog is growing in all cities and few cities emulate Singapore's brilliant model. As early as 1992, Agenda 21 called for such reforms, but few States or local authorities have done so.

(2) In addition, "green" buildings would be required and older buildings be retrofitted. No building codes now require this vision, and indeed most of the nations do not even have building codes. Maximum use of solar energy photovoltaic electricity would be required, and all fossil fuel electrical generating facilities retrofitted and converted to natural gas or to non-carbon fuels or closed down. To meet growing energy market demands, however, the opposite is taking place. New plants are built burning fossil fuels, oil shale and secondary recovery is used in once abandoned oil wells in order to increase the supply of oil; very few old plants have been closed, and companies seek life-extension permits to get longer use out of their sunk costs and existing electrical generating capacity. Use of nuclear power is growing slowly, but still lacks the highly trained personnel and the security and the fuel cycle waste and reprocessing management systems to make it politically acceptable in many regions. Few developed or developing nations possess domestic capacity to build a new nuclear power plant, for instance, even the USA would have to contract with Japanese firms to build a new plant, if it could be licensed, since no current firm in the USA has built a nuclear plant or has the technological capacity to do so.

(3) Finally, current clear-cutting of forests for timber or for agricultural expansion or crop production for biomass fuel production is eliminating forest sequestration areas. Wetlands, also rich as biotic sequestration areas, are being filled or eroded or engulfed so that their loss is rapid world-wide. Some modest afforestation is underway, but the world's once leading nation-wide afforestation programs in China are less popular today as its economic development booms and social obligations to plant trees recede. No nation is conducting afforestation at the scale needed to provide sequestration. Moreover, in conversion of forest or grasslands to agricultural uses and biofuel production, up to one-half of soil

carbon is lost, as decomposition increased by aerating organic matter during plowing and cultivation activities. Conservation tillage and soils management is needed world-wide, and yet there is no international treaty on soils, although Iceland and the IUCN Commission on Environmental Law have called for a soils agreement to be negotiated.

3. The Lagging Pace of International Law-Making

What is to be done?

Nations are still fixated on the an agenda of the past. They concentrate on material and manpower in military security, not on ecological security. The argue about how the developed states are responsible for the crisis and so should pay for the solution, just as the New International Economic Order[13] was to require the former colonial states to pay for the state of under-development in once new nations. While the developing states can win the ethical and rhetorical battle, the debate poses little likelihood that it will help cope with the effects of climate disruption.

Consider the slow pace of decision-making:

1972 UN Stockholm Conference[14] class for creating the UN Environment Programme, to monitor environmental trends and help developing nations build environmental stewardship capacity.

1979 The World Meteorological Organization (WMO)[15] convenes First World Climate Conference.

1980s Canada convenes conferences to study a Convention on the Atmosphere, like one on the Law of the Sea — States decline to move ahead this idea.

---

[13] http://lexnet.bravepages.com/NIEO.htm

[14] http://www.unep.org/Documents.Multilingual/Default.asp?DocumentID=97&ArticleID=1503

[15] www.icsu.org/Gestion/img/ICSU_DOC_DOWNLOAD/863_DD_FILE_ICSU_Strategic_Plan.pdf

1988	WMO and the United Nations Emergency Redistribution Program (UNERP)[16] convene first IPCC Session.
1990	United Nation General Assembly (UNGA)[17] Convenes the Intergovernmental Negotiating Committee for a UN Framework Convention on Climate Change.
1992	UN FCCC signed at Rio de Janeiro UN Conference on Environment & Development (Chaired by Ambassador, Prof. Tommy Koh).
1994	UNFCCC enters into force.
1995	First UNFCCC Conference of the Parties adopts the Berlin Mandate,[18] aiming to have Parties use best efforts to roll back greenhouse gas emissions to the year 1990 in order to stabilize the changing climatic conditions, with negotiation of a time-table to achieve "quantifiable emission limitation and reduction objectives."
1995	Second IPCC report[19] — demonstrates that climate change is man-made.
1997	Kyoto Protocol negotiated, with its complicated, compromise(d) process to induce reductions in emissions.
2000	Third IPCC Report[20] — examines actual effects on environment of man-made climate change.
2005	Kyoto Protocol enters into force, with Russia vesting its emission reductions from the closing of inefficient Soviet factories and the USA refusing to participate.
2005	European Union inaugurates European Trading System.
2006	States of New York and New England in USA agree on a Northeast trading system; California establishes a trading system.

[16] http://nooventures.edublogs.org/2007-09-03-a-basic-plan-for-sharing-the-worlds-resources-from-wwwstwrnet/

[17] http://en.wikipedia.org/wiki/United_Nations_General_Assembly

[18] http://unfccc.int/resource/docs/cop1/07a01.pdf

[19] http://www.ipcc.ch/ipccreports/assessments-reports.htm

[20] http://www.google.com.sg/search?hl=en&q=Third+IPCC+Report++&btnG=Search&meta=

2007 Fourth IPCC report[21] — demonstrates climate disruption is well underway.

2008 UNFCCC Conference of the Parties in Bali in December, 2007, for the agreement of a negotiation road-map to 2012 and shaping post-Kyoto new agreements.

By the time major agreements are made on climate change, and barely implemented, if at all, in many States, nearly one generation of human beings will be have been born. There are reasons why nations are so slow to respond to climate disruption. This pace of negotiations reflects the fact that most national leaders do not yet take climate change as a serious threat to their economic or social or environmental well-being. Although the IPCC has changed many opinions, the slow pace also reflects a lack of knowledge or study of the problem by economic and political leaders, interestingly, the leaders of multinational corporations and international financial institutions, including insurance and reinsurance companies, are responding information of climate change faster than governments. It also reflects a deep unwillingness for individuals (whether human, business or government agencies) to incur hardships in their immediate wealth in order to help resolve the common and diffuse hardship that the public, or others are suffering from climate disruption. This latter intransigence is seen the callous disregard that many nations show toward the pleas of the Association of Small Island States (AOSIS) in the UN General Assembly or in the UNFCCC COP. Finally, the climate and the environment are interdependent and complex systems, and virtually every human activity has positive or negative implications; the techniques for interdisciplinary planning and decision-making are rudimentary, despite global experience with legal methodologies such as environmental impact assessment (EIA), which have developed since 1970 to cope with a wide array of environmental problems and build sustainability into development.

[21] http://www.google.com.sg/search?hl=en&q=Fourth+IPCC+report&btnG=Search&meta=

Although States have the technology to implement their obligations under Article 4 of the UNFCCC to stabilize atmospheric conditions from all sectors, inertia, however, and a Pollyannaish hope that "business as usual" will allow us to muddle through, prevents rapid deployment of needed technological practices and the cessation of existing practices. We have now lost nearly 30 years of response time, and locked in the concomitant changes to the climate.

4. Precaution and the Scale of Responding to Climate Change

In 1992, at Rio de Janeiro, it was expected that rapid actions might be taken to implement the recommendations of Agenda 21 and the obligations of Article 4 of the UNFCCC. In chapter 9 of Agenda 21, entitled "Protection of the Atmosphere," States agreed to coordinate their measures to cope with climate change in an integrated way with their work for socio-economic development, "taking into account the needs of developing countries for the achievement of sustained economic growth and the eradication of poverty" (Para 9.3). In Chapter 9, many of the technological and managerial measures needed to cope with the effects of climate change were set forth.[22] However, few nations have implemented these recommendations.

[22] States agreed in Chapter 9 that "the need to control atmospheric emissions of greenhouse and other gases and substances will increasingly need to be based on efficiency of energy production, transmission distribution and consumption and on growing reliance on environmentally sound energy systems, particularly new and renewable sources of energy" (Para, 9.11), and agreed that "more efficient design and management of traffic and transport systems" is essential (Para. 9.13), and further agreed that "Land-use and resource policies will both affect and be affected by changes in the atmosphere. Certain practices related to terrestrial and marine resources and land use can decrease greenhouse gas sinks and increase atmospheric emissions. The loss of biological diversity may reduce the resilience of ecosystems to climatic variations and air pollution damage." (Para. 9.19). States emphasized the importance of the control measures in the Montreal Protocol to prevent stratospheric ozone depletion (Para. 9.22) and agreed to build capacity to prevent transboundary air pollution as priorities in the protection of Earth's atmosphere. (Para. 9.26). See N.A. Robinson, Agenda 21: Earth's Action Plan (Oceana Publications, 1992).

As the evidence has mounted regarding climate disruption, States have chosen to discount it. This is contrary to the norm adopted at the Rio de Janeiro Conference on Environment and Development, in Principle 15: "the precautionary approach shall be widely applied by States according to their capabilities. Where there are threats pf serious or irreversible damage, lack of full scientific certainty shall not be used as a reason for postponing cost-effective measures to prevent environmental degradation." If States had taken the Precautionary Principle seriously, all States would now have enacted legal and other measures to anticipate, prevent or minimize the causes of climate change. Clearly announcing a principle is not enough. There must be tools put in place to make the principle operational. EIA and emission trading systems represent two variations of such tools. Among the wide array of tools that environmental law has crafted since its inception as a field of law over the past 4 decades,[23] is it not extraordinary that only ETS is being promoted today to address climate change? Where is integrated coastal zone management, or provision of urban mass transit?

If we consider the myriad sources of green-house gases, what should be done first? The rush to deploy biofuels and to create a new global market to cap and trade emission levels is an instinctive business as usual approach. As Pacala and Socolow demonstrate, these measures alone cannot stabilize greenhouse gases at levels that prevent further climate disruption. Environmental legal principles can offer some guidance. If nations were to deploy the precautionary principle, they would at once phase out all practices that exacerbate the greenhouse effect. In 1992, former Secretary of State James Baker called this the "no regrets policy." His policy could have had the USA implement its UNFCCC obligations by all measures that could eliminate greenhouse gases without negative impacts on socio-economic conditions, such requiring high fuel efficiency standards in motor vehicles, or green building codes, or shifting to natural gas and out of

[23] See the 2 volumes on Capacity Building in Environmental Law in the Asian and Pacific Region (Asian Development Bank), edited by Donna Craig, Koh Kheng Lian and Nicholas A. Robinson.

coal or oil combustion. These are measures that save money and are good for society, as in reducing urban smog and safeguarding public health. Taken further, the "no regrets policies" could have removed perverse tax incentives that encourage more oil and gas production, and shifting to use of solar energy or other renewable sources. However, the new Clinton Administration rejected the "no regrets" policy, was unable to secure Congressional support for the Kyoto Protocol, and thereafter the current President and Vice President regressed to once again promoting oil, gas and coal.

Embracing the precautionary principles would draw us back to at least the "no regrets policy", and probably to the sort of "technology forcing" requirements that Congress one mandated in the Clean Air Act. In 1970, Congress adopted a law that required polluting electrical generating facilities to protect the public health by either cleaning up air emissions or shutting down operations. Industry challenged enforcement, but the Supreme Court upheld its constitutionality.[24] Industry then developed the new technologies, rather than relying on the "business as usual approach" which would continue to profit from extending the use of obsolete technologies just because they were paid for.

What is needed? There are a set of clear rules that oblige society to change for its own good. World economy addicted to oil will not give it up soon, but it can be persuaded to ban used cars with internal combustion engines, and replace the world's fleet of motor vehicles with hybrid or non-polluting vehicles and mass transit. Markets will not require this change, but the parliaments can. Nations could within a year frame a global green building code, and all new construction could "go green." Nations could at once deploy air force and army capacity to fighting forest fires, and not wasting money preparing to fight some presumed "enemy" or hostile nations. Nations could at once eliminate use of virgin forests and plant new

---

[24] *Union Electric v. EPA*, in which a St. Louis Missouri electrical utility argued that its mandate to supply electricity came before any duty to clean the air, and that shutting it down could not have been a serious intent of Congress just to protect the diffuse public health concerns.

tree plantations, to produce renewable timber and sequester carbon dioxide. These and scores of readily available measures can be mandated; the environmental law techniques of "technology forcing" can be used to realize the Pacala and Socolow vision.

What prevents nations from embracing the Precautionary Principle? They often need procedures to do so. Again, environmental law provides procedures, such as the technique of Environmental Impact Assessment (EIA). Rio Principle 17 provides that "Environmental Impact Assessment, as a national instrument, shall be undertaken for proposed activities that are likely to have a significant adverse impact on the environment and are subject to a decision of a competent national authority." The EU, Canada and USA widely use advanced EIA techniques, but none yet apply them to reducing greenhouse gas emissions or sequestering gases. All States need to enact EIA laws and apply them with rigor. New York State is now preparing regulations to require EIA under the State Environmental Quality Review Act (SEQRA)[25] to address climate change. State such as those in the ASEAN region should take the lead in developing advanced EIA to cope with sea level rise and ensuring stable flows of energy, potable water, and other environmental infrastructure, given the region's vast coastlines and the needs of its growing population. Can ASEAN afford to be less developed in its legal infrastructure than the EU or North America in this regard? ASEAN member States have the present capacity to enact and apply EIA far more extensively than is today done.

5. Legal Precedents for More Effective Climate Change Regimes

The climate change debates, and Kyoto Protocol negotiations, are less effective than they might because they are premised on a false assumption that nations can address climate change slowly, and as just one of their many priorities. Indeed, as a new and little understood priority (high level officials who should know better still like jokes about

[25] Article 8 of the NY Environmental Conservation Law.

"global warming"), few States have cabinet level offices of climate change just as few have cabinet level offices for environmental quality assurance. Lacking a focal point and a way to access the highest levels of government, the response to climate disruption is relegated to low levels of government. Even with its continuing disasters in New Orleans and the Coast of the Gulf of Mexico, for instance, the USA federal government has made no effective effort to better prepare to cope with the effects of climate disruption.

If States were to take their UNFCCC obligations seriously, they would model the next post-Kyoto agreements one or two precedents. The first is Agenda 21, adopted at the Rio Earth Summit. Its many prescriptions would both advance sustainable development and reduce greenhouse gases. However, as the Johannesburg Plan of Implementation[26] sadly showed in 2002, far too little is done to implement Agenda 21 by most UN agencies and most States. Since the States are responsible for the UN agencies, the deficit in action returns to the leadership in each national capitol. If States embraced Rio Principle 10, on public participation, the public could press leaders to do more to cope with climate change; lacking means for public participation in environmental decision-making, this does not happen. Above all, the multi-sectoral and inter-dependent design of sustainable development objectives in Agenda 21 needs to be brought to the climate change negotiations. As Pacala and Socolow demonstrate, one sector (e.g. greenhouse gas trading systems, even with clean development mechanisms attached) is too limited to have meet the obligations of Article 4 of the UNFCCC.

Second, and more focused, is the model of the Vienna Convention for the Protection of the Stratospheric Ozone Layer,[27] with its Montreal Protocol and the London,[28] Copenhagen and Beijing accords. Under this treaty system, one potent class of greenhouse gases, the "ozone depleting substances" such as chlorofluorocarbons (CFCs), have been

---

[26] http://www.environment.gov.za/nssd_2005/Web/Introduction%20and%20 Background/JPOI_Response_Strategy_2003.pdf

[27] http://sedac.ciesin.org/entri/texts/vienna.ozone.layer.protection.1985.html

[28] http://sedac.ciesin.org/entri/texts/montreal.protocol.ozone.amend.1990.html

reduced enormously across all developing and developed nations. By banning uses and banning trade (indeed, if a nation trades in CFCs outside the treaty system, all States agree to cut of all trade with such a nation). Nations have created national focal points and agreed in a reporting system and established a non-compliance review to help States come into compliance with targets and time-tables. Capacity building measures have been employed, and international cooperation is the fundamental norm informing the system, rather than assessing blame and taxing for costs.

Why is the Vienna Convention so helpful? It shows that States can reduce and replace greenhouse gas emissions effectively, albeit in one sector. More must be done to protect the stratospheric ozone, which is still degrading, and even when legacy CFCs are no longer being released into the atmosphere it will take a decade for the CFCs to reach the stratosphere, so we must wait and hope that Earth's natural systems will re-stabilize the stratospheric ozone layer that protects life on earth from the harmful extreme ultraviolet solar radiation.

Nations need to adopt a model such as that for ODS. There will be immediate cut-backs by everyone; the larger emitters will need to do more than the smaller emitters, but everyone must engage in a pervasive retooling of how human economic activity works. A model to be adopted is the ISO-14000 pattern, where industry designs systems to do so and is certified as meeting those systems. This approach should apply to agriculture, forestry, urban transportation, etc. An analogue would be the use of design specifications; all cities should adopt green building codes, and nations manufacturing or importing new motor vehicles should require that more than fuel efficiency standards. They should follow the precedent of the Vienna Convention and restructure the product entirely. States should mandate that the product be built without internal combustion engines, by a target year, and ban all use or export of used cars. International aid to assist developing nations import and use only clean motel vehicles would follow.

Most national leaders, whether parliamentary, executive or diplomatic, do not believe that there is the sort of urgency that States brought to the stratospheric ozone hole, as there is to climate disruption.

The road to the UNFCCC COP in Bali in December 2007, needs to set in place a multi-sectoral approach to coping with climate disruption. Every sector needs to form taskforces and teams to redesign the activities and products of each sector. This will, of course, unleash an era of rich creativity and innovation, and human ingenuity may indeed rise to the occasion. Certainly many will make new profits, and one need not assume that this will be negative to the overall economy. There will be losers and winners. But governments must maximize the space for the winners and keep the level playing field for all. This will mean helping the developing nations, who must import most of the technology and the products and practices.

At the recent UN General Assembly thematic debate on the Challenge of Climate Change,[29] Robert Socolow outlined a way to keep the emissions as of 2005 constant through 2055, and not increasing. Without stabilizing emissions, current expansion will let emissions grow from 7 billion tons of carbon emitted per year (GtC/y) to 14 GtC/y. He would replace carbon as a fuel, and supply energy from other sources. To hold global emissions flat, the OECD today must cut-back emissions on a magnitude of 4 GtC/y, and this would allow the non-OECD nations to continue to emit as they develop economically and reduce their own emissions on the magnitude of 2 GtC/y.[30]

For the UN General Assembly debate, Socolow computed the per-capita carbon distributions of five classifications of nations. The OECD accounted for 45% of carbon emissions, with its 3 billion resident people. The non-OECD nations had 5 billion people, and 55% of the emissions. However, the per capital emissions from the least developed non-OECD nations were negligible; to require them to

---

[29] See the debate at www.un.org/ga/president/61 and the Program for the Thematic Debate on the Challenge of Climate Change, 31 July 2007. Socolow's statement, and that of other experts, is found on the first day; the statements of UN Member States and Observer Missions are on the next two days, 1–2 August 2007.

[30] See Socolow and Pacala, "A Plan to Keep Carbon in Check," *Scientific American*, September 2006, pp. 50–57, at p. 56.

contribute to large cut-backs when they have urgent development needs just in order to cope with the effects of climate change, would not be productive nor workable.

Socolow suggested that to avoid the coming "environmental havoc" from the effects of climate change, the developed nations should focus on their adaptation and mitigation and help the developing states with their adaptation. A division of labor is needed, and keeping with the broad reading of "common but differentiated" responsibilities. Socolow's per capital approach, rather than an approached based in sovereign equality of states, makes sense. He re-characterizes nations in terms of those who must take the lead and those that need some capacity building. He suggests that for the least developed nations, there is no gain to be had in requiring cut-backs in climate emissions because they are so small, and these states need to meet urgent sustainable development objectives if their civilizations are to continue. For these states, all other states should lend as helping hand.

If the negotiations between 2007 and 2012 take this broader view, there are many ways that States could promote an Agenda 21/Climate Adaptation program for all nations, while promoting new economic models that the developed nations would have to fashion for themselves and for the benefit of all economies. In a "flat earth" these benefits can be exported and imported rapidly. For instance, the imposition of a small transfer fee in the movement of oil, gas and coal across borders, paid into an international fund for adaptation, could finance extensive investments in building climate adaptation systems for the poorest nations. This would allow pervasive macro-economic reform. The problem with ETS and the clean development mechanisms is that they are micro-reform techniques and only states with well developed governmental infrastructures can take advantage of them. Most small island states might never see such CDM projects.

The United Nations system also needs to do more for climate change. Every specialized agency and every MEA needs to develop a climate change model, and assume some of the adaptation burden. This cannot be done through the UNFCCC COP alone. States made clear this year that they do not want climate changed assigned to the Security Council, since every State wants to be a decision-maker. They accept the UNFCCC COP as a forum, but they also accept the UN General Assembly as a forum. Some consensus on the focal point for international cooperation on climate change will be needed.

Indeed, a system-wide approach to climate change is needed. Perhaps the UN should establish a Climate Council, taking over the chamber of the Trusteeship Council; this is where the August 2007 UNGA thematic debate on climate change was held, rather than in the UNGA Hall. All inter-agency cooperation needs to have a climate change focus also. Without a system-wide approach, it is likely that needed reforms cannot be mobilized in time to avert a worsening of climate disruption. Chief among the MEAs in this regard is the Convention on Biological Diversity (CBD). The CBD needs to be given the task to marshal the human support to sustain and expand sequestration systems for forests and wetlands and marine biota. If this is done with integrity, the transparent rigor of scientific disciplines, and with respect for the other ecosystems and their life forms, then the CBD can help fashion ways to ensure both continuation of biodiversity and a maximization of green-house gas sequestration.

While carbon trading will surely become a global system,[31] it is essential that it not be seen as a panacea or even a dispositive step toward coping with climate change. It is a fledgling preliminary expression of concern that action must now come. Its weaknesses need to be understood. We need not only compensate for the less developed nations, which cannot easily reduce emissions; we must also compensate for the releases of greenhouse gases from melting tundra, and other natural sources, such as forest fires. OECD nations must compensate for far more than just their own past, present or future emissions.

---

[31] See, for instance, the robust systems already inn place. J. Robinson, J. Barton, C. Dodwell, M. Heydon and L. Milton, Climate Change Law: Emission Trading in the EU and UK (Cameron May, 2006).

The global economy requires a major roll-back in emissions and a major effort to sequester carbon. Instead of making this the highest priority, the tentative post-Kyoto negotiations defer tough decisions, and nations domestically are at work on increasing supply side with manufacture and trading in biofuels or developing trading in emission reduction objectives, or "credits," in order to use the market to achieve mitigation of emissions. Some attraction of the credit trading is to divert funding into socio-economic development or even efforts at elimination of poverty. Yet the lion's share of all work goes into design of emission trading; major financial centers vie to become the trading capitals of these credits, and secure the wealth from the transactions that will take place. The euphoria of a consumer society attracts us to emission trading schemes; we would do well to go back to school with the IPCC and look at what else we need to address. We invest time and concern for carbon markets, but both the public and the private sector are investing vastly less time and concern in examining how the Convention on Biological Diversity can ensure that sequestration will be sustainable.

6. Institutional Dysfunctionality

What this paper outlines is dysfunction in the international and national legal systems. International Law — and all legal systems and their norms — evolved at times in which natural conditions around the Earth were relatively stable. We assumed that the Earth would accommodate us, and paid little attention to understanding how Earth's systems worked. Our assumption of the endless bounty of nature was naïve and uninformed. As natural resources became compromised through over use, the conservation movement was founded, the first wildlife laws were enacted, and in 1948 the International Union for the Conservation of Nature and Natural Resources was established. As pollution grew, the environmental movement was born and pollution control and environmental laws emerged, and UNEP was established. At the end of the 20th century, States came to recognize that carbon based fuels (coal, oil and gas) were becoming scarce and thus more expensive as energy sources, and laws for energy efficiency were enacted. However, at the foundation of each of these efforts was the motivation to continue business as usual, to sustain a yield in flora and fauna that could be harvested and to trade and consume what humans wished to use.

These traditional conservation and environmental objectives are no longer sufficient. Adaptation and building robust capacity for Earth's natural systems to adapt to the changing conditions of life on Earth is the new mission. Human activity has constrained natural systems and weakened the capability of flora and fauna to adapt. Evolution and natural selection is a powerful force in nature and humans need to give it its widest latitude to sustain biological diversity and conditions amicable to the human species.

If one basic objective of law is to stabilize conditions and ensure that traditional and established practices of commerce, culture, social order can function "as usual," then we need to build into this strong legal order the procedures to accommodate change and evolution. We need to re-establish law reform commissions, which many nations have allowed to fall into disuse or have outright abolished. We need to re-establish technology assessment and risk-assessment procedures, and above all widen the use of EIA.

Another basic objective of law is to attain justice. Climate disruption will harm rich and poor alike, and create new conditions of social injustice. Law reform will be needed to address the resulting social inequities or harms. Historically, law has done more to bolster the established order than it has functioned to redress unjust conditions. The abolition of slavery or the extension of suffrage to women or the labor reforms still being advanced by the UN International Labour Office, are examples of how the law has redressed broad social wrongs. Those of us who claim to understand the ameliorative functions of justice and law must now build on these precedents to align a new era of adaptive environmental laws to cope with climatic disruptions.

This task is not easy, but it is the over-arching task for the coming several generations of humans, and must become the over-arching mission of legal education. Perhaps we need to take a leaf from the International Labour Organization's (ILO) experience; the ILO delegations are made up of labor union, industrial and commerce and government ministerial representatives. For climate negotiations, we might want to consider shaping our national delegations to climate negotiations to have at least 4 sectors represented on each delegation: a qualified scientific delegate, an executive branch focal point delegate, an economic sector delegate (commerce, industry or agriculture, as appropriate), and delegate representing social and environmental concerns NGOs (Non-Governmental Organizations). These stakeholders are the ones who will help make adaptation to climate change effective, or not.

7. Epilogue

The task for Asia and the world on the road to Bali in December is to make adaptation to climate disruption the high and all-encompassing priority that it must become. If States address short term and narrowly focused interests, such as who gets what benefit from what emission trading or development project, all they do is delay the day when effective international cooperation will begin in the effort to stabilize the Earth's atmospheric and climatic conditions.

It is likely, in the short term, that every social reform movement will try to gain some leverage or tactical advantage by embracing a role in coping with climate change. Women's movements will (rightly) say women are at risk and if their needs are met, they can do much to help adapt to climate change. The military will show how energy efficiency can be built into aircraft, ships or tanks. Lobbyists for each commodity will position their economic clients to come out ahead in the jockeying for a climate change advantage. As existing sectors behave in this way, all sectors tend to discount the real urgency of helping the small island states such as, Bangladesh and delta states, whose territory is vanishing. These tragic developments are not merely a competing for resources or reforms; they are of a different magnitude. The effects of climate disruption rob people in some places of their shelter, their livelihood, their culture and ultimately their lives.

ASEAN and other regions need to unite to build their regional systems for climate adaptation, and to position their economies to participate in climate mitigation by removing greenhouse gases from the atmosphere. The road to Bali and 2012 should not mask the need to take urgent action today. A nation, or regional group of nations, that takes the lead in adapting to the conditions of climate disruption will be a winner; those who wait for the UN or for the international community to attain a final consensus on a course of action, will be losers.

Earth and human civilization are at the beginning of a new era, a kind of paradigm shift, whether we like it or not. We are all New Orleans, not yet back on its feet two years after the hurricanes and disaster in 2005. We are all a small island state, for which climate change is an existential challenge: adapt and cope and strive to sustain a civilization or it will be one with the fabled *Atlantis.* Local authorities are the "front lines" in coping with climate disruption, and they need to deploy local environmental law tools such as Integrated Coastal Management (ICZM) now. New wetlands must be designed and accommodated as old ones are flooded and eroded; coastal infrastructure needs to be relocated and buffered. While each such step seems mundane and not worthy of being a high matter of states, in fact helping each wetland adapt or sustaining each old growth forest or each tree for afforestation is of transcendental importance.

In short, the common concern of the atmosphere, whether in the stratospheric ozone or the climate dimensions, require all states to act in solidarity. This is not a choice, it is a necessity. To act in solidarity, however, requires each state to mobilize all of its own domestic economic and social sectors to address adaptation to the effects of climate disruption, and the removal of billions of tons of gaseous carbon from the atmosphere for the next 3–4 decades. If local authorities in many areas can eliminate cigarette smoking in one generation, individuals can embrace reuse, recycle and restore activities in their personal lives. To facilitate community-wide reforms, national action plans with the force of law will be needed. While acting locally, states must also act globally. New measures to cooperate on assisting migration of ecological refugees must be built, as must means to prevent and combat forest fires. All the global sectors of this "flat Earth" need to begin to work in tandem to retool, reinvent and re-imagine their roles. The profits from this retooling will be real, unlike the ephemeral and transitory fees gained from the transaction costs in managing transfers of emission credits.

These efforts will open a new kind of political space in international cooperation. The calls by France for a new UN environmental agency may enter this space, and a new system-wide climate focus can reinvigorate international cooperation. Without a world-wide effort addressed to removing carbon from the atmosphere, one can easily envision a range of environmental tragedies and human rights disasters and economic defaults. Sustainable development will be set back. To avert such a future, intergovernmental cooperation should facilitate a mobilization of public/private activity on the scale of the efforts of World War II, only this time for civilian and peaceful objectives.

Mobilization of such resources has been advocated before, as in the UN Millennium Objectives to eradicate poverty. The response to such calls is dismal. States thought that they could choose to ignore or use the recommendations of Agenda 21; development would come sooner or later. Climate disruption renders this "business as usual" approach obsolete. The effects of climate change are not a classic "externality" in which a polluting state transfers the cost of its waste to the receiving state. Earth has only one atmosphere, and we shall all experience climate disruption at one time or another. Historic pathways for hurricanes and typhoons can change; traditional patterns of rainfall relied upon for generations can change. Human mistakes, such as the desiccation of the Aral Sea, can get worse as the remnant waters in the Sean's tributaries dry up, as Central Asia's glaciers melt. Endangered fish stocks fall as already threatened coral reefs are covered with deeper waters.

Either we mobilize reforms across all sectors, or all locations and sectors will pay the costs of this neglect. Past generations did not understand what harm they did when they released carbon into the atmosphere. Although all too gradually they came to understand the harm of clear-cutting the forests, those who continue this vice think of themselves as creating wealth, not causing a problem. The present generation is the age that for the first time understands what humans have done to the atmosphere of the Earth, to its oceans to its flora and fauna and ecosystems. It is the future generations that will judge how we rise to this challenge. Our generations cradle babes in our arms, and needs to muster all the talent for parenting we can, to be worth of the child's trust in us.

CHAPTER 2

REFRAMING GLOBAL WARMING: TOWARD A STRATEGIC NATIONAL PLANNING FRAMEWORK

Scott Victor Valentine*

National University of Singapore
469C Bukit Timah Road, Singapore 259772
scott.valentine@nus.edu.sg

Are these the shadows of the things that will be,
or are they shadows of things that may be?

Ebineezer Scrooge in *A Christmas Carol* by Charles Dickens

There is now compelling evidence that in the absence of an expedient and rigorous international effort to abate greenhouse gas emissions, the economic and social costs associated with global warming will be historically unprecedented. Yet despite such evidence, the international community is slow to rally behind a cohesive strategy to mitigate such costs. This paper argues that the prevalence of an overly-narrow conceptualization which treats climate change as a cost to be mitigated is a key barrier to enhanced international cooperation. Initiatives in respect to alternative energy endeavours in Singapore are put forth as examples of how a centrally orchestrated, strategic approach to climate change can uncover hidden economic benefits that are overlooked when the prevalent cost-mitigation perspective is employed by country planners. Extending lessons from the Singapore case study, this paper adapts principles from corporate strategic management to create a national

* Tel. +65-6516-7501 Fax: +65-6775-6471.

strategic planning framework which will help policy makers identify and exploit national core competencies in order to capitalize on opportunities which will arise amidst the inevitable global transition to alternative energy. Reframing the response to climate change in this strategic manner may provide the impetus necessary to encourage more proactive behaviour from leading greenhouse gas emitting states such as the United States and China which are currently resisting pressure to embrace leadership roles in addressing climate change.

1. Introduction

In the immortal Dickens' [1843] tale, *A Christmas Carol*, Ebenezer Scrooge — a bitter, miserly financier — is visited on Christmas Eve by three apparitions. These ghostly visitations force Scrooge to reflect on the adverse impact that his miserly, selfish ways have had on his life and will have on his future. The visitations catalyze a metamorphosis in old Scrooge. Overnight, he transforms from a miserly, sour-puss to a man of unparalleled generosity and goodwill toward others [Dickens (1843)].

The enduring popularity of this tale can be partly attributed to the underlying theme which conveys the hopeful observation that it is not too late to realize the error of one's past ways and to make amends. In short, it is a timeless tale of optimism. However, less optimistically, *A Christmas Carol* also conveys an age-old truth about paradigm change — paradigms do not change without a significant catalyst which causes a person (or a society) to rethink established tenets [Kuhn (1996)].

These two sociological nuances of *A Christmas Carol* combine to form an appropriate anecdote when reflecting upon the threats posed by global warming. It isn't too late to realize the error of our past ways and make amends; however, to do so, a global paradigm shift is required.

The intention of this paper is to contribute to efforts to expedite an international response to abate global warming by creating an impetus to speed up the transition to alternative energy. The paper argues that this can be accomplished by reframing the paradigm through which states view global warming. In turn, this can be facilitated by adopting a strategic approach which emphasizes a search for opportunities that arise as a result of the inevitable changes that will occur in society as a transition to alternative energy takes place.

An inductive approach has been adopted in developing a strategic model for national climate change planning. The paper first introduces a case study which outlines a number of initiatives in which Singapore's government is ostensibly embracing a strategic economic perspective in regard to alternative energy transition and climate change policy. Then, lessons from Singapore's strategic initiatives will be extracted to demonstrate compatibility with a strategic framework which has been borrowed and adapted from a resource-based view of corporate strategic management. Therefore, the main contribution this paper seeks to make is to introduce and validate how a strategic framework can be used by country-level planners to identify and prioritize market opportunities which are emerging as states strive to abate global warming.

2. The Emergence of Consensus on Climate Change

Historical accounts of climate change may refer to the period between the summer of 2006 and the Spring of 2007 as the period during which the world woke up to the significance of the threats posed by climate change.

In the summer of 2006, the release of Al Gore's *An Inconvenient Truth* brought the issues associated with climate change to the general public. Despite controversy over the claims made in the film [See for example Lindzen (2006)], the film has become the third-highest grossing documentary in United States' history.

In October 2006, a comprehensive independent review called the Stern Review commissioned by the Chancellor of the Exchequer in the UK, presented an assessment of the anticipated impacts of climate change. As a foreboding sign of the content which would follow, the report began by describing climate change as "the greatest and widest ranging market failure ever seen" [Stern (2006, p. i)]. The conclusion of the report was that the long-term costs of climate change are expected to be so great, that early action to abate global warming is the most cost-effective alternative. It estimated that the net benefits (benefits less costs) from reducing GHG emissions to achieve a stabilization level of 550 parts per million (ppm) by 2050 would be in the neighborhood of US$2.5 trillion [Stern (2006)].

In February 2007, the first of four reports that comprise the Fourth Assessment Reports of the United Nations Intergovernmental Panel of Climate Change (IPCC) was released. The goal of this first report was to "describe progress in understanding of the human and natural drivers of climate change, observed climate change, climate processes and attribution, and estimates of projected future climate change" [Intergovernmental Panel on Climate Change (2007a, p. 2)]. Overall, the report upgraded the international consensual likelihood of global warming being attributable to human causes from *likely* (66% or greater probability) to *very likely* (90% or greater probability). The data presented in the report was unremarkable in the sense that it mirrored data already available in the public domain; however, the report was significant in that it presented a consensus view of UN member nations. Symbolically, it represented the juncture in which humanity formally accepted culpability in causing climate change.

In April 2007, the second of four reports that comprise the Fourth Assessment Reports of the IPCC was released. This second report focused on "current scientific understanding of impacts of climate change on natural, managed and human systems, the capacity of these systems to adapt and their vulnerability" [Intergovernmental Panel on Climate Change (2007b, p. 1)]. Comparatively, the report was less comprehensive than the Stern Review in its assessment of the current and anticipated economic impacts of global warming on humanity and global ecosystems. However, it did serve to solidify the emergent consensus that climate change was significantly harming hydrological, terrestrial and biological systems [Intergovernmental Panel on Climate Change (2007b)].

Given the escalation in international consensus that climate change is an immediate threat to both the social and economic well-being of humanity, the intuitive international response should be to put vested interests aside, hoist the sails of initiative and embark on rigorous greenhouse gas (GHG) abatement programs. However, such departures have not materialized. In fact, one is tempted to glibly question whether members of the international policy community have misconstrued the Stern Review's admonition — "delay in taking action on climate change would make it necessary to accept both more climate change and, eventually, higher mitigation costs" [Stern (2006, p. xv)] — as a policy recommendation.

Ideologically, a divide has emerged between a *Kyoto Camp* — Kyoto Protocol Annex B countries which support absolute reductions in GHG emissions — and an *Emission Intensity Reduction Camp* — countries such as the US, Australia and Singapore which support initiatives to reduce GHG emissions measured as a percentage of GDP. The ideological divide has arisen because countries in the Emission Intensity Reduction camp contend that initiatives to achieve the absolute GHG reductions implied under the Kyoto Protocol would adversely influence their respective economies [Institute for Global Environmental Studies (2005)].

Unfortunately, cohesion within the Kyoto camp also seems to be weakening. At the end of 2005, the Institute for Global Environmental Studies (IGES) released the results of regional consultations with government representatives in Asia which indicated that some Kyoto Protocol non-Annex B countries such as China, India, South Korea and Indonesia would be extremely reluctant to accept GHG emission targets post-2012 [Institute for Global Environmental Studies (2005)]. Furthermore, Canada which is currently in the Kyoto camp as an Annex B country has recently given indications of a shift in policy which favours ideology espoused by the Emission Intensity Reduction camp [Environment News Service (2007)].

The end result is that the Kyoto Protocol is showing signs of collapse. On May 11, 2007, the 15th annual meeting of the UN Commission on Sustainable Development concluded without reaching a consensus on a document summarizing common ground on energy and climate change. The level of international divisiveness demonstrated by this failure raises doubt over the success of international climate negotiations scheduled for Bali in December 2007 [Environment News Service (2007)]. To complicate matters, inter-country differences of opinion over how to approach GHG emissions abatement are now giving rise to calls for post-2012 agreements that would supplant the Kyoto Protocol [Knell (2007)].

For those who accept the validity of the scientific evidence presented in the Stern Review, the growing number of countries migrating to the Emission Intensity Reduction camp is worrisome. According to the Stern Review, the planet's atmosphere can assimilate 5 Gigatons of carbon-equivalent GHG emissions [Stern (2006)]. Anything over that amount leads to a build-up of GHG in the atmosphere and, thereby promotes further global warming. Stern [2006] estimates that humanity is currently emitting 25 Gigatons of carbon-equivalent GHG emissions into the atmosphere each year. Therefore, the challenge is clear — GHG emissions must be curtailed by at least 80%. It is highly unlikely that an emission intensity reduction strategy will catalyze such deep reductions when overall energy consumption is expected to increase by 71% between 2003 and 2030 [Energy Information Administration (2006)]. Clearly then, a more effective strategy is needed to (1) persuade the countries that are currently in the Emissions Intensity Reduction camp to agree to absolute abatement targets, (2) persuade Kyoto Protocol non-Annex B countries that do not currently have emission reduction targets to accept reduction targets and (3) motivate Kyoto Protocol Annex B countries that are currently bound by voluntary targets to accept deeper post-2012 reductions when the second reduction period is slated to commence.

Problematically, persuading sovereign states to play a role in achieving the three goals outlined in the previous paragraph is ostensibly hindered by one prominent barrier — concerns over the adverse affects that a forced transition to alternative fuels sources will have on a country's capacity to support economic development. Facilitating a transition to alternative energy requires time-consuming, costly changes to a country's energy provision infrastructure [See for example Johns (2006)]. Power plants would need to be upgraded or replaced by alternative energy generation sources, transportation sector reform would be necessary (vehicles would need to be modified and infrastructure put in place to provide accessibility to clean fuel sources) and policy measures to improve energy efficiency (i.e. regulations, taxes, emission caps etc.) would have to be fastidiously implemented. All of these initiatives are expensive undertakings that for the most part necessitate the diversion of resources that could be used for other purposes.

Resistance to voluntary GHG reduction targets is not surprising when viewed from the contextual understanding that politics is characterized by short-term preservation of economic national self-interest [Ball (2007)]. Predictably, when negotiation of emission reduction targets take place, the discussions inevitably digress into arguments over which countries should bear which level of responsibility in abating GHG emissions. Critics may be ethically justified in condemning politicians for approaching climate change negotiations from such a nationalistic, myopic perspective. However, if improving international collaborative efforts in addressing global warming is the goal, we must move beyond criticism and begin to consider strategic approaches which will minimize the impediments posed by such national self-interests.

It is hoped, and in many cases assumed, that eventually the international community will collaboratively agree on GHG reduction targets that will bring GHG emissions back to a level of atmospheric assimilative capacity. After all, the international response to the ozone layer degradation problem in the 1980s demonstrated that when humanity has its collective back against a wall, cooperation is indeed possible. However, the history of climate change negotiations has demonstrated a significant divide exists between good intentions and concrete commitments to action [Eizenstat (1998)]. Entrenched commercial interests and high costs associated with alternative energy transition make it likely that collective international action will materialise slower than is desirable.

It is highly probable that without new incentives to hasten international collaboration, the sluggish international response will exacerbate the ill-effects that will unquestionably intensify as the planet continues to warm. Regrettably, the parties most affected will be the parties most ill-equipped to deal with the emergent problems.

The poor will suffer disproportionately [Miller (2004)]. Furthermore, it is anticipated that a 2% increase in temperatures could place an additional 15–40% of our animal species at risk of extinction [Stern (2006)].

Given the restorative promise associated with encouraging the planet's largest GHG emitters to adopt improved abatement practices, operationalising approaches to incentivize such behavior is desirable. One possible approach is to demonstrate how the climate change challenge can be reframed in a strategic manner in order to allow countries to benefit from emergent change. Facilitating a transition to alternative energy need not be narrowly viewed by countries as a burden and a cost. There are a host of opportunities that arise in conjunction with the challenge of facilitating a transition to alternative energy.

3. Climate Change as an Opportunity?

Intuitively, an assertion that climate change can give rise to positive national economic benefits may seem naively optimistic. As outlined earlier, the vast majority of public policy research in regard to climate change centres around comparing the costs of abating global warming to the costs associated with the impact of global warming should business continue as usual. Under the current ideological perspective, the challenge is ostensibly to identify and embrace the path which gives rise to the lowest economic costs to humanity.

Interestingly, a decade ago, a similar mindset exemplified corporate thinking in regard to environmental management issues. Pollution and waste management programs were widely perceived as unavoidable costs of doing business. Few companies envisaged a role for environmental management in strategic planning. Few companies, that is, until an authority in strategic management made the suggestion that ignoring the strategic value of environmental initiatives may actually represent diminished competitive capacity. Revisiting the foundations of this strategic perspective of corporate environmental management can help to shed insight into how to reframe the climate change dilemma from a cost-based paradigm to a paradigm centred on the search for national competitive opportunity.

3.1. *Climate Change as an Opportunity*

In 1995, Michael Porter and Claas van der Linde published a paper entitled *Ending the Green Stalemate*, in which they argued that the traditional adversarial perspective from which companies approached environmental management issues was misguided. Their paper challenged this entrenched mentality by presenting evidence that proactive strategic treatment of environmental management issues could in fact enhance competitive advantage [Porter and Van der Linde (1995)].

In particular, Porter and van der Linde put forth two axioms that merit attention in the context of this paper. The first axiom is that pollution and waste are more than merely disposal burdens, they are also indicative of inefficient processes; and therefore, represent wasted resources. Applying this axiom to the climate change challenge, the implication is that the tons of CO_2 emitted from automobiles and smokestacks around the world are symbolic of inefficient use of fossil fuel resources. Thus, improving energy efficiency can potentially benefit firms or countries that undertake such initiatives because resources conserved are economically as valuable as paying to replenish resources.

The second axiom is that environmental regulations which have traditionally been seen as threats to corporations can also be exploited to gain competitive advantage. According to Porter and van der Linde [1995], regulations act as barriers to entry which prohibit inefficient firms from poaching customers through price-undercutting strategies. Regulations also force firms to innovate to improve resource usage. Since all firms must adhere to the same regulations, the firm that learns how to meet regulations most cost effectively will gain competitive advantage [Porter and Van der Linde (1995)]. Applying this axiom to the climate change challenge, the inference is that national regulatory restrictions on carbon emissions could force domestic firms to improve their products and production methods and thus, gain long-term competitive advantage over foreign firms that do not operate under such stringent regulations.

Although Porter and van der Linde's work was far from the first work to consider environmental management from a strategic perspective [See for example Schmidheiny (1992)], it was highly influential because of Michael Porter's reputation in the strategic management community. In fact, to the misfortune of Claas van der Linde, the gist of the paper became known simply as *Porter's Hypothesis.* On the heels of interest from the corporate community catalyzed by the publication of *Porter's Hypothesis,* a number of studies emerged to explore numerous aspects of strategic corporate environmental management [See for example Graedel and Allenby (2001); Hawken, Lovins and Lovins (1999); McDonough and Braumgart (1998)]. The overall contribution that these works made to improving corporate environmental governance is that they altered the conceptual lens through which business strategists approached environmental management. No longer were environmental issues viewed as merely additional costs to the firm; environmental management had become a strategic tool to generate new revenue streams and reduce costs.

Returning to the theme of climate change, one cannot help but be struck by the similarity between the adversarial manner in which the corporate world viewed environmental management in the 1970s and 1980s and the manner in which many economic players (governments, industry, economists) now view the climate change issue. The consensus is that it is *less costly* to incur the costs of abating GHG emissions now than to be faced with the spectre of huge economic and ecological costs in the future if global warming remains unabated [Intergovernmental Panel on Climate Change (2007a); Stern (2006)]. Through the current conceptual lens, global warming is a cost not an opportunity.

That is not to say that there has been a complete absence of research into opportunities that arise due to global warming. Many authors have written about the benefits of leveraging carbon credits under the Kyoto Protocol's Clean Development Mechanism (CDM) and the EU's Emissions Trading Scheme to improve the viability and profitability of alternative energy projects [See for example Glenn and Beech (2002); Sorrell and Sijm (2003)]. Furthermore, the popular press is awash with daily reports on opportunities which are emerging in the alternative energy sector. Even the Stern Review included a chapter (Chapter 12) on business opportunities and concluded that business opportunities exist across a "wide range of industries and services" which could be "worth at least US$500 billion by 2050" [Stern (2006, p. xvi)].

Unfortunately, the ongoing debate over apportioning the costs associated with global warming mitigation dilutes existing efforts to focus on strategically exploiting the emergent economic opportunities as humanity undertakes the transition to alternative energy. Given the final repercussions of a transition to alternative energy, there are remarkably few examples of countries taking a proactive approach to strategically capture national leadership in market niches associated with this transition. Instead, the overwhelming focus of international attention is ostensibly on defending national self-interests in terms of GHG abatement commitments. Organizational learning expert, Peter Senge [2006, pp. 397–398] labels this type of systematic failure a "tragedy of the commons archetype". He attributes this failure to an inability to break free from constrictive mindsets which inhibit holistic understanding of systematic effects arising from change. In short, humanity's preoccupation with estimating the adverse economic impact of climate change has precluded comprehensive consideration of economic benefits associated with the requisite shift to alternative energy sources.

The Stern Review provides as an apt illustration of how a focus on costs can subordinate the search for opportunities. It comprehensively demonstrates that the cost of abating the worst effects of global warming are estimated to be about 1% of GDP by 2050. Conversely, by not undertaking such abatement investments, global GDP may contract by 10–20% as a result of global warming [Stern (2006)]. Mention of generic business opportunities are not made until twelve chapters into the report.

Would a strategic approach which explores all costs and benefits related to climate change provide more incentive for a recalcitrant country to support more expedient efforts in facilitating a transition to alternative energy? This paper supports an affirmative perspective. However, as will be emphasized, in order to maximize the benefits from adopting a strategic perspective, a high level of political integration is needed to ensure that strategic footholds can be defended over the long-run.

In the next section, a case study of Singapore is presented firstly, to demonstrate the types of lucrative opportunities that lie on the fringe of the climate change challenge and secondly, to provide insight into how political integration supports development that is synergic with national competencies.

4. Why Singapore?

On paper, Singapore appears to have little to gain on the supply side from a transition to alternative energy. Aside from hobby farms and a few specialized agri-business ventures, agriculture in Singapore is virtually non-existent. In contrast with large countries such as the United States and China, Singapore cannot derive substantive benefits from growing biofuel crops. Moreover, the combination of lack of land in Singapore and relatively high costs of labour place Singapore at a disadvantage when competing as a site for locating major alternative energy system manufacturing plants.

On the demand side, the following passage from its *Initial National Communication* to the UNFCCC in 2000 highlights the logistical dilemma posed by alternative energy transition:

> *To date, there are no renewable energy sources that Singapore can harness to reduce its reliance on fossil fuels. It does not have hydro and geothermal energy sources. Nuclear energy cannot be used due to geopolitical and safety concerns. Calm seas and lack of sea space mean that it cannot tap wave energy. Neither does it have the land to cultivate biomass as an energy source. Wind energy has limited potential due to lack of land to establish wind farms and the relatively low wind speed... In the absence of alternative energy sources, Singapore has to rely mainly on fossil fuels for primary activities such as power generation*

and transportation. The only means available to Singapore to mitigate GHG emissions is to use energy as efficiently as is practically possible. [Singapore Ministry of the Environment and Water Resources (2000, pp. 19–20)]

Even in terms of securing carbon credits for abatement of GHG due to energy efficiency improvements, Singapore is at a disadvantage as a developed country. For example, 74% of Singapore's primary electricity comes from relatively clean burning natural gas [Singapore Ministry of Environment and Water Resources (2006)]. Similarly, in the transport sector, automobile ownership is tightly regulated and heavily taxed. Consequently, the limited numbers of cars that are on the road are newer fuel efficient models. Therefore, for Singapore to benefit from alternative energy transition, it will have to respond strategically — there are no fast profits to be derived from exploiting natural resource advantages or from replacing coal fired power plants to earn carbon credits. In short, perhaps more so than any other country in the world, Singapore truly does need to adopt a strategic approach if it wishes to benefit from alternative energy transition.

5. Economic and Political Realties in Singapore

Singapore, with a land area of just under 700 square kilometres, has been dubbed the "little red dot" in reference to its indicative marking on world maps [Koh and Chang (2005)]. It is an underdog state that is severely constrained by a lack of natural resources. However, despite a deficiency in natural resources, the country has risen to economic prominence thanks in part to a highly efficient, tightly integrated government bureaucracy and a highly competitive workforce. The 4.5 million people that call Singapore home had an average per capita GDP (PPP) of US$31,400[a] in 2006.

Domestically, economic prosperity takes precedence because a thriving economy is seen as the engine to support progress in all other initiatives. As Prime Minister Lee Hsien Loong pointed out during

[a] *Source*: CIA World Factbook: www.cia.gov

his National Rally Day speech on August 22, 2006, "the economy...is a precondition for everything we want to do" [Lee (2006)]. The government's commitment to managing national competitiveness in a methodological manner is exemplified by the engagement of national competitiveness expert, Michael Porter to provide counsel to the government. Moreover, it has delegated responsibility for facilitating research and discourse with stakeholders in this regard to the Asia Competitiveness Institute at the National University of Singapore. As a consequence of this mandate to preserve economic growth prospects, all policies which undermine Singapore's ability to compete on an international stage are subordinated and diluted to strike a balance that avoids derailing economic growth [Lye (2002)].

Singapore's governance approach has been referred to as "father-leadership" in reference to the tightly engaged approach the government takes to planning national development [Cheng and Low (2006)]. The government has clear ambitions to facilitate national competitive advantage for Singapore in high-tech manufacturing, knowledge-based industries, and service sector ventures [Singapore Economic Development Board (2006)]. In national planning, it has turned Singapore's geographic limitations into an advantage by ensuring that growth industries are well-supported and interconnected to maximise synergies. Well-focused subsidies, tax breaks, training schemes, and governmental financial support are all mechanisms used in Singapore to encourage and support growth in strategically targeted sectors [Singapore Economic Development Board (2006)].

Regarding the global transition to alternative energy, despite Singapore's supply and demand side limitations, there are signs that the government is actively encouraging Singaporean firms to move strategically toward assuming leadership roles in earmarked sectors. This coordinated effort is allowing national competitive strengths to dictate the emergence of new markets. The inherent benefit is that by leveraging national strengths to establish market footholds, the first-mover advantages gained as Singaporean firms establish leadership in emerging market niches may also be defendable [Grant (2005)].

6. Four National Strengths Related to Alternative Energy

In order to demonstrate the coordination between business and government in alternative energy sector initiatives, four national strengths and specific benefits related to these strengths are introduced below. Following a brief elaboration of these strengths, three emerging market niches related to alternative energy will be introduced to demonstrate how Singaporean firms with government assistance are exploiting these national strengths to gain first-mover advantages in these budding market niches.

(1) **Geographic Advantage:** Singapore is centrally located amidst a region where CDM project activity is burgeoning. Malaysia, Indonesia, China, Thailand, Vietnam and the Philippines are all major markets for CDM projects. Political and business ties between Singapore and these countries are extensive predominantly thanks to an integrated social web which links ethnic Chinese business interests [Lasserre and Schutte (2006)]. In East Asia where social relationships underpin business relationships, these existing social webs provide Singaporean firms with a competitive advantage over non-Asian firms.

(2) **Human Capital Advantages:** Singapore boasts a sophisticated education system which helps cultivate highly competent engineers, scientists and financial experts [Singapore Economic Development Board (2006)]. Moreover, Singaporeans are blessed with linguistic advantages. The vast majority of Singaporeans are educated in English. These combined traits produce an alluring talent pool for multinational alternative energy firms. In addition to English, the majority of Singaporeans also possess proficiency in their ethnic languages (predominantly Hokkien, Malay, Mandarin or Tamil). Thus, the cultural and linguistic diversity in Singapore implies that Singapore's workforce is capable of providing a bridging function for engaging with the culturally diverse markets around the ASEAN region [Lasserre and Schutte (2006)].

(3) **Infrastructure Advantages:** A safe, hospitable living environment, efficient commercial and residential support services,

sophisticated transport networks and modern health and education facilities enhance the allure of Singapore as a regional hub for many multinational alternative energy firms [Lasserre and Schutte (2006)].

(4) **Strategic Planning Advantage:** The Singapore government has been critically described as a "soft authoritarian regime" [Zakaria (1994)]. However, in terms of national economic planning, the cohesiveness with which Singapore is governed is arguably a national competitive advantage. Even critics of the system reluctantly acknowledge that national planning in Singapore can be more efficiently operationalised thanks to political and bureaucratic cohesiveness [Leong (2000)]. Hence, initiatives designed to capitalize on a global transition to alternative energy can potentially be more effectively coordinated when compared with other countries where bureaucratic disharmony leads to implementation delays and inefficiencies.

7. Leveraging National Strengths for Market Development

Opportunistically, with government inducements and support, domestic and foreign firms in Singapore have established strategic footholds in market niches that are synergistically supported by Singapore's national strengths. Three of the more significant niches which exploit the strengths identified earlier include alternative energy technology design and development, conference services and financial services.

(1) Singapore as a Centre for Alternative Energy Technology

One facet of Singapore's evolving economic growth strategy is to nurture the development of high-tech industry [Singapore Economic Development Board (2006)]. The government has been aggressive in enticing high-tech industry to relocate to Singapore. In 2005, high-tech business start-ups and market entries in Singapore reached an all-time high of 3908 firms [Singapore Economic Development Board (2006)].

In terms of attracting alternative energy enterprises, successes to date include Singapore's first solar photovoltaic cell manufacturing plant which was set up by Solar Energy Power in January 2006, commitments in early 2007 for a $500 million regional research facility being planned by wind energy giant, Vestas and a 200 000 tonne biodiesel plant commissioned by Peter Cremer of Germany [Iswaran (2006)]. These developments help distinguish Singapore as an emerging regional hub for high-tech alternative energy research.

A noteworthy justification for Singapore's national interest in fortifying a position in alternative energy is to provide a hedge against risk. A number of leading oil firms — Shell, Chevron, Exxon — have regional operations coordinated from Singapore. Furthermore, Singapore is a world leader in petroleum refining and oil bunkering (Lye, 2002). Accordingly, there is a substantial degree of economic reliance on foreign oil interests maintaining a presence in Singapore. Nurturing businesses operating in alternative forms of energy is a strategic hedge against downturns in the oil industry.

(2) Singapore as a Conference Hub

Singapore's conference industry is the largest in Asia and Singapore ranks within the top ten convention cities in the world with over 4000 events and over 400 000 foreign participants every year [Lew (1999)]. Recently, plans for two integrated resorts have been announced that will bolster the country's already impressive conference capabilities. Accordingly, the Singapore Exhibition and Convention Bureau (SECB), a government body, has set aggressive annual growth targets exceeding 10% per year [Singapore Tourism Board, (2006)].

Convention activity associated with climate change and alternative energy initiatives could contribute significantly to achieving the ambitious growth targets of the SECB. Aside from official UN Framework Convention on Climate Change (UNFCCC) meetings (of which there were two in Jakarta in 2006), there are a number of annual regional conferences related to the Kyoto Protocol. When one considers that in 2006 there were large regional events related to the Kyoto Protocol in Indonesia, Malaysia, China, the Philippines and

Thailand, the potential benefits to the Singapore conference industry of adopting a more proactive national approach to climate change and alternative energy transition are self-evident.

Singapore is already actively engaged in pursuit of conferences related to alternative energy, carbon credit trading and finance. For example in the latter half of 2007 alone, major Singapore conferences relating to these three areas include Sustainable Energy Asia Congress (June), Renewable Energy Finance Asia (June), Energy Efficiency Management Conference (July), ASEAN Energy Business Forum (August), Energy Risk and Trading Asia (September), Wind and Solar Asia (October), Clean Energy Asia (October), Ethanol and Biofuels Asia (October), Carbon Finance Asia (October), Carbon Forum Asia (November), and Energex (November).

(3) Singapore as a Hub for the CDM

The financial services sector is a strategically important sector in Singapore because financial services supported large scale commercial activity. Singapore considers itself to be competing with Shanghai and Hong Kong to be the dominant hub for finance in Asia. As such, Singapore development authorities proactively seek out initiatives that can give the country a competitive edge over these rival financial centers [Monetary Authority of Singapore (2006)].

Financial transactions related to the Kyoto Protocol's Clean Development Mechanism (CDM) are accelerating in Singapore and lending support to the drive to become the dominant financial hub in Asia. For example, on the CDM project financing front, prior to Singapore's accession in July 2006 to the Kyoto Protocol, Singaporean investment firms had to work harder to make inroads into the CDM project sector because government contacts help facilitate projects between sponsor and host countries. Singapore's accession to the Kyoto Protocol has invigorated the investment landscape concerning CDM projects. It draws Singapore and the country's financial institutions firmly into the CDM project development foray and it allows the Ministry of Trade and Industry to begin working on strategies to position Singapore as a market of first recourse for global CDM project developers. In another development, Singapore is now home to the only carbon emissions trading exchange in Asia — Asia Carbon. In April 2007, it conducted its first successful auction in which €7.2 million of carbon credits were traded.

8. Questions, Suppositions and Lessons

The three market niches identified above intermesh effectively with Singapore's overall national competitiveness strategy (Singapore Economic Development Board, 2006). To what degree has the commercial activity in these market niches been centrally planned and to what degree are these market niches simply natural responses to existing economic development undercurrents that exist in Singapore? Without inside knowledge which is seldom possible in Singapore, it is difficult to surmise. However, it is clear that intentional or not, the emergent clean energy initiatives in Singapore have integrated smoothly with Singapore's national development strategy.

To what degree did the prospect of facilitating such initiatives influence Singapore's decision to accede to the Kyoto Protocol? Again, it is hard to ascertain without insider knowledge. However, to imply that the Singapore government is not aware of these benefits greatly under-estimates the integration and sophistication of Singapore's governance and economic efforts. Michael Porter — who as mentioned earlier is a strategic advisor to the Singapore government — is a staunch advocate of building national competitive strategy around national core competencies [Porter (1998a)]. It is entirely plausible that Porter's principles have influenced the Singapore government's approach to alternative energy sector development.

There are two valuable lessons to be learned from Singapore's experience to date in responding to a transition to alternative energy. First, even in the absence of natural resource advantages, competitive niches can be identified and exploited. Second, in competitive situations, national strengths must dictate which emergent opportunities governments should support. These two lessons give rise to a useful admonition for country-level planners in other countries: *Integrating an alternative energy strategy into a national development plan can produce seamless economic benefits provided the country planners prioritise emergent opportunities which are best supported by existing and / or cultivatable national strengths.*

Although it must be acknowledged that Singapore's compact size permits the government to be more effective in achieving integrated development of industry, this should not be misconstrued to imply that larger countries cannot also achieve similar or greater benefits from applying similar strategic principles. Integrative efficiencies that larger countries lose due to broader scale and scope are counterbalanced by enhanced opportunities that arise due to broader scale and scope.

In the next section, insights from the Singapore case study will be merged with corporate strategic management theory to develop a national strategic planning framework that can be employed to best exploit national strengths and thereby, optimize opportunities which are emerging amidst the global transition to alternative energy.

9. A Corporate Perspective on Exploiting Strengths

To management scholars, the manner in which alternative energy market niches have evolved in Singapore to complement national strengths is not an unusual development. Exploiting market opportunities that are synergic with the core competencies of a given firm is a key tenet of the resource-based view (RBV) of strategic management [Barney (1995)]. In industry, the rationale supporting the RBV perspective is that a firm should first assess what it is good at before deciding where it can best compete. An example of the RBV approach applied in practice is the manner in which Honda has leverage excellence in motorcycle engine manufacturing to enter a number of markets which utilize gasoline-powered engines (lawnmowers, outboard motors, automobiles etc.). The corporate strength that made it possible for Honda to make successful inroads into these new markets is world-class proficiency in the engineering and design of engines.

A second key tenet of the RBV perspective is that firms can further insulate themselves from rival attacks by entering market niches in which a number of corporate strengths can be integrated to produce less assailable competitive offerings [Barney (1995)]. The rationale behind this tenet is that over-reliance on one strength may allow competitors to target their efforts at usurping that one strength. Conversely, by leveraging a host of core strengths to add value to a product or service, the end-product becomes much more difficult to replicate. 3M for example boasts a product list of over 30 000 separate products, many of which are product category leaders. Superiority in key technologies related to film and adhesive coatings which are then bolstered by a number of competitively superior systems developed to encourage innovation, engineering excellence and comprehensive market saturation is the key to 3M's success [Grant (2005)]. For example, although many competitors now make products which rival 3M's Post-It notes, none have achieved the level of market penetration that 3M has established by combining and integrating these strengths.

A third key tenet of the RBV perspective is that firms must nurture competencies which are: (1) sustainable in the face of competitive advances, (2) vital components of features or services that are important to the customer, and (3) transferable to future applications. For example, Intel leads competitors in R&D investment because by doing so it can produce computer processors that are superior to competitive offerings (sustainable), fast and reliable (important to the customer), and viable in more applications (transferable).

In summary, the essence of the RBV perspective of strategic management is that the capabilities and resources of a firm (or more broadly of a country in the adaptation that will follow) need to be analyzed and matched to available market opportunities in order for a firm (or country) to achieve lasting competitive (or national) advantage. Applying the RBV perspective to national development strategies can help countries achieve national competitive advantage in the same manner that firms for years have been using the RBV approach to gain competitive advantage within an industry context. Figure 1 demonstrates the cognitive stages through which country

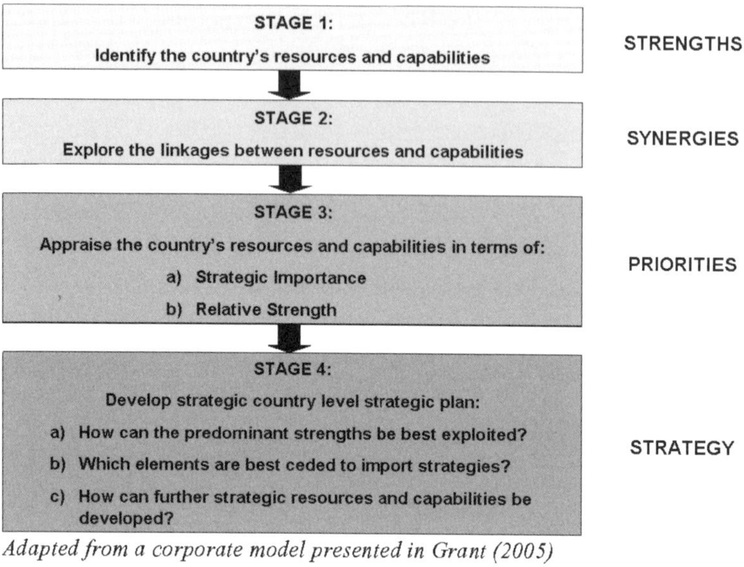

Fig. 1. Country-level RBV framework.

Adapted from a corporate model presented in Grant (2005)

planners need to proceed in order to apply the RBV approach to national planning.

10. Demonstrating Relevance

To describe how such a framework can be applied to national strategy formulation in alternative energy planning, the four stages outlined in Fig. 1 will be elaborated upon in the context of alternative energy.

Stage 1 — Identify Strengths: The Heckscher-Ohlin Theory posits that a country will benefit the most from free trade by exporting goods and services that it can produce most efficiently and importing goods and services that it produces inefficiently [Perkins, Radelet and Lindauer (2006)]. Hence, the first step in developing a national strategy for alternative energy should lie in assessing the capabilities and resources in the country that could potentially be exploited for national advantage. For example, Brazil is a major producer of sugarcane which is a source crop for the production of ethanol. Accordingly, to exploit this national strength, the national oil company, PetroBras has been charged with the task of building a US$1 billion transportation network to complement its production network and facilitate effective distribution of ethanol to world markets [Ngo (May 28, 2007)].

As outlined earlier in the Singapore case, Singapore's highly educated workforce is the strength which supports entry into the alternative energy technology development niche, its mature financial sector provides the core competency necessary to achieve success in entry to the CDM project financing niche and its sophisticated conference infrastructure facilitates an effective foray into event management for climate change policy and alternative energy conferences.

Stage 2 — Identify Synergies: In strategic management, the term "product integrity" has been used to describe the process of matching and integrating resources and capabilities in order to create competitively superior product or service packages [Clark and Fujimoto (1990)]. As the 3M example illustrated, competitors may be able to replicate one feature; however, by bundling strengths together, a firm can create total product packages that are competitively difficult to replicate.

This concept of product integrity can be extended to national initiatives to gain and defend competitive advantage in alternative energy market segments. For example, there are a number of countries that produce palm oil (Indonesia, Malaysia, Philippines etc.). However, of these countries, Malaysia has arguably done the best job of supporting the development of forward and backward linkages in order to ensure that the cost of producing Malaysian palm oil remains competitively lower. In Malaysia, government sponsored research programs support the development of agricultural techniques to maximize palm oil yields. Furthermore, government investment in transportation infrastructure for rural communities allows the product to be expediently delivered to market [Perkins, Radelet and Lindauer (2006)].

In competition between countries, the countries that develop competitive advantage in any given market niche will be the countries that can entice and nurture firms to deliver competitively superior value to the customer. Hence, national planning of alternative energy competence should focus on niches where national core competencies can deliver superior value. Once such niches are identified, national planners should seek to bolster competitive advantage by providing infrastructure, R&D and human capital support to firms to encourage clusters of excellence to develop [Porter (1998b)].

In the Singapore case study, the emergent market niches in alternative energy reflect a confluence of strengths which are fortified through government support. For example, consider the alternative energy research facilities that are being established in Singapore. They benefit from Singapore's talented work force which enhance and support R&D activities. Such talent is nurtured by elaborate training schemes developed by the Singapore government. The research facilities are also attracted to Singapore thanks to government inducements to relocate to Singapore. Furthermore, these facilities often find homes in one of Singapore's high-tech industrial parks which have been customized to appeal to such businesses.

The same pattern holds true for the emergent alternative energy conference niche. Singapore is an attractive venue for such activities because it has perhaps the most efficient airport in the world [Finch (2006)], liberal visa regulations for visitors, world class hotels, efficient transportation services and sophisticated communications infrastructure. In short, it is not one strength that allows Singapore to succeed in these initiatives; rather, successful niche markets emerge in areas where numerous national strengths are combined and continually fortified by government support to produce synergic relationships which are difficult for other countries to replicate.

Stage 3 — Set Priorities: There is clear government incentive to try and support corporate success because profitable businesses pay corporate taxes and ensure continued employment. In order to attract and nurture the development of competitively superior firms, support elements such as access to financial support, technical guidance, R&D

subsidies and suitable infrastructure influence corporate location decisions [Porter (1998a)]. Accordingly, a degree of prioritization by public authorities is necessary to ensure that firms operating domestically (homegrown and transplanted) have access to the support features necessary to succeed when competing against foreign-based firms that are often subsidized in their base countries. For example, although hundreds of countries around the world have access to the ocean, Scotland has established itself as a centre for excellence in wave energy technology research and production thanks to government support for R&D in this field (£50 million government grant) and relocation subsidies for attracting new firms [Jeffries (2005); Reina (2005)].

When prioritizing support for alternative energy industries, governments must ensure that attention is given to both (1) an assessment of the relative strengths of the capability/ resource mix (in order to determine strength of competitive advantage) and (2) an evaluation of the strategic importance of the supported industry (to ensure that resources committed in support of each industry maximise synergic strengths).

In Singapore's case, the priorities are clear: establish clusters of high-tech R&D and production, enable Singapore's formidable financial services industry to aggressively engage in alternative energy projects and entice conference planners to choose Singapore for energy conferences. Like Scotland, Singapore's government could have elected to encourage the development of a wave energy technology industry; however, such initiatives conflict with national shipping interests. There is not enough open space in Singaporean waters to facilitate and commercially support the robust development of such an industry. Accordingly, such opportunities are de-emphasized in favour of initiatives which match national competitive aspirations more effectively.

Stage 4 — Develop a Country-Level Plan: Country-level plans send signals to the market regarding which investments will be most welcome in a country and which market niches will be strategically supported in the future. By sending such signals, the government encourages entrepreneurs to respond through investment. In turn, investment inflows catalyze the creation of forward and backward linkages which enhance the synergistic impact and competitiveness of industry clusters.

Questions that should be asked when preparing a country-level plan include:

(a) *How can the country's predominant strengths be best exploited?* Answers to this question will define the level of government involvement needed to catalyze growth, the extent of financial support that will be required to entice new start-ups and foreign direct investment and the type of infrastructure that must be supplied to allow businesses to succeed in the targeted strategic areas.
(b) *Which elements are best ceded to import strategies?* Some market niches are not economically viable in certain countries. For example, in land-starved Singapore, processing biofuel may be an option but growing biofuel crops would not be competitively feasible. By only producing the products and services that a country has current competitive advantages in, countries can establish clusters of regional or international competitive superiority [Porter (1998a)].
(c) *How can further strategic resources and capabilities be developed?* A key challenge of strategic country-level planning is to anticipate future trends and to encourage development of national competencies which will place the country in a position to take advantage of strategically significant opportunities as they emerge [Grant (2005)]. The key criteria for each country should be to ensure a degree of national strategic stretch to ensure the country is developing progressive competencies that will equip the country to compete into the future [Hamel & Prahalad (1993)].

In Singapore's case, although Singapore's approach to governance is less inclusive of public participation in strategic decisions, the government signals strategic intent through consultation exercises. For example, the National Climate Change Strategy Consultation

Paper can be considered to be such a document. The paper is available on the National Environmental Agency web-site (http://www.mewr.gov.sg/nccs/index.htm) despite the fact that the consultation period has ended. Conceptually, one of the purposes this public communication serves is to convey to the corporate world what market initiatives the Singapore government plans to support and where emergent opportunities may appear. On the other hand, by communicating intentions in this low-profile manner, the government keeps some of the salient information out of the public domain and thus, shields sensitive information from foreign competitors.

11. Concluding Thoughts

The adapted RBV framework is by no means the only strategic planning tool for guiding national planning in exploiting a transition to alternative energy; however, the four stages do represent fundamentally important elements of any strategic planning process — identify strengths, search for synergies to build unassailable competitive advantage, prioritize initiatives to maximize the impact during implementation and create a detailed plan of attack so all prospective stakeholders know what to expect. This paper has demonstrated how these elements mirror Singapore's strategic response to climate change. By following a strategic path, Singapore has managed to reframe climate change issues — the cost-side threats are now supplemented by revenue-side opportunities.

Lauding Singapore's unique strategic approach to planning for energy transition by no means implies that Singapore's declared emission intensity strategy for controlling GHG emissions is an acceptable international contribution to abating global warming. To the contrary, given the critical state of affairs governing climate change, all countries in the *Emission Intensity Reduction camp* are derelict in their global responsibilities because without absolute reductions, the adverse effects of global warming will intensify [Stern (2006)]. However, this criticism of Singapore's national climate change strategy (and that of all other countries in the *Emission Intensity Reduction camp*), should not obscure the promising approach that the country is taking to play a proactive role in the *business* of facilitating a transition to alternative energy.

Since climate change is now the largest threat the planet has ever faced, it is reasonable to expect that in spite of international complacency, nations will eventually be forced to take action. Accordingly, in anticipation of this, each country has a clear choice: continue along a reactive path or adopt a proactive strategic approach which will allow the country to develop first-mover advantage in strategic alternative energy market niches.

For pioneering countries, a strategic response to capitalizing on opportunities that arise due to global warming abatement initiatives holds much promise. For example, the US possesses highly sophisticated research capabilities and an abundance of land for biofuel production. It could likely become a global centre for biofuel research which would eventually allow the country to shift from a dependence on imported oil to reliance on home-grown biofuel. As another example, China's workforce is rich in engineering competencies. China also possesses extensive experience in manufacturing the types of components that go into energy technologies such as solar cells and wind turbines. Furthermore, it has access to abundant supplies of steel and sand which are key components to wind and solar industry initiatives. Could these two countries which are the two largest emitters of CO_2 on the planet benefit economically from a global transition to alternative energy? It appears possible if a strategic lens were used to refocus national strategic efforts.

Ironically, both countries are impeded from grasping first-mover advantages in many lucrative alternative energy niches because both countries are reluctant to play a proactive role to expedite the transition to alternative energy out of fear of what this will do to their economies. This exemplifies the postulation that one of the biggest threats faced by humanity in crisis situations is breaking free from historical mind-sets which prevent us from recognizing the opportunities associated with change [Senge (2006)].

The ill-effects that humanity will undoubtedly experience as the planet warms can be likened to the visitation of the first ghost in

A Christmas Carol described at the beginning of this paper. Perhaps the question we should now be asking is: Will one shocking visitation be enough to encourage us to realize the error of our ways and change?

References

Ball, J. (2007). Global Debate Is Just Heating Up [Electronic Version]. The Wall Street Journal, June 2. Retrieved June 12, 2007, from http://online.wsj.com/article/SB118072947602021787.html?mod=googlenews_wsj.

Barney, J.B. (1995). Looking inside for competitive advantage. *The Academy of Management Executive* **9**(4), 49–61.

Cheng, K. and Low, P. (2006). Cultural obstacles in growing entrepreneurship: a study in Singapore. *The Journal of Management Development* **25**(2), 169–182.

Clark, K.B. and Fujimoto, T. (1990). The Power of Product Integrity. *Harvard Business Review, November –December*, 107–118.

Dickens, C. (1843). *A Christmas Carol.* London, UK: Chapman and Hall Publishers.

Eizenstat, S. (1998). Stick with Kyoto: A sound start on global warming. *Foreign Affairs* 77(3), 119–122.

Energy Information Administration. (2006). *International Energy Outlook 2006.* Retrieved December 11, 2006. from www.eia.doe.gov/oiaf/ieo/index.html

Environment News Service. (2007). Sustainable Development Commission Stumbles on Climate Change [Electronic Version], May 14, 2006. Retrieved May 16, 2006 from Factiva.

Finch, E. (2006). Editorial. *Facilities* **24**(3/4), 73.

Glenn, S. and Beech, J. (2002). Emissions trading: Are you ready? *CA Magazine* **135**, 24–29.

Graedel, T. and Allenby, B. (2001). *Industrial Ecology* (2nd ed.). New York, USA: Prentice-Hall Publishers.

Grant, R.M. (2005). *Contemporary Strategy Analysis* (5th ed.). London, UK, Blackwell Publishing.

Hamel, G. and Prahalad, C.K. (1993). Strategy as Stretch and Leverage. *Harvard Business Review*, 75–84.

Hawken, P., Lovins, A. and Lovins, L.H. (1999). *Natural Capitalism: Creating the Next Industrial Revolution*. USA, Little, Brown and Company.

Institute for Global Environmental Studies. (2005). *Asian Perspectives on Climate Change Regime Beyond 2012, Concerns, Interests and Priorities.* Tokyo, Japan.

Intergovernmental Panel on Climate Change (2007a). *Contribution of Working Group I to the 4th Assessment Report of the Intergovernmental Panel on Climate Change: Summary for Policy Makers.* Retrieved May 16, 2007. from http://www.ipcc.ch/WG1_SPM_17Apr07.pdf

Intergovernmental Panel on Climate Change. (2007b). *Contribution of Working Group II to the 4th Assessment Report of the Intergovernmental Panel on Climate Change: Summary for Policy Makers.* Retrieved May 16, 2007. from http://www.ipcc.ch/SPM13apr07.pdf.

Iswaran, S. (2006). Speech by S. Iswaran, Minister of State for Trade and Industry, *Financing Energy Projects in Asia Conference 2006.* Pan Pacific Hotel, Singapore.

Jeffries, E. (2005 October 28). Lost at sea. *Utility Week* **24**, 18–19.

Johns, J. (2006, June 2). Pulling power. *Utility Week* **25**, 19–21.

Knell, S. (2007). Will China and India Take the Bait in Alternative Climate Change Strategy? [Electronic Version]. *Global Insight Daily Analysis.* Retrieved May 16, 2007 from Factiva.

Koh, T. and Chang, L.L. (eds.). (2005). *The Little Red Dot: Reflections by Singapore's Diplomats.* Singapore: Institute of Policy Studies and World Scientific Publishing Co. Ltd.

Kuhn, T. (1996). *The Structure of Scientific Revolutions* (3rd ed.). Chicago, USA: University of Chicago Press.

Lasserre, P. and Schutte, H. (2006). *Strategies for Asia Pacific: Meeting New Challenges* (3rd ed.). New York, USA: Palgrave Macmillan.

Lee, H.L. (20 August 2006). *Transcript Of Prime Minister Lee Hsien Loong's National Day Rally English Speech*, University Cultural Centre, Singapore.

Leong, H.K. (2000). *The Politics of Policy-Making in Singapore.* Oxford, UK, Oxford University Press.

Lew, A. (1999). Where the World Meets: Regionalism and Globalization in Singapore's Convention Industry. *Journal of Convention & Exhibition Management* **1**(4), 17–36.

Lindzen, R. (2006). There is No Consensus on Global Warming [Electronic Version]. *Wall Street Journal*, June 26, 2006. Retrieved April 20, 2007 from Factiva.

Lye, L. H. (2002). Chapter 13: Singapore. In T. Mottershead (Ed.), *Environmental Law and Enforcement in the Asia Pacific-Rim*. Singapore: Sweet and Maxwell Publishers.

McDonough, W. and Braumgart, M. (1998). The Next Industrial Revolution. *The Atlantic Monthly, October 1998,* 82–92.

Miller, G.T. (2004). *Environmental Science: Working with the Earth* (5th ed.). Canada: Brooks Cole Publishers.

Monetary Authority of Singapore. (2006). *Annual Report 2005/2006.* Retrieved June 17, 2007 from http://www.mas.gov.sg

Ngo, P. (28 May 2007). Petrobras to begin building dedicated ethanol pipeline [Electronic Version]. *Ethanol & Biodiesel News*, 19. Retrieved 31 May 2007 from Proquest.

Perkins, D., Radelet, S. and Lindauer, D. (2006). *Economics of Development* (6th ed.). New York, USA: W.W. Norton and Company.

Porter, M.E. (1998a). Clusters and the New Economics of Competition. *Harvard Business Review* **76**(6), 77–90.

Porter, M.E. (1998b). *On Competition*. Boston, USA: Harvard Business School Press.

Porter, M.E. and Van der Linde, C. (1995). Green and Competitive: Ending the Stalemate. *Harvard Business Review*, 120–134.

Reina, P. (9 May 2005). Wind and Tides to be Tapped to Meet Scotland's Goals *ENR* **254**, 16–17.

Schmidheiny, S. (1992). *Changing Course*. Boston, USA: MIT Press.

Senge, P.M. (2006). *The Fifth Discipline: The Art and Practice of the Learning Organization*. New York, USA: Random House Publishing.

Singapore Economic Development Board. (2006). *Economic Development Board Annual Report 2006*. Singapore.

Singapore Ministry of Environment and Water Resources. (2006). *Singapore's National Climate Change Strategy: Consultation Paper*: Government of Singapore.

Singapore Ministry of the Environment and Water Resources. (2000). *Singapore's Initial National Communication: Under the United Nations*

Framework Convention on Climate Change. Retrieved May 18, 2007. from http://www.unfccc.int

Singapore Tourism Board. (2006). *BTMICE Fact Sheet*. Retrieved December 12, 2006. from http://www.visitsingapore.com/publish/stbportal/en/home/mice_home/newsroom0/fact_sheets.html

Sorrell, S. and Sijm, J. (2003). Carbon Trading in the Policy Mix. *Oxford Review of Economic Policy* **19**(3), 420–437.

Stern, N. (2006). *The Stern Review: Report on the Economics of Climate Change*. London, UK: Cabinet Office — HM Treasury.

Zakaria, F. (1994). Culture is Destiny — A Conversation with Lee Kuan Yew. *Foreign Affairs* **73**(2), 109–126.

CHAPTER 3

CLIMATE CHANGE — LIVING IN THE ANTHROPOCENE

Jeff Obbard

Division of Environmental Science & Engineering, Tropical Marine Science Institute, National University of Singapore

The scientific evidence for global climate change is now overwhelming and irrefutable. The skeptics who defy the data are now disappearing faster than most of the world's ice sheets and glaciers. Although climate scientists have recognized the risks of greenhouse gas accumulation in the atmosphere for many decades, there is now a sense of urgency and fear in the international community about the dangerous and imminent ramifications of climate change. In the last decade, the rate and extent of change has accelerated, leading to concern that the global atmosphere may now be in a state of positive feedback, heading towards an irreversible 'tipping point'. Aside from the environmental risks, leading economists have begun to acknowledge the dire threats posed to the global economy. International efforts to mitigate climate change via the imposition of greenhouse gas emissions caps are admirable, but feeble and flawed and do not address the massive socio-economic changes needed to achieve a low-carbon global economy. This chapter will cover some of the key mechanisms of the global ecosystem that lead to natural climate regulation. It will also highlight how anthropogenic activities are destabilizing these regulatory systems and pushing the atmosphere into a state of potentially irreversible, run-away climate change. Although the global community is now focused on capping atmospheric carbon emissions, such a unilateral approach underestimates the inherent risks in the degradation of ecosystems that are essential to planetary atmosphere regulation. Without the wider protection of natural ecosystems, positive feedback mechanisms will lead to a dwarfing of anthropogenic emissions from carbon sinks, and the trespass of environmental 'event horizons'. Now that we have entered the 'period of consequences', the author will examine some of the technological solutions being offered to mitigate climate change. The author will present arguments that such methods are futile without a whole scale revision of current international policy to protect the integrity of global ecosystems that provide the essential ecological services that sustain all life on the planet.

"*Friday, 2 February 2, 2007 may be remembered as the day that the question mark was removed from the question as to whether humans are to blame for climate change*"

said Achim Steiner, Executive Director of the United Nations Environment Program upon the release of the first of four reports for the 2007 Fourth Assessment by the Intergovernmental Panel on Climate Change (IPCC).[1]

The IPCC has classified the likelihood of human activities being the principal cause of atmospheric warming as "very likely", with a probability of over 90%. Some scientists wanted to classify the probability as "virtually certain" at 99%, but a few developing countries played down the risk — concerned that caps on carbon dioxide emissions would stymie their growing economies. Regardless, the higher level of certainty in the latest IPCC report — up from a 60% "likely" risk given in the 2001 report is poignant. The first 2007 report from the IPCC paints a stark picture of a warming world: it is estimated that human activity now accounts for 13 times as much atmospheric warming than can be pinned down to natural climatic variation. The American Meteorological Society[2] and the American Association for the Advancement of Science[3] have both issued statements recently stating that evidence for human modification of climate is now compelling. If humanity is to arrest, and then reverse, the growing specter of climate change then it is necessary to develop a strong understanding of the complex nature of the Earth's dynamic natural systems, the ways in which we are affecting those systems and the socio-economic consequences facing humanity.

The IPCC was established in 1988 as a partnership between the World Meteorological Organization and the United Nations Environment Program. Its purpose is to assess the risk of human-induced climate change, its potential impact and the options available for adaptation and mitigation. Although the IPCC does not conduct any research of its own, its assessments are based on peer reviewed global scientific literature. A summary of the 2007 IPCC Working Group I main predictions are given in Table 1. Overall, the report leaves little doubt that climate change is now underway and, if not addressed promptly, will result in severe and adverse impacts on human civilization.

The IPCC predicts that temperatures are most likely to rise by between 1.8 and 4°C by 2100, but the possible upper limit is greater due to the possibility of positive feedback mechanisms kicking in to exacerbate the impact. Climate predictions of the future consider different scenarios of population increase, fossil and alternative fuel use and the consequent CO_2 increases projected.

In the wider political domain, the rhetoric of climate change is also getting stronger — in the last few years the phenomenon of "catastrophic" climate change is often referred to. The use of this pejorative term and other qualifiers such as "chaotic", "irreversible" and "rapid" has led to statements referring to an irreversible tipping in the Earth's climate. The former Prime Minister of the United Kingdom,

Table 1. Summary of IPCC climate projections — Report of Working Group I, 2007.

IPCC Projections for 21st Century:

- Probable atmospheric temperature rise between 1.8°C and 4°C;
- Possible atmospheric temperature rise between 1.1°C and 6.4°C;
- Sea level most likely to rise by 28–43 cm;
- Arctic summer sea ice likely to disappear in second half of 21st century;
- Increase in heat waves very likely;
- Increase in tropical storm intensity likely.

Tony Blair, warned in an open letter to European Union that "We only have a window of only 10–15 years to take the steps necessary to avoid crossing a catastrophic tipping point. The consequences for the planet of inaction are, literally, disastrous. This disaster is not set to happen in some science fiction future many years ahead, but in our lifetime".[4] Together with Albert Gore's Oscar winning documentary 'An Inconvenient Truth',[5] the onslaught of hurricane Katrina in the summer of 2005, the shattering of summer-high temperature records across the planet, so-called peak oil and now the stronger language of the latest IPCC reports, it is hardly surprising that climate change has been catapulted to the top of the international political agenda. This emergence of climate change awareness has now been strongly underpinned by the Norwegian Nobel Committee following its decision to jointly award Gore and the IPCC the Nobel Peace Prize in October 2007. By linking climate change to peace, the committee has signaled its view that climate change is now one of global society's defining security issues. Although controversial to some, it follows the first meeting on climate change and security by the UN Security Council in April 2007.[6] The meeting was held in response to scientific predictions that land and water resources will become scarcer in the coming years, threaten access to energy, increase food scarcity and transform peaceful competition for resources into violence as floods and droughts spark massive human migrations, polarize societies and weaken the ability of countries to resolve conflicts peacefully.

Scientific data points to incontrovertible signs of climate change across the globe; eleven of the last 12 years have been the warmest on record; mountain glaciers and ice cover have declined drastically; satellites have detected a sea level rise; and more intense, longer mega-droughts are prevailing in the great continental land masses of Africa, Asia and Australia. The pace of change also seems to be quickening with the annual rate of CO_2 increase in the atmosphere from fossil fuel burning and industry almost tripling since the turn of the century.[7] Since the beginning of the industrial revolution in 1750, CO_2 levels in the atmosphere have risen by around 35% from around 280 part per million (ppm) to about 385 ppm today. Under a current 'business-as-usual' scenario a doubling of CO_2 to about 550 ppm is expected by 2050.

Although climate change rhetoric is up, action is down — annual fossil fuel derived CO_2 emissions actually increased from an average of 6.4 GtC (billion tons of carbon) in the 1990s to 7.2 GtC in 2000–2005.[7] The daily burden of CO_2 emitted into the atmosphere continues largely unabated — despite the Kyoto Protocol of 1997 which obliges industrial nations to curtail annual carbon emissions by an average by 5.2% relative to 1990 levels by 2012. Developing countries are racing up the league table of CO_2 emissions, thanks to their energy hungry, rapidly expanding economies. It has just been reported by the Netherlands Environmental Assessment Agency (June 2007)[8] that China is now building an average of two power stations per week — fuelled by its vast reserves of coal. China is now the world's largest emitter of greenhouse gases, although per capita emissions are still around four times lower than those of the US.

The now famous "hockey stick" graph endorsed by the IPCC shows the sudden surge in mean global temperatures that have been detected in the atmosphere over the last two decades. Figure 1 shows this phenomenon on two time scales, over the last 140 years and the last 1000. Temperature aberrations from the 1961–1990 average are particularly apparent since 1980.

Rising concern over climate change is no longer just the preserve of Earth scientists, it is now also firmly quartered in the economic fraternity. Lord Nicholas Stern, former chief economist of the World Bank, released his startling appraisal of the potential impacts of climate change on the global economy and human civilization in November 2006.[10] Based on latest scientific evidence, he reported that if humanity fails to act then the cost of tackling resultant economic disruption in coming decades would cost at least 5%, and as much as 20% of the world's economic output under the worst case scenario. In contrast, the cost of action to halt and reverse climate change would cost just 1% of world output. This 'no-brainer' is the essence of the precautionary principle — taking appropriate action before a situation becomes too costly to rectify — or even irretrievable. Should we rise to the climate change challenge and respond using a precautionary approach or wait until an irreversible change is upon us? A summary of the economic impacts induced by climate change predictions, as reported by Stern, are shown in Table 2.

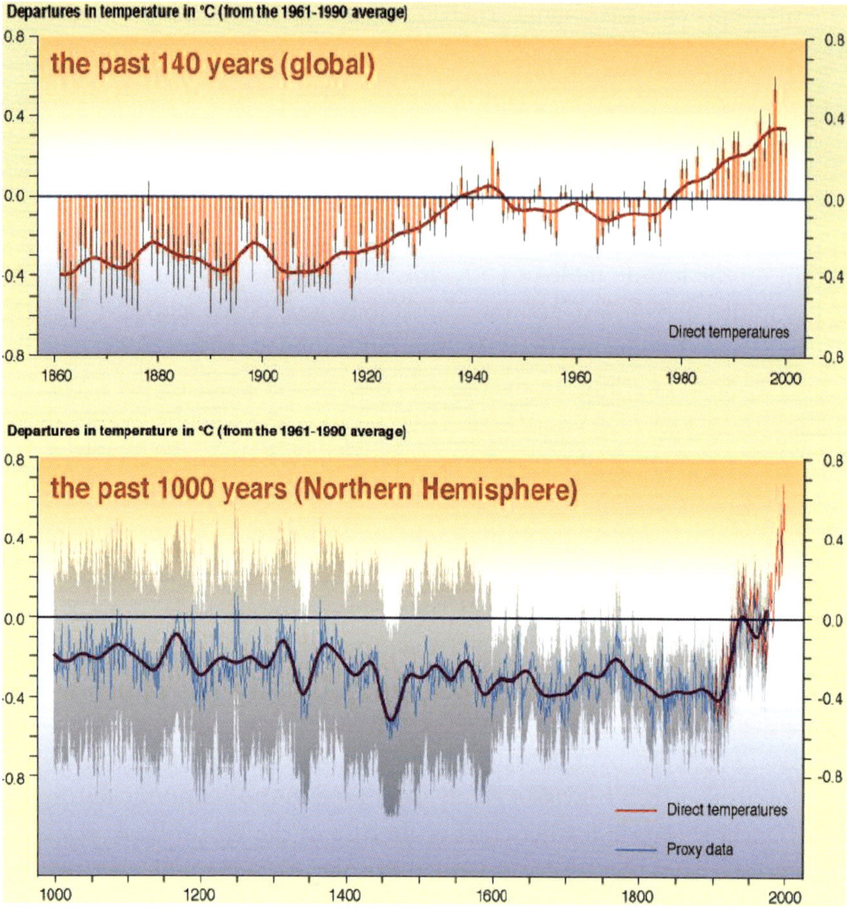

Fig. 1. Departures in atmospheric temperature from the 1961–1990 average over the past 140 years (global) and the past 1000 years (Northern Hemisphere).
Source: IPCC.

Although the scientific evidence for climate change is now largely irrefutable, the big debate is whether these observations are down to long term natural cyclical events or to anthropogenic impacts on the atmosphere. We know from a wealth of geological, paleoclimatic and cryological evidence that the planet's climate is constantly changing and has been markedly different in the past — so why all the

Table 2. Summary of the economic impacts predicted due to climate change. Stern Review, November 2006.[10]

- Extreme weather events could reduce global gross domestic product (GDP) by up to 1%;
- A 2–3°C rise in temperatures could reduce global economic output by 3%;
- If temperatures rise by 5°C, up to 10% of global output could be lost. The poorest countries would lose more than 10% of their output;
- In the worst case scenario global output would fall 20%;
- To stabilize CO_2 levels in the atmosphere to manageable levels, emissions would need to stabilize in the next 20 years and fall between 1% and 3% after that. This would cost 1% of global GDP.

fuss? Basic knowledge of climate change is now prevalent within the general public — even children in elementary school are taught the basics of atmospheric science and global warming these days. What is perhaps less well understood is the fact the Earth has a natural, sustained greenhouse effect that has rendered the planet habitable for life over the last several billion years. All of the major greenhouse gases of concern — CO_2, water vapor, methane, nitrous oxide etc. occur naturally in the atmosphere and trap sufficient longwave radiation to ensure a life-conducive warming of the planet's surface. Without it, we would be in the deep freeze with an average surface temperature of minus 18°C — making the planet uncomfortable for life at best. This natural blanket of atmospheric greenhouse gases is absolutely essential in allowing life, as we know it, to flourish on Earth.

The Earth-Atmospheric System and Climate Change

Many of the greenhouse gases in our atmosphere are artifacts of biological processes taking place in the Earth's ecosystems, or are heavily moderated by them. The atmosphere has an intimate interface with the planet's oceans and terrestrial aquatic systems (hydrosphere), the planet's soils and geology (lithosphere) as well as its flora and fauna (biosphere). Each gas has its own distinct capacity for greenhouse warming (or radiative forcing) and a distinct chemical half-life — that is the time a typical molecule of gas spends in the atmosphere.

Carbon, together with the other essential elements for life (nitrogen, phosphorus, water etc.) is in a state of flux between the various spheres. These interactions combine to form the great biogeochemical cycles on Earth — the term itself implies a close and intricate relationship between the biotic and abiotic components of the global environment. Together, these cycles work to provide the essential ecological services to sustain life — clean air, pure water and waste removal that together manifest to produce equitable conditions for life over eons of geological time. These global biogeochemical cycles interact together to represent a planetary green machine — keeping the planet healthy and habitable for its panoply of life. The malleable nature of the biosphere, via the process of evolution, allows life in whatever carbon-based form, to adapt, change and perpetuate itself within the global environment. It is, indeed, remarkable that, despite many catastrophic events in the Earth's history which have precipitated mass biological extinction events the chain of life, since its inception about 3.8 billion years ago, has never been extinguished and the boundary conditions for life have never been exceeded.

The nature and composition of the Earth's atmosphere is in stark contrast to our neighbouring planets, Venus and Mars. The Earth is sometimes referred to as the 'Goldilocks' planet of the solar system — not too hot for life, unlike Venus, and not too frigid, unlike Mars. However, this is no cosmic act of fate. The gaseous composition of our atmosphere defies conventional chemical equilibrium theory — it is in a dynamic, reactive state of *disequilibrium,* where gases such as oxygen and methane are always present, but should not theoretically co-exist. This phenomenon can only be reasonably explained by the presence of the biosphere itself which continually works with climatic and physical processes to cycle the life-essential elements. In contrast, the atmospheres of Venus and Mars are lifeless and inert — in steady-state equilibrium with no evidence of biogeochemical pathways or atmospheric regulation.

The Impacts of Climate Change

The IPCC's second report, released in April 2007 entitled Climate Change 2007: Climate Change Impacts, Adaptation and Vulnerability[11] presents clear and compelling evidence that climate change is already having a direct impact on the Earth's biosphere and hydrosphere — with profound implications for the world's natural and agricultural systems. "For the first time, we are no longer arm-waving with models; this is empirical data, we can actually measure it," said Martin Parry, co-chairman of IPCC Working Group II that produced the report.[11]

The literature reviewed by the IPCC included more than 29,000 pieces of scientific data on observed changes in physical and biological aspects of natural ecosystems. About 90% of these data confirm a warming trend - the undeniable fingerprints of global climate change on the biosphere. The key findings of the second IPCC report are shown in Table 3. The assessment underlines the action needed to reach international consensus on reducing greenhouse gas emissions, and the need for humanity to adapt to committed climate change in the near future.

Table 3. Key findings of Working Group II of the IPCC Fourth Assessment, 2007.[11]

- 20–30% of all plant and animal species at increased risk of extinction if temperatures rise between 1.5–2.5°C;
- Glaciers and snow cover expected to decline, reducing water availability in countries supplied by melt water;
- 75–250 million people across Africa could face water shortages by 2020;
- Crop yields could increase by 20% in East and Southeast Asia, but decrease by up to 30% in Central and South Asia;
- Agriculture fed by rainfall could drop by 50% in some African countries by 2020.

The Climate Change Puzzle

Knowledge about atmospheric temperature and its relationship to the Earth's energy balance is by no means recent — it was the brilliant French mathematician and physicist, Joseph Fourier who, in the 1820s, realized that incoming solar energy to the Earth's atmosphere must be balanced by energy returning to space.[12] He showed that the interception of longwave, infrared radiation by the atmosphere following its emission from the Earth-atmospheric system keeps the planet warmer than it should otherwise be. Fourier compared this heat trapping effect of the atmosphere to a greenhouse which traps heat — hence the term "greenhouse effect".

The Earth and space are in a dynamic state of radiative energy equilibrium, meaning there is no overall net gain or loss of heat. About one third of the solar radiation reaching the Earth from the sun is reflected back to space by clouds, dust or high albedo surfaces. Of the remainder that gets through, over 80% warms the lower troposphere and powers the cycling of water through the biosphere. After Fourier's great discovery, it was John Tyndall who, in the 1860's, discovered that CO_2 is highly absorbent of infrared energy at certain parts of the light spectrum.[13] CO_2 and the other naturally occurring gases interact with solar energy that has been re-emitted by the Earth-atmospheric system as infrared radiation to increase their kinetic energy, thereby inducing natural atmospheric warming. It is the ultimate re-radiation of this attenuated energy from the atmosphere to space that maintains the Earth's energy balance. It is fascinating to note that green plants, algae and certain bacteria use less then 0.1% of the incoming solar energy to fuel photosynthesis and generate biomass. Although only a tiny fraction, this captured organic carbon, as we shall see later, has a disproportionate effect on atmospheric temperature regulation and powers almost all life present on Earth.

Following Tyndall, in the 1890s, the Swedish chemist Svante Arrhenius calculated the impact of fluctuating CO_2 concentrations on atmospheric temperature.[14] He was able to deduce that a doubling of pre-industrial levels of CO_2 in the atmosphere would result in a projected warming of about 5°C. This, and later predictions by the British engineer Guy Stewart Callendar, did not however raise too much scientific concern — it was predicted that a doubling of CO_2 levels from industrial emissions would occur only very slowly over a period of several hundred years.[13]

It was only in the 1950s when Charles Keeling began his eloquent measurements of atmospheric concentrations of CO_2 at the Hawaii Mauna Loa Observatory that the harsh reality of rising levels of CO_2

in the atmosphere started ringing alarm bells.[15] The march of industrialization, human population growth and its impact upon our atmosphere became blatant for all to see — the unremitting, inexorable rise in CO_2 levels year-upon-year. This increase, superimposed upon the seasonal variation in CO_2 levels caused by the capture and release of CO_2 from the world's flora, continues its upward trend today and allows extrapolation of predicted CO_2 levels into the future. Although CO_2 levels have increased by around 35% since 1750, other gases are 'partners in crime' in global warming and have also risen as a result of human activity: methane (CH_4) and nitrous oxide (N_2O) levels, for example, are now about 145%, and 15% higher respectively than in pre-industrial times. Some greenhouse gases have a much stronger radiative forcing than CO_2, so although present at much lower concentrations they punch above their weight in terms of warming impact. The effect of this additional 'blanket' of human-derived emissions on atmospheric heating and its disruption to the global carbon cycle cannot be overlooked.

The Global Carbon Cycle

A more comprehensive understanding of the potential impacts of current and future greenhouse gas emissions and their intricate relationship with climate change requires a deeper appreciation of the global carbon cycle, its fluxes in the environment and is interaction with the biosphere. The atmosphere's heat regulating thermostat is an intricate, dynamic and finely tuned mechanism and at its heart is CO_2. However, the CO_2 present in the atmosphere is only a tiny fraction of the total carbon present on Earth, where carbon is in a state of flux with the biotic (biosphere) and abiotic components (hydrosphere, atmosphere and lithosphere) of the planet. Vast reservoirs or 'sinks' of carbon abound in these different spheres and are interconnected via numerous environmental processes and pathways. An overview of the global carbon cycle is given in Fig. 2.

The carbon cycle can be divided between the endogenic cycle where carbon turnover is slow — in the ocean sediments and soil; and the exogenic cycle where turnover is faster — in the atmosphere and

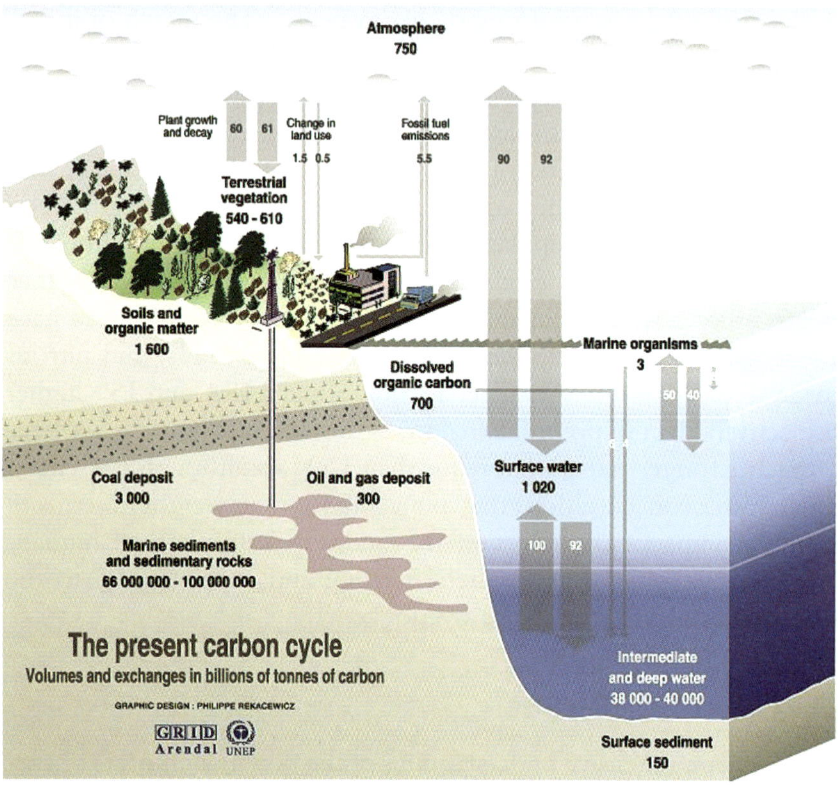

Fig. 2. The global carbon cycle.

Source: United Nations Environment Program.[16]

parts of the biosphere. Of all CO_2 emitted by human activities into the atmosphere, about half is still there;[17] the rest goes into carbon sinks that provide an essential service in regulating the atmospheric thermostat by mopping up excess CO_2 and slowing down global warming. The two major natural carbon sinks in the global environment are the planet's oceans and the terrestrial biosphere which take up approximately equal amounts of CO_2. The atmosphere contains around 800 GtC of carbon as CO_2 (a gigatonne is a billion metric tones) to which humans add around 7.2 GtC per year. The most active reservoir in terms of carbon flux and interaction with the atmosphere is the terrestrial biosphere in which around 2000 GtC of carbon are present as plants and soil humus, and the largest is the ocean with around 40 000 GtC (excluding marine methane clathrates). Both the terrestrial and ocean ecosystems each absorb about quarter of all annual anthropogenic CO_2 emissions, with the remainder staying in the atmosphere for at least 100 years. This essential means that the carbon emitted from the coal burning fires lit by our great grandparents is still warming the atmosphere today.

Carbon is essential for life, having an integral role in cell structure, biochemistry and nutrition in all life forms on the planet. Life itself plays an important intermediary role in the carbon cycle, where photosynthesis and respiration result in the fixation and release of carbon between the biosphere and the atmosphere over a relatively short time period. In the upper, sunlit 'photic zone' of the ocean biological productivity is prolific. Microscopic floating plants or phytoplankton, convert captured carbon into biomass and carbonates. Although some CO_2 returns to the atmosphere from the warm upper layer of the ocean, there is a net loss of carbon from the photic zone to the deep ocean abyss via gravity and sedimentation when cells die. This process, referred to as the 'biological pump', results in biologically fixed carbon being transferred to the deep ocean, where it essentially becomes locked away from the global carbon cycle for long periods of time. On land, the CO_2 captured by plants is returned to the atmosphere via respiration and the decomposition of organic tissues, except for a small fraction that is sequestered away to eventually form rich deposits of organic carbon — again locked away from the global carbon cycle.

This gradual trickle of carbon from the biosphere has been going on for several billion years and was on over-drive during the Carboniferous period about 360 to 290 million years ago when much of the Earth's vast oil and coal deposits were formed. This process can still be witnessed today in the rich swamplands of Sumatra, for example, where fallen trees are slowly compressed and converted to peat and then coal. This slow, incessant capture of atmospheric carbon through the biosphere is the source of the carbon-based fuels that we consume so voraciously today, and the reason why fossil fuels are sometimes referred to as "fossilized sunlight".

Another vital segment of the planetary biogeochemical cycling of carbon is the weathering of terrestrial rocks via carbonic acid reactions. This process results in the production of bicarbonate ions that are transported in the hydrosphere to the ocean where they react to form marine carbonates. This terrestrial-aquatic carbon exchange mechanism is important in controlling the hydrogen ion concentration (pH) of the ocean and the process sequesters carbon in a form that is not readily returned to the atmosphere. Animal shells made of calcium carbonate eventually become captured in ocean sediments and, under immense geological pressures, eventually become limestone — again resulting in a net loss of carbon from the cycle. Volcanic eruptions intermittently release various greenhouse gases including CO_2 to the atmosphere. The magnitude of this geologically released CO_2 is roughly equal to the amount removed by terrestrial rock weathering; so the two processes, essentially the chemical reverse of each other, sum to roughly zero.

Overall, an insight into the complex mechanics of the planetary carbon cycle leads to one over-arching conclusion — there has been a net loss of carbon from the planet's biosphere and atmosphere over geological time. The combustion of biologically-derived carbon stored in the form of coal, oil and natural gas essentially re-releases the carbon that has been stored away from the atmosphere. Therefore, net changes in organic carbon sinks as a result of human activity directly affect atmospheric concentrations of CO_2 and amplify natural global warming.

The dynamic chemistry of the 'great aerial ocean' that is our atmosphere provides clues as to the importance of CO_2 that is at the heart of the planets atmospheric thermostat. Given that the regulation of CO_2 in the atmosphere is fundamentally intimated with biological processes in the terrestrial and ocean ecosystems, what can this tell us in the context of current anthropogenic emissions? Since the last ice age ended, about 10 000 years ago, the global atmospheric thermostat has been set at an average of 14.4°C — a temperature highly conducive to the progress of human civilization.[18] The development of agriculture, cities and the global networked community, is a remarkable achievement for the human species and this, by and large, has been achieved under favourable climatic circumstances.

Despite this multi-millennia stability in temperatures, the atmosphere is influenced by complex and dynamic feedback mechanisms in the carbon cycle which may be negative i.e. serving to dampen fluctuations from the prevailing level or positive i.e. serving to amplify fluctuations. Negative feedback tend to lead to balanced, homeostatic conditions in atmospheric CO_2 levels and temperature, whereas positive feedback can lead to detrimental, exacerbating deviations from the status quo. What does all of this mean in the Industrial Age that has been largely built on energy derived from the combustion of fossil fuels?

Carbon and the Gaia Hypothesis

The Gaia hypothesis, as proposed by British scientist James Lovelock in 1979[19] offers a radical insight into how regulatory feedback mechanisms work on Earth, and how we can interpret anthropogenic impacts on the global ecosystem. Gaia is essentially an ecological hypothesis which proposes that the biotic and abiotic components of the Earth behave as an interacting system to maintain global homeostasis. Lovelock defined Gaia as "a complex entity involving the Earth's biosphere, atmosphere, oceans, and soil; the totality constituting a system which seeks an optimal physical and chemical environment for life on this planet".[19] With the hypothesis, Lovelock claimed the existence of a global control system for regulation of the mean surface temperature on the planet. The basis of the hypothesis is that the biosphere, hydrosphere, lithosphere and atmosphere all act in concert as a self-regulating entity to render global conditions suitable for life, in whatever carbon-based form, over geological time.

Broadly, Lovelock's hypothesis is premised on the fact that since life began on Earth, about 3.8 billion years ago, the incident energy impinging on the planet's atmosphere from the Sun has been increasing — despite the fact that the surface temperature of the planet has remained remarkably constant over the geological timescale. Although average temperatures fluctuate over relatively brief geological periods, they have never exceeded the boundaries of tolerance for life on Earth. The Earth has sustained an equitable temperature for life over epochs of geological time even though the Sun, like all stars, has obeyed the laws of physics and now radiates approximately 35% more energy than it did when life fist began on the planet.

After initial skeptism, the Gaia hypothesis has now been broadly accepted in the scientific community as consistent with our contemporary understanding of global ecology which embraces not only the biosphere, but also the atmosphere, hydrosphere and lithosphere. Research conducted over the last three decade has given support to the hypothesis, where the global ecosystem does indeed demonstrate behavior conducive to being a regulating entity comprised of interacting physical, chemical and biological components that work together to maintain conditions fit for life on the planet. Although the interactions and feedbacks between the component parts of the carbon cycle are highly complex and exhibit temporal and spatial variability, our knowledge of the natural dynamics of the Earth's system has advanced greatly in recent years and provides a basis for predicting the impacts and consequences of human-driven climate change.

The Biosphere and Climate Change

If the interaction of the biosphere and carbon is an integral part of the planetary wide biogeochemical cycle and that the presence of life itself supports a homeostatic mechanism that regulates surface temperature, then what is the state of the world's biodiversity and how are human activities encroaching upon the biosphere? This question is complimentary to the issue of anthropogenic greenhouse gas emissions — while CO_2 emissions are stressing the global ecosystem, in terms of regulation of the atmospheric thermostat, then damage to the biosphere are undermining its integrity to function. To gain further insight into the question, it is useful to examine ecological data collected over recent decades that serves as an indicator to the health of the Earth's global ecosystems and their biodiversity.

According to the 2006 biannual Living Planet Report of the WWF[20] the biosphere is now being degraded at a rate unprecedented in human history — to the extent that human consumption levels are likely to result in a large-scale collapse of global ecosystems by the middle of this century. The WWF report is based on two prime indicators: the Living Planet Index (LPI) which assesses the health of the planet's ecosystems; and the Ecological Footprint (EF) which measures human demand on the natural world in terms of the use of biologically productive land and water required to meet human demands. The LPI has tracked the populations of over 1300 vertebrate bioindicator species of fish, amphibians, reptiles, birds and mammals from around the world for over three decades and shows that species populations in terrestrial, marine and freshwater ecosystems have declined by over 30% since 1970. This rapid loss of biological capital has been caused by the consumption of renewable natural resources faster than the biosphere can replenish them and is incidental with the increase in human population size and growth in the world economy. The trends in the LPI and EF are illustrated in Fig. 3. Humanity's demand for resources is now outstripping biocapacity by about 25%. Meanwhile the health of the planet's ecosystems, measured by the living planet index, is falling.

Exceeding biocapacity means that the Earth's ecosystems can no longer keep up with the demands being placed upon it, and that

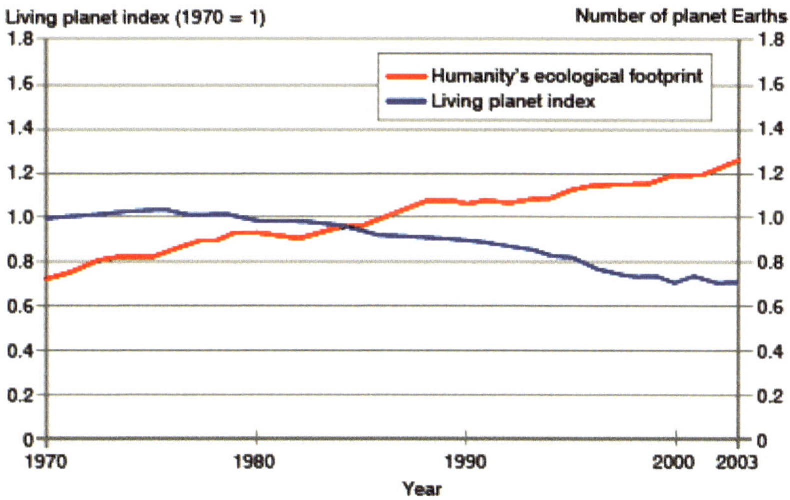

Fig. 3. Humanity's ecological footprint and the planet's health.
Source: WWF Living Planet Report, 2006.[21]

humanity is now dominating more of the natural ecosystems that are integral to the biogeochemical cycles. The findings of WWF are supported by an international team reporting in the Proceedings of the National Academy of Sciences in the US. Their manuscript 'Tracking the Ecological Overshoot of the Human Economy'[22] is based on existing data that translates human demand on the environment into the land area needed for the production of food and other goods, and for absorbing wastes. In 1961, humans were using 70% of the capacity of the global biosphere and by 1999 that had risen to 120%. The watershed was in 1986 — when 5 billion people started to consume the entire Earth's annual renewable production of resources. The overshoot at the end of the last century means that 1.2 Earth equivalents were needed to regenerate the renewable resources consumed by humanity in a single year. This overshoot is expected to increase into the 21st century due to a rising human population and developing world aspirations for western living standards. Under a 'business-as-usual' scenario, the EF is expected to rise to an equivalent of two planets worth of biocapacity by 2050.

The 20th century started with a global human population of around 1.7 billion people and closed with a population of 6 billion. The United Nations projects that global population will begin to level off at around 9 billion by the middle of this century[22] (see Fig. 4). The exponential growth in human population, particularly since the end of World War II, has been coupled with a radical growth in industrialization and a large increase in per capita energy consumption.

Projecting forward, The International Energy Agency in its World Energy Outlook report of 2004[23] has given us a glimpse of the future and its projected energy demand over the next 25 years. In this short time, world energy demand is expected to double, with fossil fuels supplying most of this need. Emissions of carbon dioxide from all sectors are likely to increase by 62%. Clearly, such a large ecological debt is not sustainable on a finite planet and represents a severe risk to the viability of the biosphere and its ability to provide critical ecological services. As Stephen Carpenter, the past President of the Ecological Society of America commented "There is a common pattern we are seeing in ecosystems around the world: gradual changes in vulnerability

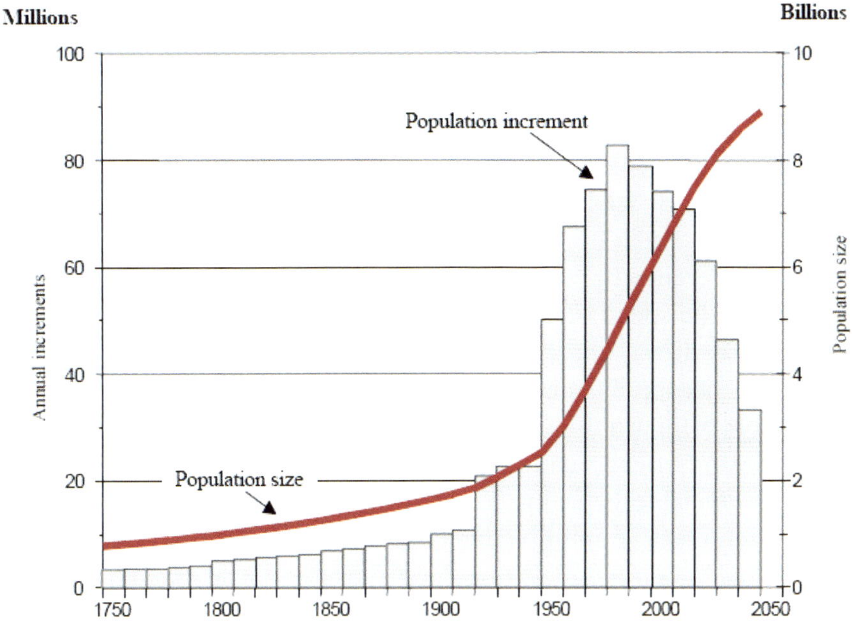

Fig. 4. Long term world population growth 1750 to 2050.
Source: United Nations Population Division.[24]

accumulate and eventually a shock to the system, perhaps a flood or a drought, knocks it into another regime".[24]

Overall, based on available biodiversity and other environmental data, ecologists have concluded that anthropogenic change imposed upon the Earth's land surface, atmosphere, biosphere and its biogeochemical cycles is now beyond the boundaries of natural variability. The plethora of changes now occurring simultaneously in the Earth's system, their magnitudes and rate are unprecedented. The Earth's ecosystems are now in unchartered waters with respect to their integrity and viability to regulate essential biogeochemical cycles. We are sailing into planetary *terra incognita*.

Many scientists are concerned that human activities have the potential to damage the Earth's CO_2 thermostat to the point that damage may become irreversible — at least in a time frame that is relevant to the human species. In his latest book, The Revenge of Gaia[25]

Lovelock argues that the lack of respect humans have for the biosphere is undermining the Earth's capacity to dampen the effects of the additional burden of atmospheric CO_2 emissions. Anthropogenic activities can be viewed as undermining the planet's homeostatic negative feedback control mechanisms, thereby increasing the likelihood of positive feedbacks and the specter of runaway global warming. Although the Earth has gone through dramatic changes in its history, the real difference between what is happening now and the geological past, is the speed of change in our natural environment. The risk of a human-driven abrupt, near-term change in the atmosphere has yet to be accurately quantified but is not negligible. The climate change problem can no longer be easily understood in terms of a simple cause and effect paradigm where change will occur according to linear, predictable behavior. Human-driven changes to the environment often result in multiple effects that cascade through the Earth's ecosystems in complex and unanticipated ways. Historical changes in our atmospheric system, at least the ones that we know of, have usually taken place over tens of thousands of years allowing organisms time to change, accommodate and evolve. Scientists are now concerned that natural climatic fluctuations and cycles have been superseded by human-induced change, and this has serious implications for the stability of the climate on which most life on the planet depends. This profound change has led the Nobel Laureate, Paul Crutzen, best known for his work on atmospheric ozone depletion, to claim that the Earth has now entered a new geological era — the 'Anthropocene', where human activities are now overwhelming natural, regulatory environmental processes on Earth.[26]

From an ecological perspective, the chief concern regarding climate change is not necessarily the scale of anthropogenic CO_2 emissions, but more the impact of human activities upon global ecosystems and associated biodiversity which are intimately involved in the planet's biogeochemical cycles.[27] We know that mass extinction events have happened at least on five occasions in the history of life on Earth, but we now appear to be in the midst of the sixth mass extinction event caused by a single species — *Homo sapiens.* It is estimated that the current biological extinction rate is now running somewhere between

100 to 1000 times greater than the natural background rate.[28] This has led the eminent biologist, Edward Wilson, to predict that projected increases in human population and resource consumption are likely to cause the premature extinction of up to half of the Earth's species by the end of this century.[30] The dire state of the planet's ecological systems has been highlighted yet again, following the release of the United Nations Global Environment Outlook Geo-4 report in October 2007 that gives the stark warning that the future of humanity itself is now at risk.[29] Twelve per cent of birds, 23% of mammals, 30% of amphibians and almost a third of primates now face extinction.[30] It is clear that the human population is living far beyond its means, where demand for resources now far exceeds what is replenishable at current consumption rates.

Feedback Mechanisms and Climate Change

The prospect of positive feedback mechanisms causing runaway climate change due to a vastly increased transfer of sequestered deposits of organic carbon to the atmosphere, coupled with widespread destruction of the biosphere, is the most disturbing aspect of climate change. A quickening of positive feedback mechanisms affecting atmospheric CO_2 levels is now evident. Crossing such environmental 'event horizons' represent triggers for accelerated human-induced global warming. A few examples are listed below:

- Human induced global warming will trigger due to melting decomposition of vast organic carbon deposits in the world's frozen tundra and permafrost at high latitudes leading to releases of greenhouse gases that could dwarf anthropogenic emissions;[31]
- The melting of the arctic ice sheet will lead to accelerated absorption of solar radiation by the ocean via loss of high surface albedo, thereby hastening the sheet's melting and demise. A summer season ice-free Arctic may be with us as early 2040 based on recent predictions.[32] The latest news on this issue is even more disturbing — there was a 22% shrinkage in the minimum summer arctic ice sheet cover between 2005 and 2007, opening up the fabled Arctic

'Northwest Passage' for the first time.[33] This rate of loss is an order of magnitude higher than even the worst predictions, but while Earth system scientists are alarmed others are already rushing in to claim the mineral wealth of the arctic seabed and valuable new shipping routes to Asia. Other climate-related impacts on the oceans include an impaired ability to absorb CO_2. This affect has been anticipated in global climate models, but appears to be happening several decades sooner than expected.[34] Again, the very latest reports show a dramatic halving of the ability of the North Atlantic to absorb atmospheric CO_2 from the mid 1990s to 2002–2005. The exact reasons for this remain unclear, but such loss in capacity of such a major carbon sink will mean an accelerated accumulation of anthropogenic carbon in the atmosphere;[35,36]

- The increased prevalence of forest fires due to higher atmospheric continental temperatures and reduced precipitation leading to large scale release of biologically captured carbon from plant biomass. Coupled with this is the widespread destruction of the planets great tropical, temperate and boreal forests that cover about one third of the planet. It was recently reported by the United Nations that Borneo and Sumatra in Southeast Asia will be denuded of their primary rainforests as early as 2022.[37] It is also speculated that an atmospheric rise in temperature of two to three degrees centigrade will decimate the ecology of the Amazon rainforest as we know it, with major droughts already evident.[38,39] Furthermore, there is the possibility of forest CO_2 sinks and soils flipping over to become net sources of emissions to the atmosphere following a modest rise in atmospheric temperature;[40,41]
- The increased prevalence of glacial earthquakes caused by basal slippage of ice masses as a result of percolating melt water and dynamic ice thinning leading to accelerated movement and collapse of ice sheets in Greenland and the West Antarctic.[42]

Perhaps most significant of all is the prospect of a gigantic release of the powerful greenhouse gas methane (with a radiative forcing about 23 times stronger than CO_2) from the huge reserves of methane clathrates found in the world's oceans due to rising sea temperatures.

This reservoir of carbon is many times larger than all of the anthropogenic emissions to the atmosphere over the last 250 years. If clathrate carbon were to find its way into the atmosphere, any meaningful reversal of CO_2 back to pre-industrial atmospheric levels would become meaningless on a human time-scale.

The risk of a break down in the planetary regulation of atmospheric temperature and the danger of positive feedback mechanisms is perhaps best summed up by Lovelock[25] "What makes global warming so serious and so urgent is that the great Earth system, Gaia, trapped in a vicious circle of positive feedback, whether from greenhouse gases, the disappearance of the Arctic ice or the Amazon forest, is amplified, and its effects are more than additive. It is almost as if we had lit a fire to keep warm, and failed to notice, as we piled on the fuel, that the fire was out of control and the furniture had ignited. When that happens, little time is left to put out the fire. Global warming, like a fire, is accelerating and almost no time is left to act".

Certainly, it is prudent to regard positive feedback mechanisms on the atmospheric thermostat as Achilles' heels which should be treated with the utmost respect. Scientists are already speculating that the Earth may have already crossed a significant event-horizon — a "magic gate" through which changes may occur more rapidly than envisaged due to a cascade effect of positive feedback mechanisms. However, to add to the complexity, it is also possible that negative feedback mechanisms may be triggered in a warming climate that may serve to mitigate atmospheric warming. For example, it is possible that plants may take more CO_2 from the atmosphere as their growth speeds up in warmer conditions, though this has been debated as recent evidence suggests that forest and vegetation become net emitters of CO_2 under warmer conditions.[38] Tropospheric aerosols (microscopic airborne particles) resulting from combustion of fossil fuels, biomass burning and volcanic sources have led to negative solar forcing, acting as a brake on atmospheric warming.[43]

Although we have barely begun to recognize the danger of setting off rapid and irreversible changes in the Earth's ecosystems, it is erroneous to regard nature as fragile, readily succumbing to human destruction. Although humans have shown an amazing ability to thrash global ecosystems, the prospect of a complete crash of the biosphere is unlikely. Lovelock refers to Gaia as "Like an old lady who has to share her house with a growing and destructive group of teenagers. Gaia grows angry, and if they do not mend their ways she will evict them." Less graciously perhaps, Professor Lynne Margulis, Lovelock's co-proponent of the Gaia hypothesis, refers to Gaia as "a tough bitch — a system that has worked for over three billion years without people". In her opinion, the planet's surface and its atmosphere and environment will continue to evolve long after humanity ceases to exist.[44]

The rich complexities of the earth's climate and biosphere mean we cannot be absolutely sure what changes will result from an increase in atmospheric CO_2 concentrations, although our ability to model the global ecosystem and predict impacts becomes ever more sophisticated. The sheer number and complexity of predictive global climate models (GCMs) is quite overwhelming. GCMs are computer programs that simulate the earth's climate, taking into account an extraordinary number of variables describing the physical and chemical properties of the atmosphere, oceans, and continents. The US National Center for Atmospheric Research (NCAR) Community Climate System Model (CCSM) is one of the world's most sophisticated and widely used climate models.[45] The graphic in Fig. 5 shows the many complex components included in the CCSM to predict future climate change, ranging from the effect of cirrus and stratus clouds, to ocean currents and soil moisture.

Although the quality and power of GCMs has improved dramatically over recent years, several weaknesses still remain that need to be corrected to improve accuracy. For example, much work remains to be done in simulating the behavior of the Earth's oceans, as well as the physics of cloud formation and their propensity to reflect and absorb radiation. Overall, the balance between positive and negative feedbacks remains a major cause of uncertainty in climate prediction and scientists are not really sure how the complexities will play out. For example, a shut down of the Atlantic Thermohaline Circulation that transfers heat from the equatorial tropics to higher latitudes may occur due to a sudden influx of melt water into the North Atlantic from the

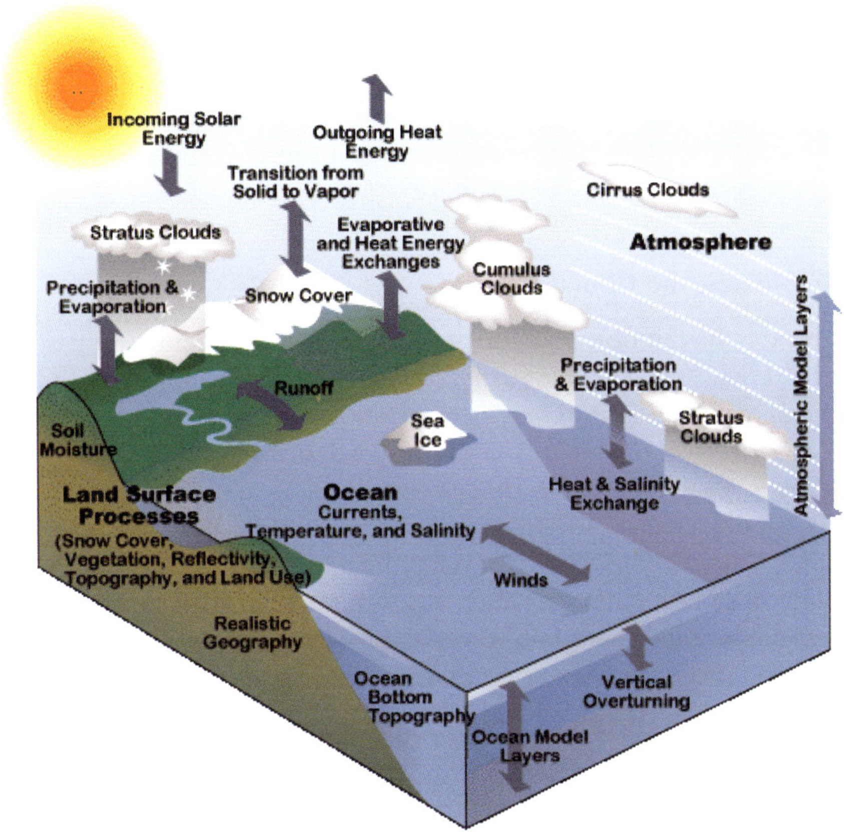

Fig. 5. Components of the NCAR-based community climate system model.
Source: National Center for Atmospheric Research.[47]

Greenland ice sheet.[46] The consequent reduction in ocean salinity would have a major impact on the Earth's heat regulation system, and could actually induce a short ice-age in Northern Europe as a result of the loss of convected heat from the ocean to the atmosphere.[47,48]

Climate modeling is an evolving science, but it is likely that continued the pressure on natural regulatory processes in the atmosphere and biosphere will render humanity increasingly vulnerable to more frequent climatic outlier events and raise the chance of crossing an atmospheric tipping point. Evidence suggests that Earth-system dynamics are characterized by critical thresholds, and that abrupt atmospheric change may occur over time periods perhaps as short as decades or centuries, the equivalent of a geological heart beat.[18] Although there is still great uncertainty over the sensitivity of the global ecosystem, the atmosphere may undergo a series of progressive, short-term jolts to reach an eventual warmer, steady-state.

Stabilizing the Climate

The UN has an objective for the stabilization of greenhouse gases at a level that would prevent dangerous anthropogenic interference with the climate system.[49] There is a scientific consensus that a warming target of no more than 2°C above pre-industrial levels is the most realistic goal to achieve climate stabilization. This is based on an assumption that positive feedback mechanisms do not kick in beforehand to exacerbate the risk. According to the IPCC, if nothing is done to reduce emissions then, according to climate models a global temperature increase of up to 6.4°C by 2100[1] is possible. It is important to recognize that this figure represents an average. The fallacy of averages means that temperature could rise much more at higher latitudes, particularly in the polar regions. Due to the longevity of carbon dioxide in our atmosphere and the natural lag in response time of ecosystems, we are already locked into climate change, meaning that if we make cuts in greenhouse gas emissions today then we are still committed to some degree of change. Given that we are already observing significant changes in the atmosphere-biosphere system at an average temperature increase of only 0.8°C, then the potential impact of future climate change on human health are manifest and complex. For example, the World Health Organization has warned that the health of millions could be threatened by an increase in malaria, water-borne disease and malnutrition as disease vectors extend their ecological range northward and upward as a result of warmer climatic conditions.[50]

Despite all of the gloomy scenarios portrayed by the IPCC and others, the uncertainty continues to fuel skeptic's comments on the significance of climate change. After all, if we find it challenging to predict what the weather will do in the next few days how can we predict changes in the climate over the coming century? Although scientific data is compelling, debate rages on as to what extent the observed warming trend is due to human activities, and what the repercussions of a warming world will be. Cynicism about global climate change falls into three main arenas: (i) those who maintain atmospheric temperatures are not rising; (ii) those who accept the climate is changing, but claim it is largely down to natural variation; and (iii) those who accept the theory of human-induced warming, but say it is not worth tackling as other global problems are more pressing. Those in the first group are now disappearing faster than some of the world's glaciers — the evidence is now irrefutable. As for the second group, it is a fact that natural variations in the Earth's climate have been occurring for millennia, but nonetheless there is now enough scientific evidence to conclude that, even on top of expected natural variability in climate, something out of the ordinary is happening and humans are to blame. For those in the third group, such as the skeptic Bjorn Lomborg,[51] it is certainly true that the world has many other pressing issues including poverty, hunger, disease, and nuclear proliferation to name but a few. However, we are also being warned by more eminent economists and scientists, that the 'do nothing', 'business-as-usual scenario' will exact a heavy price on the global economy and the very future of civilization itself. The 2006 Stern Review, made it clear that the world has to act now on climate change or face devastating economic consequences.[10] Regardless of what the skeptics may think about global warming and the causes behind it, there are still valid reasons for weaning ourselves off our addiction to fossil fuels. Much of the world would be less dependent on volatile and undemocratic regimes in hostile countries for a start, and most people would genuinely prefer to live in a less wasteful, less polluted world.

The IPCC and others are now calling for drastic cuts in CO_2 emissions. In its third report of 2007 Mitigation of Climate Change (May 2007),[52] a 50–80% global cut in CO_2 emissions is needed by 2050 to stabilize CO_2 levels at twice their pre-industrial levels to avoid 'dangerous' climate change. In November 2006, the UK governments proposed a stringent goal of a 60% reduction in CO_2 emissions by 2050 as a legally binding target, although details of interim reductions were not clearly defined.[53] These types of cuts have profound implications for the global economy and major ramifications for power generation, transportation, and industrial systems. Indeed, the change required to transform to a low-carbon global economy has led the European Union to call for a second 'industrial revolution' in the energy sector.[54] Whereas some leaders fear the disruption and impact of reducing CO_2 emissions on their fossil fuel driven economies, others advocate the abundant opportunities that exist for innovation, creativity and technology to decarbonize and transform civilization to a low or even post carbon economy. The good news is that the enormous carbon reductions needed are achievable based on current technologies, and could be introduced in a series of manageable measures across industry, transport and power generation sectors. This is basis of the ingenious 'wedge concept' of CO_2 reductions as proposed by Pacala and Socolow.[55] Scaling up from present capabilities, using a range of mitigation strategies and technologies, the wedges would stabilize CO_2 levels in the atmosphere to around 500 ppm within 50 years and cap the temperature rise to 2°C.

Overall, the broad scientific consensus is that humanity has a window of opportunity to act on climate change — a failure to do so could mean that the Earth's people and its whole panoply of life could face the most severe consequences. The Kyoto Protocol now seems largely symbolic, but it shows that the world can act to safeguard the Earth's future, and sets a precedent for future action. The post-2012 replacement for Kyoto must set-out a clear path for global development which unhooks the world's global addiction to fossil fuel and decarbonizes the world's economy. After strongly objecting to curbs on carbon emissions, even the most recalcitrant nations to the protocol, including the United States and Australia, are now calling for action on climate change. Recently, China too has announced its own curbs on CO_2 emissions, although all three nations refuse to compromise domestic economic growth and set binding targets for CO_2 reduction.[56] A continued emphasis on expansion of global GDP as an economic priority with no real political breakthrough on curbing CO_2

emissions indicates that humanity is still not moving swiftly enough. Never before has an effective multilateral political system been more necessary to address a global threat.

Together, the 2007 fourth assessment reports of the IPCC strongly emphasize that some degree of climate change is now 'locked in' — inevitable due to the carbon we have already emitted to our atmosphere. Humanity will have to adapt to this. Although there is still uncertainty about the scale of future change, it must be acknowledged that by the end of this century global temperatures could increase by up to 6.4°C. The growing accuracy of current and past climate simulations using coupled atmosphere-ocean GCMs has greatly increased our confidence in the projections of future climate change. The doubling of CO_2 from pre-industrial stable levels could happen as early as the middle of this century, and it is widely accepted that a temperature increase of more than 2°C should be avoided to avert dangerous changes in the Earth's climatic system.

Conclusion

The 2007 message from the IPCC is stark. If we are to avoid crossing a significant event horizon in our climate system by allowing atmospheric CO_2 emissions to reach 550 ppm, then it is imperative that low carbon technologies be introduced into the world economy as a matter of urgency. Avoiding 2°C of post-industrial atmospheric warming requires a decarbonization of the global economy by at least 50%, and it is now an imperative that we cut back on the burning of fossil fuels and drastically step-up international efforts to protect the biosphere. Although many details about climate interactions are not well understood, and there are ample grounds for continued research to provide a better basis for understanding climate dynamics, there is now a scientific consensus on the gravity of anthropogenically-induced climate change. Scientists and leading economists have repeatedly tried to make this clear in a succession of reports supported by an overwhelming set of field data. Now is the time for the world's political leaders to address the crises in a bold and visionary way. The real missing link in the climate change solution is universal political action. As the eminent naturalist Sir David Attenborough said "This is our planet. Since life began it has gone through extraordinary changes. Now it is being transformed not by natural events, but by the actions of one species: mankind".[57] It is the reductionist view held by mankind, treating its socio-economic activities as separate to the regulatory and life-sustaining processes of the natural world, that has brought the impending climate crisis upon us. Science led policy on climate change mitigation and biodiversity conservation, together with the full implementation of existing carbon management strategies and emerging sequestration technologies, hold the solution to climate security in the 21st century.

References

1. Climate Change 2007: The Physical Science Basis (February 2007). Working Group I Contribution to the Intergovernmental Panel on Climate Change. Fourth Assessment Report, 2007.
2. Climate Change: An Information Statement of the American Meteorological Society (2006). Bulletin of the American Meteorological Society, **88**, Issue 2.
3. AAAS Board Statement on Climate Change (2006). Approved by the AAAS Board of Directors, 9 December 2006. See: www.aaas.org/climate
4. BBC Report: Blair urges climate change action (October 2006). See: http://news.bbc.co.uk/2/hi/uk_news/politics/6068226.stm
5. Gore, A. (2006) An Inconvenient Truth. Rodale, USA.
6. United Nations Security Council SC\9000 (17 April 2007). Security Council 5663rd Meeting. Security Council Holds First-Ever Debate on Impact of Climate Change on Peace, Security. See: http://www.un.org/News/Press/docs/2007/sc9000.doc.htm
7. Raupach, M.R., Marland, G., Ciais, P., Le Quere, L., Canadell, J.G., Klepper, G. and Field, C.B. (2007). Global and regional drivers of accelerating CO_2 emissions. *Proceeding of the National Academy of Sciences.* See: http://www.pnas.org/cgi/reprint/0700609104v1
8. China now no. 1 in CO_2 emissions; USA in second position Netherlands Environmental Assessment Agency (June 2007). See: http://www.mnp.nl/en/dossiers/Climatechange/moreinfo/Chinanowno1inCO2emissionsUSAinsecondposition.html.

9. Intergovernmental Panel on climate Change (2001). Climate change 2001: The Scientific Basis. Contribution of Working Group 1 to the Third Assessment Report.
10. Stern Review. The Economics of Climate Change (November 2006). Her Majesty's Treasury, UK Government.
11. Climate Change 2007: Climate Change Impacts, Adaptation and Vulnerability (April 2007). Working Group II Contribution to the Intergovernmental Panel on Climate change Fourth Assessment Report.
12. Weart, S.R. (2003). *The Discovery of Global Warming: New Histories of Science, Technology and Medicine.* Harvard University Press, Masachusetts.
13. Fleming, J.R. (1998). *Historical Perspectives on Climate Change.* Oxford University Press, Oxford, UK.
14. Arrhenius, S. (1896). On the influence of carbonic acid in the air upon the temperature of the ground. *Philosophical Magazine* **41**, 237–276.
15. Keeling, C.D. (1978). The Influence of Mauna Loa Observatory on the Development of Atmospheric CO_2 Research. In *Mauna Loa Observatory: A 20th Anniversary Report.* (National Oceanic and Atmospheric Administration Special Report, September 1978), edited by John Miller, pp. 36–54. Boulder, CO: NOAA Environmental Research Laboratories.
16. Center for Climatic Research, Institute for Environmental Studies, University of Wisconsin; Okanagan University College in Canada, Department of Geography (1998); World Watch, Climate Change (1995); The Science of Climate Change, Contribution of Working group 1 to the Second Assessment Report of the Intergovernmental Panel on climate Change, UNEP and WMO, Cambridge University Press, UK (1996).
17. Mackenzie, F.T. and Lerman, A. (2006). *Carbon in the Geobiosphere: Earth's Outer Shell.* Springer Press, Netherlands.
18. Flannery, T. (2005). *The Weather Makers.* Penguin Books, Great Britain.
19. Lovelock, J. (1979). *Gaia: A New Look at Life on Earth.* Oxford University Press, UK
20. Living Planet Report (2006). WWF. See: http://assets.panda.org/downloads/living_planet_report.pdf

21. Wackernagel, M., Schulz, M.N., Deumling, D., Linares, A.C., Jenkins, M., Kapos, V., Monfreda, C., Loh J., Myers, M., Norgaard, R. and Randers, J. (2002). Tracking the ecological overshoot of the human economy. *Proceedings of the National Academy of Sciences* **99**, 9266–9271.
22. The World at 6 Billion. United Nations Population Division (2000). See: http://www.un.org/esa/population/publications/sixbillion/sixbilpart1.pdf
23. World Energy Outlook (2004). International Energy Symposium, IEEJ. *International Energy Agency.*
24. Scheffer, M., Carpenter, S., Foley, J.A., Folke, C. and Walker, B (2001). Catastrophic shifts in ecosystems. *Nature* **413**, 591–596.
25. Lovelock, J. (2006). *The Revenge of Gaia: Why the Earth is Fighting Back — and How We Can Still Save Humanity.* Santa Barbara, California.
26. Crutzen, P.J. and Stoermer, E.F. (2000). The Anthropocene. *International Geosphere-Biosphere Program Newsletter* **41**, 16–18.
27. Miller, G.Y. (2005). Living in the Environment. Thomson Brooke Cole.
28. Wilson, E. (2006). The Creation: An Appeal to Save Life on Earth. W. W. Norton & Company, Inc
29. Global Environment Outlook:. Environment for Development (GEO-4) (2007). United Nations Environment Program, October 2007.
30. Primates in Peril: The World's 25 most Endangered Primates (2007). The World Conservation Union, October 2007.
31. Hansen, J., Sato, M., Ruedy, R., Lo, K., Lea, D.W. and Medina-Elizade, M. (2006). Global temperature change. *Proceedings of the National Academy of Sciences* **103**, 14288–14293.
32. Holland, M.M., Bitz, C.M. and Tremblay, B. (2006). Future abrupt reductions in the summer Arctic sea ice. *Geophysical Research Letters* **33**, L23503.
33. Danish National space Centre (17 September 2007). Northwest passage opens up. See: http://www.spacecenter.dk/research/remote-sensing/northwest-passage-opens-up.
34. Sabine, C.L., Feely, R.A., Gruber, N., Key, R.M., Lee, K., Bullister, J.L., Wanninkhof, R., Wong, C.S., Wallace, D.W.R., Tilbrook, B., Millero, F.J., Peng, T.H., Kozyr, A., Ono, T. and Rios, A.F. (2004). The ocean sink for anthropogenic CO_2. *Science* **305**, 367–371.
35. Schuster, U. and Watson, A.J. (2007). A variable and decreasing sink for atmospheric CO_2 in the North Atlantic, *Journal of Geophysical*

Research, in press. See: http://news.bbc.co.uk/2/hi/uk_news/7053903.stm.

36. Canadell, G.K. and Field, C.B. (2007). Global and regional drivers of accelerating CO_2 emissions. *Proceedings of the National Academy of Science* **104**, 10288–10293.
37. United Nations Rapid response Assessment (February 2007). The last stand of the Orangutan. State of emergency illegal logging, firs and palm oil in Indonesia's National Parks.
38. Betts, R.A., Cox, P.M, Collins, M., Harris, P.P., Huntingford, C. and Jones, C.D. (2004). The role of ecosystem-atmospheric interactions in simulated Amazonian precipitation decrease and forest dieback under global climate warming. *Theoretical Applied Climatology* **78**, 157–155.
39. BBC report: 'Amazon 'faces more deadly droughts' (March 2007). See: http://news.bbc.co.uk/2/hi/americas/6484073.stm
40. Cox, P.M., Betts, R.A., Jones, C.D., Spall, S.A. and Totterdell, I.J. (2000). Acceleration of global warming due to carbon-cycle feedbacks in a coupled climate model. *Nature* **408**, 184–187.
41. Laurance, S., Oliveira, A.A., Laurance, S.G., Condit, R., Nascimento, E.M.H., Sanchez-Thorin, A.C., Lovejoy, T.E., Andrade, A., D'Angelo, S., Ribeiro, J.E. and Dick, C.W. (2004). Pervasive alteration of tree communities in undisturbed Amazonian forests. *Nature* **428**, 171–175.
42. Ekström, G., Nettles, M. and Tsai, V.C. (2006). Seasonality and increasing frequency of Greenland glacial Earthquakes. *Science* **311**, 1756–1758
43. Wild, M., Ohmura, A. and Makowski, K. (2007). Impact of global dimming and brightening on global warming. *Geophysical Research Letters* **34**, L04702.
44. Margulis, L. (undated) Gaia Is a Tough Bitch, See: http://www.edge.org/documents/ThirdCulture/n-Ch.7.html
45. US National Center Atmospheric Research Community Climate System Model. See: http://www.ucar.edu/news/features/climatechange/ccsm-illus.jsp#mediaterms
46. Curry, R., Dickson, B. and Yashaaev, I. (2003). A change in the freshwater balance of the Atlantic Ocean over the past four decades. *Nature* **426**, 826–829.
47. Swing, T.J. (2003). What future for the oceans? *Foreign Affairs* **82**, 139–152.

48. Hare, W. (2003). Assessment of knowledge on impacts of climate change: contributions to the specification of Article 2 of the UN Framework convention on climate change. WBGU Materialien, Berlin 2003.
49. Avoiding Dangerous Climate Change. International Symposium on the Stabilization of Greenhouse Gas Emission Concentration (February 2007). Report of the International Steering Committee. Hadley Centre, UK Meteorological Office, Exeter.
50. Tanser, S.C., Sharp, B. and Le Seur, D. (2003). Potential effect of climate change on Malaria transmission in Africa. Royal Society of Tropical Medicine & Hygiene **97**, 129–132.
51. Lomborg, B. (2004). *Global Crises, Global Solutions.* Cambridge University Press.
52. Climate Change 2007: Mitigation of Climate Change (May 2007). Working Group IIII Contribution to the Intergovernmental Panel on Climate Change. Fourth Assessment Report.
53. BBC report: Binding carbon targets proposed (March 2007). See: http://news.bbc.co.uk/2/hi/uk_news/politics/6444145.stm
54. News report: EU to Urge New Industrial Revolution in Energy (January 2007) See: http://www.planetark.org/dailynewsstory.cfm/newsid/39698/newsDate/5-Jan-2007/story.htm
55. Pacala, S. and Socolow, R. (2004). Stabilization Wedges: Solving the Climate Problem for the Next 50 Years with Current Technologies. *Science* **305**, 968–972.
56. BBC Report: G8 climate deal signal a breakthrough (June, 2007). See: http://news.bbc.co.uk/2/hi/europe/6732787.stm
57. Attenborough, D. (2007). See: http://www.bbc.co.uk/worldservice/specials/119_wag_climate/index.shtml

THEME II

CLEAN DEVELOPMENT MECHANSIM (CDM)

CHAPTER 4

AN OVERVIEW OF THE CLEAN DEVELOPMENT MECHANISM IN SOUTHEAST ASIA

Jolene Lin

Faculty of Law, The University of Hong Kong,
Jolene@hku.hk

This paper seeks to give an overview of the performance of the Clean Development Mechanism (CDM) in Southeast Asia. The author argues that the modest performance of Southeast Asia in the CDM world is largely due to demand-side factors that are related to the structure of the CDM market. Addressing these demand-side factors will require time and international agreement. In the meanwhile, a supply-side initiative that should be developed is greater regional cooperation amongst the Southeast Asian countries to achieve the economies of scale that come from pooling of resources to develop technical know-how and information systems that will aid project developers, financial institutions and governments. The Association of Southeast Asian Nations (ASEAN) provides an established regional framework to pursue such cooperation. This paper also examines the role that Singapore, an international financial centre that has been attracting global "green" investors and has plans to be the ASEAN region's environmental financing hub, can play to facilitate the implementation of the CDM in Southeast Asia.

1. Introduction

This paper seeks to give an overview of the performance of the Clean Development Mechanism (CDM) in Southeast Asia. It is worth having this macro-perspective because the various Southeast Asian countries have significant CDM potential, which may be better harnessed if a more coordinated regional approach is adopted in terms of developing technical know-how, information-sharing and capacity-building by donor governments and non-governmental organizations. So far, the performance of the CDM in the various Southeast Asian countries has been evaluated on a country-by-country basis.[1] It is beyond the scope of this paper to do so. Further, it is not my intention to regurgitate the existing literature that has already so eloquently identified the existing barriers to the full realization of the CDM's potential, including its development dividend.[2] Instead, what I would like to do in this paper is to explain the modest performance of Southeast Asia in the CDM world, which I suggest, is largely due to demand-side factors that are related to the structure of the compliance market that the CDM has created. These demand-side factors are international and, therefore, beyond the control of any one country. I will propose that a supply-side initiative that should be developed is greater regional cooperation amongst the Southeast Asian countries to achieve the

[1] See, for example, "CDM Development in Indonesia-Enabling Policies, Institutions and Programmes, Issues and Challenges 2006" (Second Edition), compiled by "CDM Development in Indonesia" Editorial Team in Cooperation with Yayasan Pelangi Indonesia, under the Supervision of New Energy and Industrial Technology Development Organization (NEDO) Representative Office in Jakarta; Dang Hanh, "Assessment of CDM Capacity Building Activities in Cambodia, Lao PDR and Vietnam — Lessons Learned", Hamburg Institute of International Economics (HWWA) Discussion Paper 351. This study was conducted under the project "EU-Asia Institutional co-operation and Multinational Dialogues on Enabling the Meaningful Participation of Cambodia, Lao PDR and Vietnam in the CDM" funded by "European Union under Asia-Wide Programmes"; Development Bank of Japan (Representative Office in Singapore), "Climate Change & CDM Projects in Southeast Asia", DBJ Research Report Vol. 41, April 21 2006, http://www.dbj.go.jp/singapore/english/ (accessed on 11 August 2007).

[2] The "development dividend" refers to the socio-economic and environmental benefits that should be generated by the CDM. Source: A. Cosbey *et al.*, *Realising the Development Dividend: Making the CDM Work for Developing Countries (Phase I Report — Prepublication Version)*, (Canada: International Institute for Sustainable Development, May 2005), online: www.iisd.org

economies of scale that come from pooling of resources to develop technical know-how and information systems that will aid project developers, financial institutions and governments. The Association of Southeast Asian Nations (ASEAN) provides an established regional framework to pursue such cooperation, which is in line with the general direction that ASEAN is taking towards greater cooperation in areas such as the environment and energy. Indeed, energy, the environment, climate change and sustainable development is a key theme of the 13th ASEAN Summit (and related summits) in November 2007.[3] The ASEAN Leaders' Declaration on Environmental Sustainability will also be signed at this Summit.[4]

Section 2 of this paper provides a brief introduction to Southeast Asia. Section 3 analyses the CDM performance of the Southeast Asian countries. It will be argued that, apart from the domestic factors that render Southeast Asia countries (in general) less attractive as CDM host countries, the modest performance of these countries is attributable to structural market forces. Section 4 proposes more regional cooperation amongst the Southeast Asian countries, which may be pursued within the established regional framework provided by the ASEAN. In Section 5 I suggest that as an international financial centre that has been attracting global "green" investors and has plans to be the ASEAN region's environmental financing hub, Singapore has a unique facilitative role to play in the implementation of the CDM in Southeast Asia. Section 6 concludes the paper.

2. Introduction to Southeast Asia and ASEAN

2.1. *Southeast Asia*

Southeast Asia comprises of Brunei, Cambodia, Indonesia, Laos PDR, Malaysia, Myanmar, the Philippines, Singapore, Thailand and Vietnam. These countries also make up the Association of Southeast

---

[3] "Environment to take centre stage at Asean summit in Nov", *Business Times*, reproduced on the ASEAN 13th Summit (Singapore) website http://www.13thaseansummit.org.sg/asean/index.php/web/summit_overview/overview (accessed on 15 August 2007).

[4] *Ibid.*

Asian Nations (ASEAN).[5] However, Brunei is not a party to the United Nations Framework Convention on Climate Change as well as the Kyoto Protocol. It will therefore be excluded from discussion in this paper.

The countries that make up the region are at very different stages of nation-building and economic development. For example, Cambodia, Lao PDR and Myanmar are classified as Least Developed Countries. Singapore has nearly reached the status of "developed country" while Malaysia is on the way to reaching such a level.[6] Vietnam has a rapidly growing market economy, but still faces concerns related to poverty alleviation. The information provided in Table 1 below provides some useful data on key selected indicators. Therefore, when discussing Southeast Asia and ASEAN, one must bear in mind that this region is highly disparate. The various countries have different capacities to participate in the CDM and in the wider global dialogue about climate change. However, what is evident is that rapid economic development in Asia for the past few decades has put severe strain on the natural environment. To the extent that the CDM can contribute to bringing about sustainable development in the region, and bend the carbon trajectory of these developing economies downwards, it is important that efforts be made to fulfill the potential of the CDM in Southeast Asia.

2.2. *ASEAN*[7]

The Association of Southeast Asian Nations or ASEAN was established on 8 August 1967 in Bangkok by the five original Member Countries, namely, Indonesia, Malaysia, Philippines, Singapore, and Thailand. Brunei Darussalam joined on 8 January 1984, Vietnam on 28 July 1995, Lao PDR and Myanmar on 23 July 1997, and Cambodia on 30 April 1999. The original objective of ASEAN was to

---

[5] More information on ASEAN is available on its website www.aseansec.org

[6] World Bank, *World Development Indicators 2005*, Washington, DC.

[7] The source for most information in this section is the ASEAN official website http://www.aseansec.org

Table 1.

Table 1

Selected basic ASEAN indicators

as of 25 July 2007

ASEAN Statistics

Country	Total land area	Total population[3/]	Population density[3/]	Annual population growth[3/]	Gross domestic product[4/] at current prices	Gross domestic product per capita at current prices		Merchandise trade[6/]			Foreign direct investments inflow[7/]	
								Exports	Imports	Total trade		
	km²	thousand	persons per km²	percent	US$ million	US$	US$ PPP[5/]	US$ million	US$ million	US$ million	US$ million	US$ million
	2006	2006	2006	2006	2006	2006	2006	2006	2006	2006	2005	2006
Brunei Darussalam	5 765	383	66	3.5	11 571.8	30 213.6	25 094.1	7 619.4	1 488.9	9 108.3	288.5	433.5
Cambodia[1/]	181 035	14 163	78	2.5	7 256.3	512.3	3 226.0	3 514.4	2 923.0	6 437.4	381.2	482.6
Indonesia	1 890 754	222 051	117	1.3	364 258.8	1 640.4	4 321.3	100 798.6	61 065.5	161 864.1	8 336.0	7 513.7
Lao PDR	236 800	6 135	26	2.5	3 521.8	574.0	2 332.1	402.7	587.5	990.2	27.7	187.4
Malaysia	330 252	26 686	81	2.1	156 924.2	5 880.4	12 184.9	157 226.9	128 316.1	285 543.0	3 964.8	6 059.7
Myanmar[2/]	676 577	57 289	85	2.3	11 951.0	208.6	1 958.8	3 514.8	2 115.5	5 630.3	235.9	143.0
The Philippines	300 000	86 910	290	2.0	117 457.1	1 351.5	5 332.7	47 410.1	51 773.7	99 183.8	1 854.0	2 345.0
Singapore	704	4 484	6 369	3.3	132 273.4	29 499.6	32 379.6	271 607.9	238 482.0	510 089.9	15 001.9	24 055.6
Thailand	513 254	65 233	127	0.7	206 645.1	3 167.8	9 163.5	121 579.5	127 108.8	248 688.3	8 957.0	9 750.7
Viet Nam	329 315	84 222	256	1.3	60 965.2	723.9	3 373.3	37 033.7	40 236.8	77 270.5	2 020.8	2 360.0
ASEAN	**4 464 456**	**567 557**	**127**	**1.5**	**1 072 824.7**	**1 890.3**	**5 210.2**	**750 708.0**	**654 097.8**	**1 404 805.8**	**41 067.8**	**53 331.2**

Sources: ASEAN Finance and Macro-economic Surveillance Unit Database and ASEAN Statistical Yearbook 2006 (compiled/computed from data submission and/or websites of ASEAN Member Countries' national statistical offices, central banks, and other relevant government agencies)
ASEAN Trade Database as of 18 July 2007
IMF World Economic Outlook Database as of April 2007

Symbols used
- not available as of publication time
x not available/not compiled
Data in italics means it has been revised from the previous posting.

Notes
1/ GDP per capita is preliminary figure computed using the latest/revised GDP estimate per Cambodia's submission dated 16 July 2007 and estimated population from the ASEAN Statistical Yearbook 2006.
2/ Myanmar's GDP is based on fiscal year from April to March of the following year, and computed using derived foreign exchange rate based on IMF WEO database April 2007.
3/ Refers to/based on mid-year total population based on the latest updates from AMCs as of 25 July 2007
4/ Brunei Darussalam data is based on the official revised GDP data per Brunei Darussalam submission dated 19 June 2007 and average annual foreign exchange rate, using official country data for Q1-Q3 2006 and Singapore's data as proxy for Q4 2006.
5/ Recomputed based on IMF WEO Database April 2007 estimates and actual country data
6/ Based on country updates as of 18 July 2007
7/ Refers to net inflow of foreign direct investments as measured in the balance of payments; includes equity capital, reinvested earnings, and inter-company loans, unless otherwise indicated; data is preliminary as of 14 May 2007; data for Myanmar covers the fiscal year from April to March of the following year; data for the Philippines covers reinvested earnings of local banks only; data for Thailand is based on new series, which now includes estimated reinvested earnings; detailed information, such as intra- and extra-ASEAN, country of origin, etc., on the revised 2005 data and preliminary 2006 data is not yet available as of posting date.

maintain peace and stability in a politically turbulent region. Since then, the regional association has grown and evolved. It now cooperates in areas ranging from public health, the environment to economic integration. Launched in 1992, steps towards creating an ASEAN Free Trade Area (AFTA) have been apace. It aims to promote the region's competitive advantage as a single production unit, and there has already been elimination of tariff and non-tariff barriers among Member Countries. Other economic integration activities within ASEAN include the Roadmap for Financial and Monetary Integration of ASEAN in four areas, namely, capital market development, capital account liberalization, liberalization of financial services and currency cooperation, trans-ASEAN transportation network consisting of major inter-state highway and railway networks, sea lanes for maritime traffic, and major civil aviation links; and the private sector-led ASEAN Hip-Hop Pass to promote intra-ASEAN tourism.

There is also ASEAN cooperation in the area of energy. The ASEAN Plan of Action for Energy Cooperation 1999–2004 was adopted by the 17th ASEAN Ministers on Energy Meeting on 3 July 1999 in Bangkok, Thailand. The six programme areas of the Plan of Action are:

(1) The ASEAN Power Grid (the pursuance of bilateral and cross-border interconnections that will later develop and form part of the regional grid);
(2) The Trans-ASEAN Gas Pipeline Infrastructure Project (the aim of which is to ensure greater security of gas supply, to reduce the region's dependence on imported energy as well as shifting from coal and petroleum to greater use of natural gas);
(3) Promoting clean coal technology;
(4) Energy efficiency and conservation;
(5) Renewable energy;
(6) Regional energy outlook, energy policy and environmental analysis.

Energy efficiency and renewable energy projects are suitable CDM opportunities. Given that both types of projects have been identified as focus areas in the ASEAN Plan of Action for Energy Cooperation, there are strong arguments to urge ASEAN to play a bigger role in developing regional cooperation for the CDM.

In the area of the environment, ASEAN cooperation goes back to 1987 initiating the ASEAN Subregional Environment Programme (ASEP) Phase I.[8] The Manila Declaration on the ASEAN Environment 1981, has as its objective "[t]o ensure the protection of the ASEAN environment and the sustainability of its natural resources so that it can sustain continued development with the aim of eradicating poverty and attaining the highest possible quality of life for the people of the ASEAN countries" and the adoption of the ASEP. This was followed by the ASEAN Declaration on Heritage Parks and Reserves (1984), the Agreement on the Conservation of Nature and Natural Resources (1985), the ASEAN Agreement on Transboundary Haze Pollution (2002), the ASEAN Declaration on Heritage Parks (2003), to name a few. The two main areas on which ASEAN cooperation is focused are therefore management of shared natural resources (biodiversity) and pollution control (transboundary haze). However, as has been argued by Koh and Robinson, the "ASEAN way" which emphasizes non-interference in ASEAN members' domestic affairs, and consensual decision-making, favors regional policy and "soft law" formulations (such as declarations). The general lack of concrete legal instruments for translating regional commitments into national level action has hindered implementation of environmental governance programs.[9] With regard to climate change, a future ASEAN Climate Change Initiative is currently under discussion. Climate change is also likely to receive greater attention in the run-up to the 13th ASEAN Summit in Singapore in November 2007 and the COP13/MOP3 in Bali.

---

[8] Koh, K. L. (compiler), *Selected ASEAN Documents on the Environment* (APCEL), pp. 90–113 (Singapore: 1996).

[9] Koh, K. L. & Nicholas A. Robinson, "Regional Environment Governance: Examining the Association of Southeast Asian Nations (ASEAN) Model" in Daniel C. Esty and Maria H. Ivanova (eds), *Global Environmental Governance* (Yale, USA: 2002). See also Koh, K. L. and N. A. Robinson, "Strengthening Sustainable Development in Regional Inter-Governmental Governance: Lessons from the 'ASEAN Way'", *Singapore Journal of International and Comparative Law* (2002) 6 pp. 640–682.

3. The Clean Development Mechanism in Southeast Asia

3.1. *The global context*

As of 24 July 2007, the CDM Executive Board has registered 742 projects. As can be seen from Table 2, there is 1 registered CDM project in Cambodia, 9 in Indonesia, 10 in the Philippines, 2 in Thailand and 2 in Vietnam. Malaysia tops the chart with 16 projects. There are none in Singapore. The CDM projects in the Southeast Asian countries are mostly small-scale ones which are not expected to deliver large quantities of CERs. The projects fall into four main categories: renewable energy, energy efficiency, biomass, and methane recovery and utilization.

In comparison, there are 258 registered CDM projects in India, making it the top CDM "destination" in the world (in terms of numbers of projects), followed by Brazil, China, Mexico and South Korea (see Table 3). However, India will not be the larger producer of Certified Emission Reductions (CERs) — China is expected to produce 44.29% of the expected yearly CERs. Eight Chinese hydroflurocarbon (HFC) reduction projects alone account for about 35% of the total expected credits from registered CDM projects globally. China's lead position in the CDM market is largely attributable to these projects based on the destruction of gases with high global warming potential (GWP). HFC-23 is generated as a by-product of HCFC22 manufacture, which is used as a refrigerant or as a feedstock, for example, in Teflon manufacture.[10] The 100-year GWP of HFC-23 is 11700. This means that, under the rules of the CDM which converts the other six Kyoto Protocol regulated gases to carbon dioxide and hence CERs based on their GWPs, a ton of HFC-23 abated is considered equivalent to 11700 tons of carbon dioxide.[11]

---

[10] J. Ellis & S. Kamel, "Overcoming Barriers to Clean Development Mechanism Projects" (Paris: Organisation for Economic Co-operation and Development (OECD)/ International Energy Agency, May 2007) at p. 10, footnote 4.

[11] Michael Wara, "Measuring the Clean Development Mechanism's Performance and Potential", Working Paper #56, July 2006, Program on Energy and Sustainable Development at the Centre for Environmental Science and Policy, Stanford University, at p. 26.

Table 2. CDM Projects in the Southeast Asian countries.

Registered	Title	Host Parties	Other Parties	Scale	Reductions**	Ref
10 Aug 06	Angkor Bio Cogen Rice Husk Power Project	Cambodia	Japan	Small	51620	0363
31 Aug 06	Methane Capture and Combustion from Swine Manure Treatment Project at PT Indotirta Suaka Bulan Farm in Indonesia	Indonesia	Japan	Large	166000	0450
09 Dec 06	Lampung Bekri Biogas Project	Indonesia	Switzerland United Kingdom of Great Britain and Northern Ireland	Small	18826	0616
11 Dec 06	Darajat Unit III Geothermal Project	Indonesia	United Kingdom of Great Britain and Northern Ireland	Consolidated Methodology	652173	0673

(*Continued*)

Table 2. (*Continued*)

Registered	Title	Host Parties	Other Parties	Scale	Reductions**	Ref
06 Feb 06	CDM SOLAR COOKER PROJECT Aceh 1	Indonesia	Germany	Small	3500	0218
17 Jun 06	MSS Biomass 9.7 MWe Condensing Steam Turbine Project	Indonesia		Small	56116	0406
31 Aug 06	MNA Biomass 9.7 MWe Condensing Steam Turbine Project	Indonesia		Small	46322	0407
20 May 07	PT Navigat Organic Energy Indonesia Integrated Solid Waste Management (GALFAD) Project in Bali, Indonesia	Indonesia	Japan	Large/ Consolidated/ Small	123423	0938
29 Sep 06	Indocement Alternative Fuels Project	Indonesia	Netherlands Finland	Consolidated	144414	0493

(*Continued*)

Table 2. (*Continued*)

Registered	Title	Host Parties	Other Parties	Scale	Reductions**	Ref
27 Oct 06	Indocement Blended Cement Project	Indonesia	Netherlands Finland	Consolidated	469750	0526
07 Apr 07	Energy Efficiency Improvement Project at A Beer Brewery in Lao PDR	Lao People's Democratic Republic	Japan	Small	3338	0930
07 Apr 06	Replacement of Fossil Fuel by Palm Kernel Shell Biomass in the production of Portland Cement	Malaysia	France	Consolidated	61946	0247
21 Jul 06	Kina Biopower 11.5MW EFB Power Plant	Malaysia	Japan	Small	230019	0385
21 Jul 06	Seguntor Bioenergy 11.5MW EFB Power Plant	Malaysia	Japan	Small	230019	0386

(*Continued*)

Table 2. (*Continued*)

Registered	Title	Host Parties	Other Parties	Scale	Reductions**	Ref
02 Sep 06	Bentong Biomass Energy Plant in Malaysia	Malaysia	Canada	Small	380934	0501
04 Sep 06	ENCO Biomass Energy Plant in Malaysia	Malaysia	Canada	Small	70316	0502
02 Sep 06	Johor Bundled Biomass Steam Plant in Malaysia	Malaysia	Canada	Small	130505	0503
24 Sep 06	Jendarata Steam & Power Plant	Malaysia	Denmark	Small	8851	0558
29 Sep 06	Krubong Melaka LFG Collection & Energy Recovery CDM Project	Malaysia	Japan	Consolidated	57830	0323
08 Apr 07	Kim Loong Methane Recovery for Onsite Utilization Project at Kota Tinggi, Johor, Malaysia.	Malaysia	Switzerland	Large	57656	0867

(*Continued*)

Table 2. (*Continued*)

Registered	Title	Host Parties	Other Parties	Scale	Reductions**	Ref
10 Jun 06	LDEO Biomass Steam and Power Plant in Malaysia	Malaysia	Canada	Small	208871	0395
10 Jun 06	SEO Biomass Steam and Power Plant in Malaysia	Malaysia	Canada	Small	216831	0402
23 Apr 06	Sahabat Empty Fruit Bunch Biomass Project	Malaysia	United Kingdom of Great Britain and Northern Ireland	Small	53986	0288
03 Mar 07	Factory energy-efficiency improvement project in Malaysia (MAPREC, PRDM, PSCDDM, PAVCJM, PCM)	Malaysia	Japan	Small	1312	0757

(*Continued*)

Table 2. (*Continued*)

Registered	Title	Host Parties	Other Parties	Scale	Reductions**	Ref
20 May 07	Landfill Gas utilization at Seelong Sanitary Landfill, Malaysia	Malaysia	Denmark	Consolidated	108335	0927
03 Mar 07	Factory energy-efficiency improvement project in Malaysia (PHAAM, PCOM (PJ), PCOM (SA), PEDMA, MEDEM)	Malaysia	Japan	Small	6474	0759
24 Feb 06	Biomass Energy Plant-Lumut	Malaysia	Denmark	Small	32545	0249
01 Oct 06	Wastewater treatment using a Thermophilic Anaerobic Digestor at an ethanol plant in the Philippines	Philippines	Japan	Large	95896	0504

(*Continued*)

Table 2. (*Continued*)

Registered	Title	Host Parties	Other Parties	Scale	Reductions**	Ref
10 Dec 06	20 MW Nasulo Geothermal Project	Philippines	Netherlands	Consolidated	74975	0590
31 Jan 07	Paramount Integrated Corporation Methane Recovery and Electricity Generation	Philippines	United Kingdom of Great Britain and Northern Ireland	Small	7582	0605
23 Oct 06	Joliza Farms Inc. Methane Recovery	Philippines	United Kingdom of Great Britain and Northern Ireland	Small	3656	0607
28 Oct 06	Uni-Rich Agro-Industrial Corporation Methane Recovery and Electricity Generation	Philippines	United Kingdom of Great Britain and Northern Ireland	Small	2929	0609

(*Continued*)

Table 2. (*Continued*)

Registered	Title	Host Parties	Other Parties	Scale	Reductions**	Ref
30 Oct 06	Gaya Lim Farm Inc. Methane Recovery	Philippines	United Kingdom of Great Britain and Northern Ireland	Small	3130	0611
21 Oct 06	Gold Farm Livestocks Corporation Methane Recovery and Electricity Generation	Philippines	United Kingdom of Great Britain and Northern Ireland	Small	2929	0612
13 Apr 07	San Carlos Renewable Energy Project	Philippines		Small	37658	0931
10 Sep 06	NorthWind Bangui Bay Project	Philippines	Netherlands Finland	Consolidated	56788	0453
05 May 07	Philippine Sinter Corporation Sinter Cooler Waste Heat Recovery Power Generation Project	Philippines	Japan	Consolidated	61702	0963

(*Continued*)

Table 2. (*Continued*)

Registered	Title	Host Parties	Other Parties	Scale	Reductions**	Ref
18 Jun 07	A.T. Biopower Rice Husk Power Project in Pichit, Thailand	Thailand	Japan	Consolidated	70772	1026
16 Jun 07	Korat Waste To Energy	Thailand	United Kingdom of Great Britain and Northern Ireland	Large	310843	1040
26 Jun 06	Song Muc Hydro Power Station Regeneration Project in Vietnam	Vietnam	Japan	Small	4306	0435
04 Feb 06	Rang Dong Oil Field Associated Gas Recovery and Utilization Project	Vietnam	Japan United Kingdom of Great Britain and Northern Ireland	Large	677000	0152

** Estimated emission reductions in metric tonnes of CO_2 equivalent per annum (as stated by the project participants).
Source: UNFCCC website (as of 24 July 2007).

Table 3. The Major CDM "Destinations".

Country	Number of Registered Projects	% of Total Number of Registered Projects Worldwide	Expected Average Annual Certified Emission Reductions (CERs) From Registered Projects (% of Total Global)
China	101	13.61	44.29
India	258	34.77	14.75
Brazil	104	14.02	10.77
South Korea	14	1.89	8.85
Mexico	90	12.13	3.96

Source: UNFCCC website (as of 24 July 2007).

One can see why reducing HFC-23 is an economically attractive option for CDM project developers. Thus, these projects which reduce end-of-pipe emissions of HFC are popular, as they have a short lead-time, offer large volumes of credits for a low capital investment and mitigation cost, and additionality assessments are relatively straightforward.[12] The original objective of the CDM was to promote the development of low-carbon energy infrastructure in the developing world, to achieve sustainable development goals as well as substitute for early retirement of expensive carbon-intensive energy infrastructure in the developed world.[13] It is therefore very disappointing that the CDM pipeline bears little resemblance to this vision. Instead, the subsidy provided by Annex I countries (via their CERs purchases) are going towards insuring that high Global Warming Potential (GWP) industrial gases such as HFC-23, nitrous oxide and methane (CH4) emitted in non-Annex I nations are captured and destroyed.[14]

[12] *Ibid.* at p. 10.

[13] *Ibid.*

[14] *Ibid.* This is a poor use of Annex I party resources because to abate all developing world HFC-23 emissions would cost approximately $31 million per year. Instead, through participation in the CDM, the Annex I nations will likely pay between €250 and €750 million to abate 67% of Non-Annex I HFC-23 emissions.

As most of the HFC-23 abatement opportunities are found in China (and to a lesser extent, India), it is not surprising that China is trail-blazing the CDM market in terms of the potential number of CERs it is expected to generate and has gained a major proportion of the CDM investment pie.

As rational economic actors, CERs purchasers and project developers are placing their dollars on the projects that will yield the most CERs for the least amount of money. In this regard, the CDM is achieving one of its key objectives, that is, the creation of least-cost greenhouse gas abatement opportunities for Annex I parties. However, the CDM has not been able to address the unequal global distribution of CDM projects. Southeast Asia, with its absence of high-yield CDM opportunities, is a case in point. Nonetheless, these are but early days. It is possible that once the "cheap and easy to achieve" CDM opportunities have been fully utilized, the market will then readjust towards exploiting other CDM opportunities. This hinges, in part, on the outcome of the Subsidiary Body for Scientific and Technological Advice (SBSTA)'s deliberations on whether to allow HFC destruction from new HCFC22 production to qualify under the CDM rules. This issue has caused great controversy because of the fear of creating a "perverse incentive" in that the CDM could encourage industrial facilities to increase production of HCFC22 in order to generate more HFC23 to be destroyed so as to create more CERs. HCFC22 is already to be phased out under the 1987 Montreal Protocol but developing countries (including China) are allowed to expand production until 2015 and only stop production in 2040.[15] Should the SBSTA recommend to the CDM Executive Board that HFC destruction from new HCFC22 production should be allowed to qualify under the CDM rules, the CDM will be taking steps backwards from its objective of assisting developing countries in

[15] Robin Lancaster, "The Credit for destruction", *Carbon Finance*, 15 February 2007. Also see "Relationship between Efforts to Protect the Stratospheric Ozone Layer and Efforts to Safeguard the Global Climate System: Issues Relating to Hydrofluorocarbons and Perfluorocarbons", UNFCCC website, http://unfccc.int/methods_and_science/other_methodological_issues/items/2311.php (accessed on 12 August 2007).

their quest for sustainable development (as these HFC23 projects contribute little to sustainable development), and its environmental integrity will suffer. For Southeast Asia, it will mean that CDM investment will continue to flow towards HFC23 projects and less likely to be directed to renewable energy and energy efficiency projects, which are the types of projects that Southeast Asian countries are able and interested to implement.

3.2. *Renewable energy*

Renewable energy projects face some hurdles in the operation of the CDM. These barriers are not unique to Southeast Asia at all, but certainly form part of the explanation of the low number of CDM projects in Southeast Asia.

Despite renewable energy equipment having become cheaper over the years (prompting some commentators to declare renewable energies as the new "sunrise industry"[16]), equipment costs of renewable energy projects remain significant higher per emission reduction than the costs of other types of CDM projects, such as methane flaring. The contribution of the CERs to the overall revenue stream is therefore relatively smaller for renewable energy projects than for other types of projects. This places renewable energy projects, especially the small-scale types which tend to be the sort of projects implemented in Southeast Asia so far, at a disadvantage. In addition, renewable energy projects such as geothermal facilities have a long operation lifespan, which will extend beyond 2012, the First Commitment Period under the Kyoto Protocol. As there has been uncertainty surrounding the future of the Kyoto Protocol after 2012 till recently when member states' commitment towards negotiating the continuation of theclimate change treaty regime post-2012 was announced at COP12/MOP2 in Nairobi, project developers have been reluctant to invest in

[16] See, for example, *The Economist*, "Sunrise for Renewable Energy?", 8 December 2005, Technology Quarterly.

long-term projects, preferring projects that would yield credits before 2012. In this regard, the lack of confidence in the market has led to a skew in the types of projects that have been undertaken for purposes of fulfilling the legal obligations under the First Commitment Period of the Kyoto Protocol. Also, the lack of market confidence in the future of the Kyoto Protocol regime has made it difficult for CDM projects in general to attract project financing. CER purchasers have tended to restrict their involvement in CDM projects to a commitment to pay for CERs upon delivery, rather than fund the underlying project. In developing countries like Cambodia, Laos, Myanmar and the Philippines, where it is difficult to raise capital, the absence of funding for the underlying projects by potential CERs purchasers is likely to ensure the non-viability of these projects from the outset. Finally, the transaction costs of CDM projects are high (including the costs of consultants and legal fees for the negotiation and drafting of contracts, registration fees, etc), amounting to another hurdle that must be negotiated.

3.3. *Domestic factors*

It is beyond the scope of this paper to analyze the domestic factors that affect the performance of the CDM in Southeast Asia on a country basis. In any event, there is a voluminous literature that does precisely this sort of country-specific analysis, largely the result of capacity-building initiatives by foreign governments and inter-governmental organizations that are working to further the implementation of the CDM in Southeast Asia. Some of the domestic factors that have been identified include

- the general lack of public awareness of climate change;
- concerns that addressing climate change will divert public resources away from other pressing needs such as education, healthcare and public housing;

Table 4. CDM volume-risk matrix for Asia.[17]

	Low Political Risk	High Political Risk
Low Volume of executable CDM opportunities	Singapore Taiwan Brunei Bhutan	Pakistan Afghanistan North Korea Cambodia Nepal Laos Myanmar
High Volume of executable CDM opportunities	Thailand The Philippines India Malaysia South Korea	China Indonesia Bangladesh Vietnam Sri Lanka

- convoluted bureaucratic approval procedures within the Designated National Authority or the lack of clear and transparent procedures, information of which is publicly available;
- limited domestic institutional capacity to effectively engage in market-based environmental mechanisms;
- political risk in certain Southeast Asian countries (see Table 4 below); and
- difficulties in obtaining project financing for CDM projects.

It should be noted that these domestic barriers to the successful implementation of the CDM are not unique to Southeast Asia. In fact, the Nairobi Framework which seeks to address the CDM capacity-building needs in Africa, may well be a "template" which

[17] Reproduced from page 10 of "Improving the Competitiveness of Southeast Asia on the Global CDM Market: Regional Cooperation in ASEAN on CDM in the Energy Sector", Discussion Paper, February 18 2005, http://www.capsd.nl/pages/documents/CDMASEANreport_000.pdf (accessed on 15 August 2007).

should be implemented in Southeast Asia as well (subject to modifications to suit local conditions).[18]

4. Regional Cooperation Within ASEAN

While ASEAN members may see themselves as natural competitors for CDM investment, it is arguable that there are overall benefits from cooperating at a regional level to boost the individual capacity of each country as a CDM host destination. As discussed earlier in this paper, ASEAN provides an established institutional framework for such regional cooperation. Further, cooperation to enhance the CDM capacity of ASEAN countries fits within the existing ASEAN cooperation programmes on energy and the environment. It should be made clear that I am not advocating the promotion of ASEAN as some sort of unified supplier of CERs. Pursuing this sort of integration will be taking too many steps ahead of ourselves. It is also likely that ASEAN member states will not support such integration because it arguably hints at the possibility of a cap-and-trade system akin to the European Union Emissions Trading Scheme sometime in the future. Putting aside the non-viability of such a pan-ASEAN emissions trading scheme, for economic, political and environmental efficacy reasons, ASEAN member states will not want to risk the possibility of setting the wheels in motion. All that is being suggested here is more coordination amongst the Southeast Asian countries in terms of information sharing, increasing technical and legal expertise, and regional capacity-building.

Steps in this direction have already been taken. The CDM-ASEAN Project, concluded in February 2005, aimed to support the implementation of the CDM in ASEAN countries through regional

[18] The "Possible Elements of the Nairobi Framework" contains a possible basic structure for the Nairobi Framework. The main objectives of the Nairobi Framework are (1) enhance Designated National Authorities/Establish Legal Framework; (2) Build CDM Capacity (in both public and private sector); (3) Promote Investment; (4) Information Exchange; (5) Coordination. www.unfccc.int (accessed on 1 August 2007).

capacity-building and improving the competitive position of ASEAN for investments in the energy sector. The project focused on the energy sector because of the large number of potential CDM projects and the regional relevance of the sector. The project culminated in the organization of two regional CDM seminars, the publishing of discussion papers on topics of ASEAN-wide interest, and concrete recommendations on the co-ordination of CDM implementation in ASEAN.[19] The project evaluated aspects of the role of ASEAN in CDM implementation, which are worth setting out here in some detail[20]:

(1) Improving Baseline Determination and Data Availability

A problem that CDM project developers frequently encounter is the inadequacy of data to calculate baseline emission factors. A centralised collection of such data and its continuous updating and publication would help address this problem. The ASEAN Secretariat and the ASEAN Centre for Energy were identified as "obvious candidates to coordinate these activities". It was also suggested that ASEAN could reduce transaction costs further by providing grid data for the region. As it has to collect this data anyway, there is little additional burden and consistent quality standards can be achieved. It was argued that with an increasing integration of the electricity grid in ASEAN, this would be of increasing importance and utility.

(2) Regional Capacity Building

The following areas were identified as requiring further capacity-building: (1) operational skills of the DNAs; (2) technical CDM project assessment, including baselines; (3) establishment of local operational entities; (4) project development, including the CDM life cycle,

---

[19] See "Improving the Competitiveness of Southeast Asia on the Global CDM Market: Regional Cooperation in ASEAN on CDM in the Energy Sector", Discussion Paper, 18 February 2005, http://www.capsd.nl/pages/documents/CDMASEANreport_000.pdf (accessed on 15 August 2007).
[20] *Ibid*, Chapter 8.

preparation of proposals and financing/transactional issues. It was pointed out that there have already been many donor-funded CDM capacity-building projects in the ASEAN region. However, the effectiveness of these projects was limited by, *inter alia*, overlap in the scope and content of the projects, the lack of follow-up, the lack of on-the-job training, and the failure to address the role of local government in the CDM.

(3) Financing CDM Projects

Finally, a range of possible actions at a regional level to address financing issues was recommended. These include:

- Developing CER risk mitigation mechanisms, which are important in most ASEAN countries;
- Improving the understanding of the CDM within local debt and equity markets and increasing willingness to invest;
- Establishing local support infrastructure such as Operational Entities, lowering project development costs and increasing local stakeholder involvement in project execution;
- As a longer — term option, establishing an ASEAN based fund to support CDM projects through either the purchase of credits or the supply of equity/debt to CDM projects;
- Increasing Singapore's involvement in the CDM as the country has the most mature capital markets in the region. It has the potential to meet CDM project financing requirements and be a regional trading platform for the CER commodity.

These detailed recommendations are sound and their implementation should be followed up in a consistent fashion as the initial steps towards regional cooperation within ASEAN on the CDM and the broader issue of climate change.

5. The Role of Singapore

Picking up on one of the recommendations under the category "Financing CDM Projects" above, I thought it would be useful to say a few words on the role that Singapore can play to facilitate the implementation of the CDM in the ASEAN region. A small city-state which does not have natural resources, Singapore has few, if any, CDM opportunities. However, upon its accession to the Kyoto Protocol,[21] Singapore has plans to make itself a regional centre for backing renewable energy projects and to become an environmental financing hub for the Asian region in general.[22] The launch of the Renewable Energy Exchange in Singapore by the Austrian-based non-governmental organization, the Renewable Energy and Energy Efficiency Partnership (REEEP) and Singapore's signing up to membership of this organization were steps in this direction.[23] The Renewable Energy Exchange acts as a "matchmaker" between potential investors and potential project developers. Potential investors are often on the lookout for profitable and viable investment opportunities. However, apart from the specialized few with a specific interest in socially responsible investment or sustainable development/ "green" investment, the majority of potential investors tend to shy away from renewable energy projects which are perceived to be riskier and to have less predictable revenue returns. The REEEP seeks to address this concern by conducting basic due diligence on potential projects so that, at least on a preliminary basis, the projects that are introduced to potential investors have investment potential. On the other hand, project developers are often in search for equity investors and joint venture partners. The REEEP plays the "matchmaker" by introducing potential investors to these project developers.[24]

---

[21] Singapore deposited its instruments of accession with the Secretary-General of the United Nations in New York on 12 April 2006 (*morning of 13 April, Singapore Time*). See "Singapore Accedes to the Kyoto Protocol", Ministry of Environment and Water Resources, 13 April 2006, online: Sprinter http://app.sprinter.gov.sg/data/pr/20060413999.htm. The Protocol came into force for Singapore on the 90th day after the date of deposit of the instruments.

[22] Choong Meng-Yew, "Singapore chosen as Base for Renewable Energy Projects", 23 June 2006 (the *Straits Times* (Singapore)).

[23] "Singapore Joins International Energy Partnership", Press Release (Singapore, 21 June 2006), REEEP.

[24] Interview with Dr. Mike Allen, Director, the Renewable Energy Exchange (Asia) Private Ltd (on file with author).

5.1. *Singapore as an environmental financing hub*

"Singapore, with its inherent expertise in financial services, is well positioned as a hub for environmental financing".[25]

Singapore has long been a key financial centre in the Asian region. There are more than 500 local and foreign financial institutions offering a wide array of financial products and services.[26] The local equity and debt capital markets are fairly mature and many global companies have operations in Singapore to take advantage of the depth and liquidity that these markets offer. Singapore is the fourth largest foreign exchange trading centre in the world. The Singapore Exchange (SGX) is Asia-Pacific's first de-mutualised and integrated securities and derivatives exchange and many of Asia's largest companies are listed on the SGX. The Monetary Authority of Singapore (MAS) is the regulatory authority that has oversight of all aspects of monetary, banking and financial activities in Singapore and is reputed for its prudential supervision which has, amongst other things, been credited as one of the reasons behind Singapore's ability to weather the Asian economic crisis in the 1990s.[27] Singapore is therefore well-placed to provide financial support services for CDM projects. It is also a natural candidate as a source of equity and debt financing for projects in the region. International companies, such as New York-listed energy firm AES Corp and London-listed Ecosecurities (which specialized in sourcing and trading of carbon credits), are already setting up base in Singapore.[28] A possible scenario is that Singapore will

---

[25] Speech by Dr Yaacob Ibrahim, Minister for the Environment and Water Resources, at "Emissions Trading — Economic Opportunities in Asia" Seminar and Launch of the Sustainable Energy Association of Singapore, 12 July 2006.

[26] A complete list of financial institutions with presence in Singapore is available on the Monetary Authority of Singapore website www.mas.gov.sg

[27] See "Singapore as a Regional Financial Centre", Denis Hew, Institute of Southeast Asian Studies, presented at "The Role of Capital Markets in Asian Economic Development", 7–8 March 2002, organized by the Tokyo Club Foundation for Global Studies.

[28] Matthew Phan, "S'pore draws global green investors", *The Business Times* (28 March 2007).

continue to attract such investors to establish their presence, from which they will be able to tap on the CDM opportunities in the wider Southeast Asian region.

6. Conclusion

This paper has sought to provide an overview of the performance of the CDM in Southeast Asia. In brief, the modest performance of Southeast Asia is attributable to structural forces in the CDM market as well as domestic factors that hinder the realization of the CDM's potential. This paper argues that one of the ways by which the individual Southeast Asian countries can enhance its CDM capacity is through regional cooperation, particularly on the fronts of information exchange and institutional capacity-building. ASEAN provides an established institutional framework for the pursuit of such cooperation. Finally, the paper also analyzes the role that Singapore can play to facilitate the implementation of the CDM in the Southeast Asian region.

CHAPTER 5

CDM IN CHINA

Yang Xing
Hunan Normal University, Changsha, Hunan Province, China
Wang Xi
Law School, Shanghai Jiao Tong University, Shanghai, China
wangxi@sjtu.edu.cn

CDM is one of the most important mechanisms stipulated in the Kyoto Protocol. In order to implement the mechanism and realize the GHG emissions reduction target of the Protocol, China has taken a lot of actions for implementing CDM. Based on these actions, China will make a great contribution to world-wide GHG emissions reduction. China is working hard to change to a low carbon economy.

Introduction

The 1997 Kyoto Protocol (hereinafter referred to as "the Protocol") entered into force on 16 February 2005. The Protocol shares the objective, principles and institutions of the Framework Convention on Climate change (hereinafter referred to as "the Convention"), but it significantly strengthens the Convention by committing Annex I Parties to individual, legally-binding targets to limit or reduce their greenhouse gas (GHG) emissions. Only Parties to the Convention that have also become Parties to the Protocol (i.e. by ratifying, accepting, approving, or acceding to it) will be bound by the Protocol's commitments. 175 Parties have ratified the Protocol to date. Of these, 36 industrialized countries and the European Economic Community (EEC) are required to reduce greenhouse gas emissions to levels below those specified for each of them in the Protocol. The individual targets for Annex I Parties are listed in Annex B of the Kyoto Protocol. These add up to a total cut in greenhouse-gas emissions of at least 5% from the 1990 levels in the first commitment period of 2008–2012. In order to assist Annex I Parties to realize their GHG emission reduction targets, the Protocol provides three different flexible mechanisms.

The Clean Development Mechanism (CDM) is one of the three flexible mechanisms provided by the Protocol. CDM allows industrialized countries to acquire emissions credits (the right to emit greenhouse gases) by paying for emissions reduction measures in developing countries that do not have emissions targets. These measures must also contribute to the sustainable development of the recipient country. CDM is the only mechanism that is carried out between developed and developing countries.

The purpose of the CDM is to assist Parties not included in Annex I in achieving sustainable development and contributing to the ultimate objective of the Convention. It is also to assist Parties included in Annex I in achieving compliance with their quantified emission limitation and reduction commitments under Article 3.[1] With this stated purpose, CDM projects will offer a lot of opportunities for China's sustainable development. CDM is accordingly regarded as a project-based, win-win mechanism — that is to say, on the one hand, CDM can reduce the total cost of GHG emissions reduction in developed countries; on the other hand, developing countries can obtain additional funds and technologies from developed countries. As the largest developing country in the world, China has taken an active part in the implementation of CDM projects. CDM has some direct or indirect impacts on China. This paper will make a brief introduction to the implementation of CDM in China. Four examples of CDM projects in China will be introduced.

---

[1] Item 2 of Article 12, the Kyoto Protocol.

1. The Implementation of CDM in China

1.1. *Regulations of CDM in China*

Enacting legislation is one of the basic means to comply with international obligations. The "Interim Measures for Operation and Management of Clean Development Mechanism Projects In China" was promulgated by the National Development and Reform Commission (NDRC) and the National Scientific Technology Ministry (NSTM) on 30 June 2004. This is the first administrative regulation concerning CDM. On 12 October 2005, NDRC and NSTM amended The "Interim Measures for Operation and Management of Clean Development Mechanism Projects". The name of the regulation was modified to become "Measures for Operation and Management of Clean Development Mechanism Projects in China". This regulation deals with some important issues concerning the CDM.

This regulation is composed of five parts: I General Provisions; II Permission Requirements; III Institutional Arrangement for Project Management and Implementation; IV Project Procedures; V Other Provisions. This regulation will be examined in detail later.

1.1.1. *General provisions*

Under this part, Articles 2 and 4 are the most important provisions. Article 2 stipulates the nature and legislative purpose of CDM. Article 4 stipulates the priority areas for CDM projects.

As for the nature of CDM, Article 2 stipulates that CDM is a project-based mechanism under which developed country Parties are to co-operate with developing country Parties in order to meet part of the GHG emission reduction obligations of the former.

Article 2 also provides that the purpose of this mechanism is to assist developing country Parties in achieving sustainable development and contributing to the realization of the ultimate objective of the Convention; as well as to assist developed country Parties in achieving compliance with their quantified GHG emission limitation and reduction commitments. The core of CDM is to allow developed country Parties, in cooperation with developing country Parties, to acquire "certified emission reductions" (CERs) generated by the projects implemented in developing countries.[2]

As for the priority areas of CDM projects, Article 4 provides that the priority areas for CDM projects in China are energy efficiency improvement, development and utilization of new and renewable energy, and methane recovery and utilization.

1.1.2. *Permission requirements*

CDM projects must be in conformity with the Permission Requirements. The regulation stipulates seven Permission Requirements as follows.

Firstly, CDM project activities shall be consistent with China's laws and regulations, sustainable development strategies and policies, and the overall requirements for national economic and social development planning.[3]

Secondly, the implementation of CDM project activities shall conform to the requirements of the Convention, the Protocol and relevant decisions by the Conference of the Parties.[4]

Thirdly, the implementation of CDM project activities shall not introduce any new obligations for China other than those under the Convention and the Protocol.[5]

Fourthly, funding for CDM projects from developed country Parties shall be additional to their current official development assistance and their financial obligations under the Convention.[6]

---

[2] Article 2 of Measures for Operation and Management of Clean Development Mechanism Projects in China.

[3] Article 6 of Measures for Operation and Management of Clean Development Mechanism Projects in China.

[4] Article 7 of Measures for Operation and Management of Clean Development Mechanism Projects in China.

[5] Article 8 of Measures for Operation and Management of Clean Development Mechanism Projects in China.

[6] Article 9 of Measures for Operation and Management of Clean Development Mechanism Projects in China.

Fifthly, CDM project activities should promote the transfer of environmentally sound technology to China.[7]

Sixthly, Chinese corporate entities within the territory of China are eligible to conduct CDM projects with foreign partners.[8]

Seventhly, CDM project owners shall submit to the Designated National CDM Authority the following documents: CDM project design document, certificate of enterprise status, general information of the project and a description of the project financing.[9]

1.1.3. *Institutional arrangement for project management and implementation*

The management institutions will be set out in part 1.2. This part will only lay down the implementation institutions. According to Article 17, Project owners, which refer to the enterprises funded or held by Chinese individuals or corporate entities, are responsible for implementing CDM projects. In the course of implementation, these enterprises shall[10]:

(1) Undertake CDM project negotiations with foreign partners;
(2) Be responsible for the construction of the project and report periodically to NDRC on the progress;
(3) Implement the CDM project activity, develop and implement a project monitoring plan to ensure that the emission reductions are real, measurable, long-term and additional, and subject the project to the supervision of NDRC;
(4) Contract with designated operational entities to validate the proposed project activity and verify emission reductions of the

[7] Article 10 of Measures for Operation and Management of Clean Development Mechanism Projects in China.

[8] Article 11 of Measures for Operation and Management of Clean Development Mechanism Projects in China.

[9] Article 12 of Measures for Operation and Management of Clean Development Mechanism Projects in China.

[10] Article 17 of Measures for Operation and Management of Clean Development Mechanism Projects in China.

project activity; provide necessary information and monitoring record, submit the information to NDRC for record purposes; and protect confidential information of the state and any businesses in accordance with the relevant laws and regulations;
(5) Report to NDRC on the CERs issued;
(6) Assist NDRC and the Board in investigating relevant issues and respond to any inquiries; and
(7) Undertake other necessary obligations.

1.1.4. *Project procedures*

Firstly, this regulation stipulates the Procedures for the application and approval of CDM projects. Article 18 provides as follows:

(1) Project owner, or together with its foreign partner, submits to NDRC its project application, and documents required by Article 12 above. The relevant departments and local governments may facilitate such project application;
(2) NDRC entrusts relevant organizations for expert review of the project submitted. The review shall be concluded within 30 days;
(3) NDRC submits those project applications reviewed by the experts to the Board;
(4) NDRC approves, jointly with the Ministry of Science and Technology (MOST) and the Ministry of Foreign Affairs (MFA), projects based on the conclusions made by the Board, and issues approval letters accordingly;
(5) NDRC will make a decision on a project application within 20 days (excluding the expert review time) from the date of accepting the application. The time limit for decision-making may be extended to 30 days, with the approval of the Chair or the Vice-chair of NDRC if a decision could not be made within 20 days. The project applicant should be informed of such a decision and its reasons.
(6) Project owner invites designated operational entity to validate the project for registration; and

(7) Project owner shall report to NDRC the approval decision by the CDM Executive Board within 10 days from the date of receiving the notice from the Executive Board.

Secondly, the regulation provides the procedures for project implementation, monitoring and verification. Article 20 provides the procedure as follows:

(1) Project owner is responsible for presenting to NDRC and the designated operational entity its project implementation and monitoring reports;
(2) NDRC is authorized to supervise the implementation of the project to ensure the quality of the activity;
(3) The contracted designated operational entity verifies the emission reductions of the project activity and submits a certification report to the CDM Executive Board, which will then issue CERs for the project and inform the project participants of its decision; and
(4) NDRC or other organizations entrusted by NDRC will put the CERs issued by the CDM Executive Board in file for record.

Domestic Procedures for CDM Project Approval

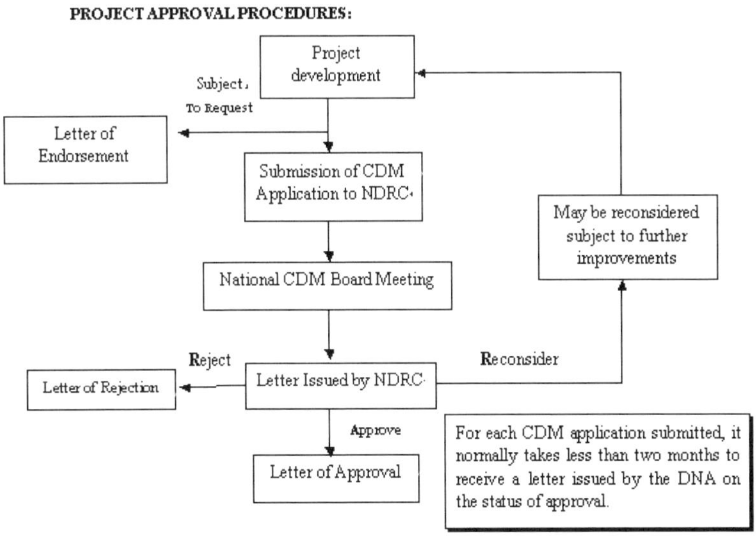

1.1.5. *Other provisions*

Firstly, Part V of the regulation defines some important legal terms. For example, developed country Parties are defined as the parties included in Annex I of the Convention. CDM Executive Board refers to the board as defined in Article 12 of the Protocol for the purpose of supervising CDM. As for Operational entity, Article 23 provides that it refers to the entity as defined in Article 12 of the Protocol for the purpose of validation as well as verification and certification of CDM project activities.

Secondly, Part V provides the allocation ratio of the proceeds from the transfer of CERs. The allocation ratio is defined as follows[11]:

(1) the Government of China takes 65% of the CER transfer benefit from HFC and PFC projects;
(2) the Government of China takes 30% of the CER transfer benefit from N_2O project;
(3) the Government of China takes 2% of the CER transfer benefit from CDM projects in the priority areas defined in Article 4 and forestation projects.

1.2. *National project management institutions of CDM in China*

1.2.1. *National climate change coordination committee (hereinafter referred to as "the Committee")*

China attaches great importance to addressing climate change issues. As early as 1990, the Committee was established under the Environmental Protection Committee of the State Council at that time. Mr. Song Jian, State Councilor, chaired the Committee. The Committee was reorganized in 1998. In the past few years, the Committee provided guidance to central government departments and local governments on addressing climate change issues.

[11] Article 24 of Measures for Operation and Management of Clean Development Mechanism Projects in China.

The Committee is an inter-ministerial body and is responsible for the deliberation and coordination of climate change-related policy issues and activities, and the negotiations with foreign parties. The Committee has the following responsibilities: (1) To review national CDM policies, rules and standards; (2) To approve members of the Board; (3) To review other issues that are deemed necessary.[12]

1.2.2. *National CDM Board (hereinafter referred to as "the Board")*

The Board is established under the Committee and it has the following responsibilities: (1) To review CDM project activities mainly in the following aspects: Participation qualification; Project design document; Baseline methodology and emission reductions; Price of CERs; Terms relating to funding and technology transfer; Crediting period; Monitoring plan; Expected sustainable development effectiveness. (2) If no foreign buyer is determined by the time a project is submitted for approval, and as a result information as to the price as requested in the project design document is not available, it must be indicated in the project design document that the emission reductions generated by the project will be transferred into China's national account in the CDM registry and can only be transferred out with the authorization of China's Designated National Authority for CDM. (3) To report to the Committee on the overall progress of CDM project activities, any issues that have emerged, and further recommen dations; (4) To make recommendations on amendments to these measures.[13]

The NDRC and MOST serve as co-chairs of the Board, and the MFA serves as the vice chair of the Board. The other Board members are the State Environmental Protection Administration, China Meteorological Administration, Ministry of Finance, and Ministry of Agriculture.

[12] Article 14 of Measures for Operation and Management of Clean Development Mechanism Projects in China.

[13] Article 15 of Measures for Operation and Management of Clean Development Mechanism Projects in China.

1.2.3. *National authority for CDM*

NDRC is China's Designated National Authority for CDM, with the following responsibilities[14]: (1) To accept CDM project applications; (2) To approve CDM project activities jointly with MOST and MFA, on the basis of conclusions made by the Board; (3) To issue written approval letters on behalf of the Government of China; (4) To supervise the implementation of CDM project activities; (5) To establish the CDM project management institute in consultation with other departments; (6) To deal with other CDM issues concerning foreign affairs.

In the meantime, the provincial development and reform commissions in different provinces and autonomous regions also have some responsibilities for regional CDM projects based on NDRC's regulations and rules.

1.3. *CDM projects in China*

CDM is a project-based mechanism. China has implemented a lot of CDM projects with some developed countries. In November 2004 the first CDM project — "Land-fill Gas Recovery and Utilization Project of An-ding Rubbish Treatment Plant" — was approved by the Chinese government. This is a CDM project implemented in Beijing, the capital of China. It is expected that this project will reduce 800,000 tons of carbon dioxide in the next ten years. China cooperated with Holland in completing this project.[15] According to the statistics of NDRC as of 12 June 2007, China has approved 524 CDM projects.[16] Up to the said date, 13 Projects have had CERs issued.

---

[14] Article 16 of Measures for Operation and Management of Clean Development Mechanism Projects in China.

[15] An-ding Landfill Plant will reduce 800,000 tons of carbon dioxide in the next ten year, available at http://news.sina.com.cn/c/2005-01-19/04005592154.shtml, 2008-04-09

[16] Office of National Coordination Committee on Climate Change, Projects Approved By DNA of China, available at http://cdm.ccchina.gov.cn/WebSite/CDM/UpFile/File1322.pdf, 2007-06-20

Projects Obtaining Issued CERs (Total 13) (Up to 12 June 2007)

No	Project Name	Project Type	Project Owner	CER Buyer	CERs Issued (tCO_2e/y)
13	Jilin Taonan Wind Power Project	Renewable energy	Jilin Mingmen Wind Power Co. Ltd.	Austrian JI/CDM Programme	71385
12	Liaoning Zhangwu 24.65MW Wind Farm Project	Renewable energy	Liaoning Zhangwu Jinshan Wind Power Ltd.	Carbon Asset Management Sweden AB	58934
11	Liaoning Kangping 24.65MW Wind Farm Project	Renewable energy	Liaoning Kangping Jinshan Wind Power Ltd.	Carbon Asset Management Sweden AB	71331
10	Project for HFC23 Decomposition at Changshu 3F Zhonghao New Chemical Materials Co. Ltd, Changshu, Jiangsu Province, China	Chemical pollutants reduction	Changshu 3F Zhonghao New Chemical Materials Co. Ltd.	World Bank	2966145
9	Fujian Zhangpu Liuao 30.6 MW Wind Power Project	Renewable Energy	Datang Zhangzhou Wind Power Co., Ltd.	Unilateral project	22202

(*Continued*)

(*Continued*)

No	Project Name	Project Type	Project Owner	CER Buyer	CERs Issued (tCO_2e/y)
8	Ningxia Tianjing Shenzhou 30.6MW Wind-farm Project	Renewable Energy	Ningxia Tianjing Shenzhou Wind Power Ltd.	Trading Emission Limited	56930
7	Project for GHG Emission Reduction by Thermal Oxidation of HFC23 in Jiangsu Meilan Chemical Co. Ltd., Jiangsu Province, China	Chemical pollutants reduction	Jiangsu Meilan Chemical Co. Ltd.	World Bank	1102461
6	Ningxia Helanshan Wind-farm Project, Ningxia Autonomous Region, China	Renewable Energy	Ningxia Electric Power Group Co. Ltd.	Trading Emission Limited	121396

(*Continued*)

(Continued)

No	Project Name	Project Type	Project Owner	CER Buyer	CERs Issued (tCO_2e/y)
5	Zhangbei Manjing Windfarm Project	Renewable Energy	Beijing Guotou Energy conservation Company (BJGT)	First Carbon Fund Ltd. (UK)	30968
4	Shandong Dongyue HFC23 Decomposition Project	Chemical pollutants reduction	Shandong Dongyue Chemical Co., Ltd	Mitsubishi Corporation, Nippon Steel Corporation, and Natsource Europe Limited	596803
3	Meizhou Landfills Gas Recovery and Utilization as Energy	Methane recovery & utilization	Shenzhen PhasCon Technologies Co., Ltd.	Austrian JI/CDM Programme, Kommunalkredit Public Consulting Gmbh	48840

(Continued)

(Continued)

No	Project Name	Project Type	Project Owner	CER Buyer	CERs Issued (tCO_2e/y)
2	HFC23 Decomposition Project of Zhejiang Juhua Co., Ltd., P. R. China	Chemical pollutants reduction	Zhejiang Juhua Co., Ltd.	JMD Greenhouse-Gas Reduction Co. Ltd.	3501022
1	Nanjing Tianjingwa Landfill Gas to Electricity Project	Methane recovery & utilization	Nanjing Green Waste Recovery Engineering Co., Ltd.	EcoSecurities Ltd. (UK)	42444

All these projects are implemented by Chinese project owners and CER buyers of developed countries. According to statistics from the website of the CDM Executive Board (EB), China has altogether 87 registered projects up to now. The total amount of the expected average annual CERs is 65014859 tons, amounting to 42.96% of global expected production.[17]

It should be emphasized that the Renewable Energy Law and the Law of Energy Conservation have not provided for the CDM, though China has implemented a lot of CDM projects with some foreign countries. In order to promote further development of CDM projects, all these laws should stipulate CDM. It is well known that a lot of CDM projects can be categorized as renewable energy projects. The provisions of the Renewable Energy Law and the Law of Energy Conservation will further facilitate the development of CDM projects.

1.4. *Provincial CDM service institutions in China*

CDM projects require high technological support from experts and scientific research institutions. A lot of provinces in China have established all kinds of CDM service institutions. Generally speaking, these institutions can provide technological consultation and service for CDM project owners in different sectors, conduct market investigations on CDM projects, contact foreign CER buyers, and help CDM project owners to apply for auditing and monitoring from or registration with Chinese authorities and EB, etc. The paper will make a brief introduction to some typical provincial CDM service institutions in China.

There is no doubt that CDM service institutions have played important roles in implementing CDM projects. It should be pointed out that more and more CDM service institutions will be launched in other provinces in the future.

---

[17] Office of National Coordination Committee on Climate Change, Projects Approved By DNA of China, available at http://cdm.ccchina.gov.cn/english/main.asp?ColumnId=26, 2007-06-21

Institution	Province	Year of Establishment
Hunan CDM Project Service Center	Hunan province	2005
Ningxia CDM Service Center	Ningxia municipality	2003
Shandong CDM Project Center	Shangdong province	2005
Hebei CDM Project Office	Hebei province	2005
Hubei CDM Service Center	Hubei province	2006
Guizhou CDM Technological Service Center	Guizhou province	2006
Shanxi CDM Research Center	Shanxi province	2006
Sichuang CDM Center	Sichuan province	2006

1.5. *Other legal instruments concerning CDM*

In June 2007, the State Council issued a notice to implement China's National Climate Change Programme. At the same time, China's "Scientific & Technological Actions on Climate Change" was jointly issued by MOST, NDRC and MFA, etc. Under these two legal instruments, CDM is regarded as one of the most important measures to mitigate GHG emissions.

2. Examples of CDM Projects in China

This part of the paper will give a brief introduction to four examples of CDM projects in China. All the following CDM projects have been registered with the CDM Executive Board, and have obtained issued CERs accordingly.

2.1. *Meizhou landfills gas recovery and utilization as energy*

2.1.1. *Project participants*

The Meizhou Landfills Gas Recovery and Utilization as Energy Project involves the following three participants:

The developer, owner and operator of the Meizhou Landfills Gas Recovery and Utilization as Energy Project is the Shenzhen PhasCon

Technologies Co., Ltd. (hereinafter as "PhasCon"). PhasCon is in charge of the design, financing, investment, implementation, management and operation of the project.

The owner of the landfill sites is the Meizhou Environment and Sanitation Administration Bureau, Guangdong, China (MESAB). MESAB is in charge of the solid waste collection, landfill disposal and daily operation of the city municipality. MESAB entered into contracts in May and July 2004 to entrust PhasCon with the collection and utilization of landfill gas (LFG) from the eight landfills in Meizhou. MESAB will provide the necessary construction tools for landfill ground cover works.

The Austria JI/CDM Programme has entered into an Emissions Reduction Purchase Agreement (ERPA) with PhasCon to purchase the CERs from this project.

2.1.2. *A brief description of the project*

Meizhou City is located in the northeast corner of Guangdong province in the People's Republic of China. The center of government for Meizhou is located in Meijiang District, together with the seven counties of Fengshun, Meixian, Xingning, Wuhua, Jiaoling, Dapu, and Pingyuan. The City has a total population of over 5 million. There is a big Longfeng landfill in Meijiang District, and there are other seven (7) landfills in the counties of Meizhou, all of the above eight (8) landfills were badly polluted by nauseous odors gas emission and leachate water.

PhasCon is a Chinese-Canadian joint venture company established in the city of Shenzhen, China. Being an energy and environmental company which specializes in energy efficiency service, renewable energy utilization and GHG mitigation issues, PhasCon is a professional project investor for landfill gas recovery and utilization in China. It is also a standing member of the China Power Supply Association, and a National Demonstration Base Enterprise certificated by the China Government for Advanced Science and Technology Applications.

Comcor Environmental Services of Canada (hereinafter as "Comcor") is a famous North American company in the landfill gas recovery and utilization business. Comcor is invited by PhasCon to provide technical assistance in design engineering, training of PhasCon personnel, and conducting safety and operational reviews for landfill gas projects. These lead to the development of a strong local Chinese capability to conduct all aspects of landfill gas recovery and utilization activities.

In 2004, PhasCon was granted a 30-year contract by the MESAB — a government agency responsible for waste collection and disposal — to introduce 2R recycling utilization technologies to collect landfill gas, explore potential utilization and treat leachate water by recycling in all of the abovementioned eight landfills in Meizhou. The specific objectives of the contract are to implement landfill gas collection and the utilization activities of the eight landfills. These will involve investing in a gas collection system, leachate recycling system, flaring equipment and a modular electricity generation plant at each landfill site. A generator compound is also necessary at each site, with an approved total capacity of up to 9 MW within the contract period. PhasCon plans to develop at least 2 MW in the first crediting period. The generators will combust the methane in the landfill gas to produce electricity for export to the grid or landfill local sites for the purpose of gas collection. Excess landfill gas, and all gas collected during periods when electricity is not produced, will be flared.

Total Municipal Solid Waste (MSW) collected in 2004 by the eight landfills averaged 1160 t/d, with an increasing annual rate of 9%. A technical analysis was conducted in order to quantify the potential volume of emissions reductions that the project can generate. The analysis was conducted based on the US EPA first order decay model for the projections of carbon emissions for the project's baseline. It was found that, by combining combustion and flaring, the project has the capacity to reduce emissions by about 5.9 million tons of CO_2e over the next 21-year lifetime. In addition, the project will lead to emissions reduction which is attributable to the displacement of grid electricity, although this is not claimed by PhasCon for the first seven years of the contract period.

The main social and environmental impact of this project will be a positive effect on the health and living conditions of the local area. Contaminated leachate and surface run-off from landfills can affect down-gradient ground and surface water quality; consequently affecting the local environment. PhasCon proposes to use its own patent in China to return the leachate that is collected at the bottom of the landfills to a distribution system at the top of the landfill. This process of returning contaminated water will serve two purposes: the first is to secure the contaminants on the landfill solids as a result of water evaporation from the recycled stream, the second is to promote the generation of LFG by the addition of water which is distributed over the waste in place to evenly supply water to the microbes that generate methane from the organic content of the waste. Once LFG recovery operations are instituted, more water will be removed from the landfill material and the biological process will be slowed down unless water is returned to the waste in place. This leachate recycling will also accelerate the production of LFG and stabilize the site earlier as far as the production of gas is concerned.

The uncontrolled release of landfill gas can also impact negatively on the health of the local environment and population and lead to risks of explosion on the landfill sites. By managing the Meizhou landfill sites gas recovery operations properly, the environmental health risks and the risks of explosion are greatly reduced. Nauseous odors from the site will also disappear, benefiting the local population.

The project will also have a small but positive impact on employment in the local area as a number of staff will need to be recruited to manage the landfill gas operations. Social benefits will include the project acting as a clean energy demonstration project, encouraging less dependency on grid-supplied electricity and better management of landfills throughout China. The project could be replicated across the region.[18]

---

[18] Meizhou Landfills Gas Recovery and Utilization as Energy, available at http://cdm.unfccc.int/Projects/Validation/index.html, 2007-08-01

2.2. HFC23 decomposition project of Zhejiang Juhua Co., Ltd.

2.2.1. Project participants

This project is carried out by Chinese and Japanese participants.

The Chinese participant is the Zhejiang Juhua Co., Ltd. It is the project owner and operator. It is a listed company founded exclusively by the Juhua Group in 1998 and is currently the largest fluoro-chemical production base in China. It consists of six production plants, including Zhejiang Quhua Fluoro-Chemical Co., Ltd., the Electrochemical Plant, the Synthetic Ammonia Plant, the Sulfuric Acid Plant, Lanxi Pesticide Plant and the Fluoropolymer Plant; and eleven subsidiary companies. These companies are mainly engaged in the production and sale of fluoro-chemical feedstock and relevant subsequent products, basic chemical feedstock and relevant subsequent products, as well as chemical fertilizers and pesticides which have already been sold to more than 30 provinces/cities in China and exported to more than 30 countries and regions around the world. The Zhejiang Juhua Co., Ltd. has an area of 1.265 m^2 and hires about 8000 employees.

The Japanese participant is the JMD Greenhouse-Gas Reduction Co., Ltd.

The JMD Greenhouse-Gas Reduction Co., Ltd. is the project sponsor, technology supplier and CER buyer. It was set up as a project company to develop this CDM project, buy accruing CERs, and provide technology, monitoring and other necessary services. The parent companies of the JMD Greenhouse-Gas Reduction Co., Ltd. are the JGC Corporation, the Marubeni Corporation and the Daioh Construction Co., Ltd.

The JGC Corporation (JGC) has accumulated rich experience from the implementation of over 20 000 projects in 50 countries including China, and is recognized internationally as a first-class energy company with strong technical capacity and strong project management capacity.

The Marubeni Corporation is one of the biggest commercial companies in Japan, with an international business network and 13 business groups. It provides professional commercial, financial and project development services. It has carried out about 150 projects through joint ventures or sole investment in China, with an excellent financial performance.

The Daioh Construction Co., Ltd. is a comprehensive construction company which also has business activities in the environmental field. It provides superheated steam decomposition technology for decomposing fluorocarbons (CFC, HCFC and HFC) and has sold more than 30 sets of decomposition facilities. The Daioh Construction Co., Ltd. has established several wind power plants with a total capacity of 35 MW.

2.2.2. *A brief description of the project*

The project will install a set of superheated steam decomposition facility which is imported from Japan. The most advanced technology of Japan will be utilized to conduct non-hazardous treatment of the HFC23 gas, the by-product generated in the production process of the fluoric refrigerant HCFC22. The HFC23 will be decomposed to hydrogen fluoride (HF) and carbon dioxide (CO_2). HFC23 is one of the six categories of greenhouse gases (GHGs) defined in the Kyoto Protocol. Its GWP of 11 700 is comparatively big. By decomposing HFC23 to CO_2, the proposed project is estimated to reduce GHG emissions by at least 5.79 MtCO2e per year, thus improving the global environment and mitigating the impacts of climate change.

China has no mandatory limits on HFC23 emissions at present. Furthermore, the decomposition of HFC23 requires the installation of facilities with high technology and a large amount of investment and has no economic benefit. The owner of the proposed project is the Zhejiang Juhua Co., Ltd. of China, which has not sold HFC23 as a commodity previously and has no such market plan at present. This project activity will decompose HFC23 released from a HCFC22 production line which was put into production in February 2001 with an annual production capacity of 16 000 tons. The proposed project activity is to decompose the HFC23 generated by the second production line. Without the proposed project, the HFC23 generated in this company will be emitted directly to the atmosphere as waste gas.

The proposed project will contribute both directly and indirectly to the sustainable development of China.

Direct contribution: Implementation of the proposed project will increase the net annual benefits of the project owner. It will also increase the annual financial revenue of the local government through sales tax, and thus may increase the investment of the local government in public areas such as education, poverty alleviation, etc. Moreover, the proposed project will create 15 new jobs.

Indirect contribution: The project owner has committed to submit a certain percentage of its economic revenue from selling CERs to the Government of China, as required by China's CDM rules, for the purpose of establishing a National Climate Change Fund or supporting the sustainable development of China.

In brief, the proposed project activity will not only bring significant GHG emission reduction benefits but also promote the transfer of HFC23 decomposition technology to China. It will thus contribute to the mitigation of the impacts of climate change and promote the sustainable development of the local area and China.[19]

2.3. *Fujian Zhangpu Liuao 30.6 MW Wind power project*

2.3.1. *Project participants*

This is a unilateral project. The Datang Zhangzhou Wind Power Co., Ltd. is the investor, constructor, operator, and owner of the project.

2.3.2. *A brief description of the project*

The project site is located in the east of Liuao peninsula, Zhangpu County in Fujian Province. It has the Taiwan Strait to the east, Gulei Peninsula in the southwest and Jiugang town in the north. The site is 48 km from the Zhangpu county government in the northwest, 100 km from Zhangzhou Municipality in the north.

---

[19] HFC23 Decomposition Project of Zhejiang Juhua Co., Ltd., P. R. China, available at http://cdm.unfccc.int/Projects/Validation/index.html, 2007-08-01

The project involves the installation of 36 turbines, each of which has a rated output of 850 kW, providing a total capacity of 30.6 MW. The project activity is expected to reduce a total estimated amount of emissions of 1 075 998 tCO2e over the third seven-year crediting period by avoiding CO_2 emissions from electricity generation in fossil fuel power plants that presently supply to the East China Grid.

The project will help reduce GHG emissions generated from the high-growth, coal-dominated power generation. Also, it will contribute to the sustainable development in the region by reducing pollution, creating employment opportunities, promoting the local tourism industry and improving the livelihoods of local people. On a larger scale, the project will assist China in stimulating and accelerating the commercialization of grid-connected renewable energy technologies and markets.[20]

2.4. *Liaoning Zhangwu 24.65 MW Wind-farm project*

2.4.1. *Project participants*

The Liaoning Zhangwu Jinshan Wind Power Ltd. (ZJWP) is the project owner. ZJWP's enrolment was approved by the Zhangwu Industrial and Commercial Bureau in August 2002. The address of ZJWP is Shaoguo village, Houxinqiu town, Zhangwu County.

The CDM project developer is the Beijing Keji Consulting Ltd. The Beijing Keji Consulting Ltd is a service institution whose enrolment was approved by the Beijing Industrial and Commercial Bureau in 2003. At the time of project registration, the CER buyer was not decided.

2.4.2. *A brief description of the project*

The objective of Liaoning Zhangwu 24.65 MW Wind-farm Project (hereafter "the Project"), a grid-connected renewable power project,

[20] Fujian Zhangpu Liuao 30.6 MW Wind Power Project, available at http://cdm.unfccc.int/Projects/Validation/index.html, 2007-08-01

is to utilize wind resources for generating renewable electricity to sell to the North East Power Grid via a power purchase agreement (PPA).

The Project is located in Zhangwu County, Shenyang City, Liaoning Province of the People's Republic of China. The Project involves the installation of 29 turbines, which have a capacity of 850 kW, providing a total installed capacity of 24.65 MW. The project is expected to generate 56 695 MWh electricity annually and to sell 54 427.2 MWh power to the grid.

The Project clearly accords with the development priority of the PRC, and will support China in stimulating and accelerating the commercialization of grid-connected renewable energy technologies and the green-power market developments. It will therefore help reduce GHG emissions resulting from the high-growth, coal-dominated business-as-usual scenario. Furthermore, the project will lead to improved energy security, improved air quality, improved livelihoods of people dependant on coal, support alternative sustainable energy futures, and contribute to the overall sustainable development of the renewable energy industry.

The objectives of the Project are to:

- reduce GHG emissions in China compared to a business-as-usual scenario;
- create local employment opportunities during the period of project construction and operation;
- provide training in the employment and management of advanced wind power technologies;
- help to stimulate the growth of the wind power industry in China.

The Liaoning Government is supportive of the Project because the development of wind power is in accordance with the national criteria for sustainable development and the national policies relating to energy resources and the environment, which will push forward the use of renewable and clean energy across China.[21]

---

[21] Liaoning Zhangwu 24.65 MW Wind-farm Project, available at http://cdm.unfccc.int/Projects/Validation/index.html, 2007-08-01

3. Conclusion

Based on the above discussion, China has seriously implemented the CDM of the Kyoto Protocol. China's CDM projects focus on energy efficiency improvement, development and utilization of new and renewable energy, and methane recovery and utilization. There is no doubt that China's implementation of the CDM will make a great contribution to the tasks of worldwide GHG emissions reduction and global environmental protection. China is a responsible country in the implementation of the CDM. The principle of common but differentiated responsibilities has also been fully complied with in China's implementation of the CDM. In order to realize the final goal of building up a world-wide Low Carbon Economy, developed countries and developing countries should make much greater progress in implementing the CDM. But all efforts to implement the CDM must be based on the principle of common but differentiated responsibilities.[22]

[22] The principle of common but differentiated responsibilities is one of the basic principles in international environmental law, the core idea of this principle is that the developed countries and the developing countries should bear different environmental protection responsibilities in all kinds of international environmental protections issues.

CHAPTER 6

EMPIRICAL CONSIDERATIONS IN THE DEVELOPMENT OF CDM PROJECTS IN ASIA

William I.Y. Byun and Felix H.C. Chan

The determinative measure for the exigency of a market based climate change regulatory framework is as a tautology, practical. Considerations for structuring the regime must be rooted in how the actual features and elements are utilized by the markets and policy adjustments must grow outwards from such market directions. With respect to the Kyoto Protocol framework, as its CDM pipeline mechanism is already operating, the market is also generating its "feedback signals" on its issues within the present system and actual market viability can be assessed. How does the scoreboard appear thus far? These signals have identified an overall positive growth and deepening of the participation into the market from a much broader corporate range. Rather than a narrower appeal to sustainable-growth orientation entities, one measure of the Kyoto framework's success has been the significantly broader participation by pure investment funds, large industrial corporations, service providers, etc., into climate change as a viable commercial market in itself.

On the other hand, such enthusiastic participation by the market has also identified potential bottlenecks in both practical implementation and aggravating cracks in the underlying policy frameworks. For financial funds, a greater market liquidity and predictive certainty is critical to its pricing and transactional needs; for the large industrial corporations, the development of a continuous volume of sophisticated technical methodologies is necessary to keep up with evolving industrial practices; for service providers, greater mechanisms of input and adjustments and stricter timelines are necessary for the system to not become bureaucratically sclerotic.

As the policy focus is now shifting towards the possible post-Kyoto landscapes, mobilizing such an enthusiastic subscription by the broader market is essential. Especially since the debate about climate change has shifted from arguments about its science to debate on the viability of the structure of the market cap-and-trade system, the climate change policy regime must accommodate this broader societal participation or risk becoming a niche system of limited impact. What may well be needed is a greater decoupling from the narrower orientation of a growth tied to sustainability, to a shift towards a more volume based "climate-positive" and more "energy efficiency-positive" growth becoming commercially sustainable.

1. A Market Based Climate Change Framework

Since the entry into force of the Clean Development Mechanism (CDM) under the Kyoto Protocol, there has been a disparity between the results of actual implementation of CDM and the expected generation of environmental, economic and social benefits that were supposed to arise from this economic mechanism. There are several Asian countries which are without any emissions reduction targets under the Kyoto Protocol. These countries have been slow to convert to more climate-friendly methods that minimize greenhouse gas emissions, since there is no short term economic benefit in sight. Fortunately, under CDM, a way forward has been provided for these emerging economies of the world to participate in the world wide efforts against climate change. The CDM allows for a financial incentive to participate in emissions reduction projects whereby "carbon credits" can be generated for an equivalent reduction in greenhouse gas emissions. These credits can then be sold to developed countries with reduction targets under the Protocol, adding a revenue stream to improve the economic attractiveness of the various power projects available in these emerging economies. Based on the potential of

CDM projects which have not yet been tapped into, the economic mechanism has not fully accomplished its purpose.

As will be explained in this paper, there are issues that need to be addressed within the current framework, in order to allow for a more successful implementation of CDM and to maximize the possible benefits of this system. Thus, the determinative measure for the exigency of a market based climate change regulatory framework is practical and necessary. The original purpose of the CDM to allow the 36 Annex 1 countries to reduce emissions at a lower cost, whilst amassing the participation of the developing countries, has not been followed through completely due to a lack of proper collaboration and partnership between the two of industrial and financial developers. Currently, there is still a high demand for carbon credits as countries are not certain of their abilities to meet their targets. This can be corrected through a more market-based framework for climate change, whereby investment in these projects will prove to be attractive enough to be considered from a purely economic standpoint. Hopefully, this can lead to better integration between the two markets and provide better benefits for both.

The necessity for a market based framework also arises from the need to involve a wider range of participation in the market. Currently, the beginnings of a broad range can be seen, from the national to corporate entities. However, to encourage a deeper commitment from more parties, the market based mechanism should be set up to allow for greater facilitation of all "climate-positive" projects, such that any available contribution to the global environment in terms of emissions reduction should be made possible, which is currently, to some extent, restricted by the UN framework. By improving the possible revenue stream from Certified Emissions Reductions (CERs) to make these projects more financially appealing through market mechanisms, and by modifying the regulations that at present make project registration difficult, the number of CDM projects registered and implemented can be increased.

The main philosophy of the mechanism is that whilst the developed countries were the ones assigned an emissions reduction target, the market-based mechanism actually allows for the emissions reduction targets to be re-assigned in the least-cost manner, whilst still benefiting the environment. Assuming that a country with an assigned target is unable to fulfill its goal, the global environment eventually loses out when the emissions reduction is not made. However, now that this target can be partly re-assigned, the country can now pay for a unit of emissions reduction while this unit of real emissions reduction occurs elsewhere. The end result is more climate positive than a simple cap on the developed countries.

2. Considerations for Structuring Market Growth

Considerations for structuring the regime must be rooted in how the actual features and elements are utilized by the markets, and policy adjustments must grow outwards from such market directions. Clearly, such structures are intended to facilitate the functioning of the regime rather than stifle it. In this regard, it must be a structure that encourages market participation and also precipitates the fundamental effects that indicated a need for linkage to a market framework. Such fundamental effects linked to a market framework must arise from the market response in an appropriate manner — that is, the structure should not cause the market bias that defeats the underlying objectives of the market exercise.

In the context of climate change, the market framework should be structured so as to avoid excessive focus on the market funding projects with limited impact, to prevent a bubble from surrounding such projects and drawing funds away from other climate change projects with merit. Instead, the market should be conditioned to focus on those projects that contribute the most benefit to the global environment. Conversely from an excessive financial speculation supply-demand caution, such market framework should not stifle commercial efforts with scalability, eschewing them in favour of smaller but more "social development" oriented projects — not only smaller in size but with limited or minimal scalability. The challenge of finding a balance has to be met from both sides of the project, those who are driven by profitability and those who seek the reversals of anthropogenic climate change.

The advantage of a flexible, market-oriented mechanism is obvious. Relatively speaking, for developed countries, implementation of CDM projects is cheaper in developing countries than in local territory because of the varying economic levels. CDM provides the pipelines for them to meet their obligations in a more economically prudent way, by offering the option of developing new projects and improving technologies in other countries, while having this translate into lower costs for the developed countries. Furthermore, with the comparatively lower level of economic and infrastructural development in developing countries, there are many more opportunities present for the implementation of CDM projects. This benefit becomes even more apparent when under a market-based framework, high levels of competition will lead to an augmented importance of the cost effectiveness of each project proposed. The hunt for these projects involves not only the efficiency on the energy side, but also that on the economic side.

For developing countries, CDM provides a way in which they can attract considerable project investment from developed countries needing to achieve greenhouse gas reductions. Many of these developing countries have intentions of streamlining their development to fit in with the UNFCCC goals as much as possible, but even so, some might not have the economic ability to do so because of the lack of capital and foreign direct investment. CDM allows these countries to participate by removing or minimizing these investment barriers that prevent them from actively participating in emissions reductions through the proposed flow of foreign direct investment. Moreover, such projects are also good samples for future development under the Kyoto Protocol after the first reduction period in 2012, whereby new protocol regulations might necessitate non-Annex I countries to begin mandatory emissions reduction as well. Though it is unclear as to when the developing countries will be eventually assigned their own emissions reduction targets, there is no doubt that in due time, such targets would be necessary for all countries in order for the atmosphere to contain a sustainable level of greenhouse gasses. In some of the countries without obligations to reduce greenhouse gas emissions, economy stimulus might be the only incentive to participate in CDM. This highlights the extent of carbon trading's influence. The market based CDM essentially exchanges economic benefits for carbon credits, making it advantageous for both parties. Hence, widespread efforts in voluntary emissions reduction are rather unrealistic without the immediate benefits to profit-seeking business entities, as they either are not interested if profits are not involved, or they might not have the capabilities by themselves to make such investments into the environment.

However, the disadvantages have also been significant. In the case of CDM, the developed country would have to rely on another government to ensure its international commitment, as approval by the Designated National Authority (DNA) of the host country is required for the project to be implemented under CDM regulations so as to have CERs approved and verified. Uncertainties exist as governments are replaced and policies are changed. If the government chooses to cater to certain interests that conflict with those of the CDM projects, potential CDM developers may find that they are forced to modify their projects with other resources, thus possibly diminishing or jeopardizing CDM projects in the pipeline. From the point of view of the host countries, it is reasonable for them to pursue more lucrative opportunities since they are not bound by protocol regulations for obligatory reduction. Thus, one of the main factors for hosting CDM projects would almost purely be the profits from CER trading, despite the fact that CDM projects contribute essential benefits to the host country's sustainable development in the long run.

In this frame of mind, the enthusiasm with which a host country will enforce the agreements pertaining to the CDM projects is vague. Indeed, developing countries could have significant economic incentives to abrogate CDM projects, particularly if they are viewed as a constraint to crucial development in the host country, or as a restriction to the employment of a natural resource that the host country would like to exploit. Thus, considerations about how to eliminate this uncertainty should be based upon making the structure of a market based climate change mechanism more secure and favourable for host countries.

Since the emissions of carbon are indirectly created through market practices and political decisions, the corresponding market structure should eventually be framed in law as have been done for other markets, in order to provide security and accountability to the market participants. Therefore, similar to other commodity markets, such as the oil, gas and power markets, policy adjustments could potentially have a key impact on market and price developments. For those who want to analyze and forecast these developments, there is a need to carefully assess the potential impact of policy direction. For the carbon market in particular, this signifies that the upcoming developments of the Kyoto Protocol are more pertinent than ever, particularly due to the lack of any concrete plans for a new protocol beyond 2012. It is extremely difficult for any entity to regulate or monitor the progress of the CDM market under such conditions, whereby the continuation of the carbon market is not established. In general, there is a fair amount of confidence that the carbon market will not cease to exist, due to the nature of the underlying problem of climate change. Rather, the difficulty lies in the uncertainty of the market structure.

3. "Feedback Signals"

With the CDM being functional for some time now, there should be enough activity to indicate how well the system is performing, and whether or not the carbon market has been viable so far. In this sense, it might be more relevant to consider the feedback from the EU cap and trade system, since it has been designed to address issues in a framework specifically intended to bring to fruition these project activities. These project activities are pro-active attempts at reducing fuel-related greenhouse gas emissions and basic industrial process efficiency upgrades that were hitherto neglected because there was no relevant price attached to the emissions to make them valuable. In the EU cap and trade system, there was no assistance to the developing world to help them get up to speed with the development of the rest of the world in a climate-positive manner. The CDM pipeline mechanism does this, in addition to offering the incentive for the reduction of greenhouse gas emissions from existing facilities.

As mentioned, carbon trading benefits those who are able to reduce emissions at the least cost possible. In a global system, this should benefit developing and emerging markets. The success of the Kyoto Protocol mechanisms in allowing for emissions reduction, whilst stimulating technology transfer to developing countries such as China, India and South America, has bolstered support for the carbon trading approach. China, in particular, has become a key supporter of CDM in the past two years due to the opportunities for local companies to profit from selling low cost CERs to buyers in the European and Japanese markets. The extension of these mechanisms to more countries and a further diffusion of state-of-the-art technologies is a key factor in designing the second stage for any global agreement in climate change. Without CDM, it is unlikely that developed countries would have allowed these technologies to cross the borders, since it is considered a direct threat to economic competition.

CDM, by way of attaching an economic cost to CO_2 reductions, has improved the viability of renewable energy technologies multi-fold. Considering the high initial cost of using renewable energy, CDM provides the project developers with extra leverage to overcome the investment barriers. The CDM could also be used as one of the means to finance performance-linked incentives for renewable energy. In this case, the additional incentive would reflect the avoided greenhouse gas emissions due to renewable electricity generation, since there is certainly no cap on the amount of greenhouse gas emissions that can be prevented. Since anthropogenic climate change is a global concern, the more emissions reduction that occurs anywhere in any form, the more of an improvement it is to the current situation for the global environment.

Given that the energy sector constitutes by far, the largest emitter of greenhouse gases, there have already been purposeful efforts to reduce emissions and cut production costs. The relatively concentrated nature of the energy sector and the extent to which it is already reviewed, regulated and supported in various respects by governments makes it likely to offer more traditional and effective targets for near term government policy intervention than the disparate and complex areas of deforestation, land use practices or population growth. The volatility of oil has also pushed many in the energy industry to search for cheaper or more stable alternatives. The solution in renewable energy has shown that it is progressing towards becoming a more mainstream industry solution, regardless of its potential for CER generation. Thus, the trends in CDM projects so far have shown concentration in the energy industry, whereby the endeavours to modify inefficient processes or increase the use of renewable energy are more feasible than the more exigent or time-consuming changes necessary in other industries.

However, with the increasing importance of CDM and the awareness level of the corporate sector, there will undoubtedly be more inquiries about climate positive changes in other industries. The continuous opportunities for increased CER generation will convince many companies and industries to push on for developing innovative technology to meet these demands. The possibility for an even more integrated and active market is clear. The question lies not in the viability of the market, but more in the planning of a structure and organization such that carbon trading is made more accessible.

4. Scoreboard

To consider the success of this framework so far, the results of the CDM process point to a rather telling conclusion. Despite a vast number of potential projects with a very plausible sounding basis, the scoreboard puts the number of CERs issued at approximately 50 million. Even so, the number of projects in the CDM pipeline proves to be on the rise. With the first commitment period of the Kyoto Protocol coming up, the demand for CERs is forecasted to increase. The projected number of CERs to be generated by 2012 already exceeds 2 billion. As the expected increase of projects in the pipeline as the CDM process solidifies, the number can only continue to rise.

Additionally, even without a surge in the number of CERs needed for the market, the availability of CDM project opportunities in various countries also holds its own appeal. The rapid pace of growth has demonstrated the capacity of the market. In just two years, the CDM has resulted in more than 700 registered projects in more than 50 countries of the developing world, stimulating North–South investment and considerable emission reductions in the process. The energy industry has already been thrust into the midst of this revolutionary movement. Even so, there are many more sites whereby simple modifications can be made to achieve efficiencies for the least greenhouse gas emissions possible. With the pressure on Annex I countries to meet their targets, they are being pressed to innovate on these projects. If not for the prospects of CDM, many of these innovations would not have come together on their own at such a swift pace.

In general, the significant involvement of Asian countries in this climate change effort can be seen. Asia is generating the largest amount of CERs and has been projected to generate even more, with the substantial number of projects already in the pipeline. The two countries with the most projects generated so far have been China and India.

India, host to the largest number of CDM projects in the world, currently has 340 registered projects, which accounts for 40% of the total number of projects. India remains a strong market for power with a potential of 100 000 MW with its expansion by 2012. It has a huge potential for the upgrade of old power generation capacities and bio-fuels with energy efficiency methods. Due to this potential, there needs to be increasing regulatory changes and standardizations to allow India to achieve its full capacity with CDM. Although the number of CDM projects is currently at the greatest within India, the quality of some projects might be lacking. Oftentimes, projects with real qualities of real sustainable development might not get the funding needed, because of the different economic agenda of the government. The variation in regional or state markets in India with regard to their regulatory framework also increases the difficulty in compliance efforts. As such, there needs to be a concentrated effort in changing the fundamental structure of the economy to fill the gaps.

The disparities in regulatory support for CDM projects has been lacking so far. This is probably due to the fact that the CDM efforts are sometimes seen as still rather premature, and the relevant authorities have not had an opportunity to get up to speed with the rush of investments that are trying to profit off the inefficiencies of the system. Furthermore, there needs to be a centralization or standardization of some kind, to ensure that all projects implemented provide real reductions, which has still been a problem in India so far. The instruments for risk assessment are also deficient. Buyers and developers of the CERs generated in India are faced with a wide variety of projects to choose from, but because of the relative youth of this industry, many of the project participants might not have the experience to properly assess the probability of success of the projects. As with all other investments, the risk factor cannot be mitigated. However, having a good grasp on the result is still better than taking blind guesses, as it will raise confidence levels in the quality of Indian CDM projects. The development of such service providers has yet to be explored fully.

China is in a similar position. As one of the biggest hosts of CDM projects, China has begun its active involvement in the climate change process. While it still does not have an emissions reduction target under the Kyoto Protocol, it is already well on its way, even compared to other more advanced countries. As one of the top polluters of the world, these efforts are especially important, in order for China to account for and counter its high emission levels. Already, the UN is targeting to have 41% of the emissions reductions for 2012 to be sourced from China, under the component of CDM, showing a still wide gap on projects to be implemented. Even so, some of the successes in Chinese projects have been part of the HFC23 debate. Another of the greenhouse gases, HFC23 causes eleven thousand times more damage than its equivalent in carbon dioxide by weight. However, the cost of solving the HFC emission problem is much cheaper, causing an imbalance in the cost-benefit equation. The number of CERs generated has caused a booming carbon trading market sourced from China. The Chinese government has already entered discussions to set up another carbon trading scheme, the first in this region and outside of the United States and Europe, as well as the developing world. The involvement of the Chinese government in developing its own projects without foreign aid has also shown the level of involvement in CDM. China could quickly become competent in managing its own emissions, although a complete balance would take a while. Even if the benefits to the environment are not enough to cover its polluting end, the number of jobs created through CDM projects has already been a stimulant to the economy. The positive effects that CDM has had on China is not subject to a slowdown any time soon.

In discussing the achievements of other Asia-Pacific countries in CDM efforts, there also needs to be consideration over the types of industries explored. So far, the following industries have seen some accomplishments: Energy efficiency, Landfill gas, HFCs and a variety of renewable energy such as wind, solar, hydro, biogas, etc. As can be seen, the bulk of development has occurred in China and India, because of the vast opportunities available. The rest of the region has to gain an awareness of the problem at hand, in order to rouse additional support from both the public and private sector. Nonetheless, as the region with the stronghold on CDM projects and CERs generated, Asia has had a fairly successful framework to support the activities, and this will only develop into a strong support mechanism for more CDM projects.

5. Growth of Corporate Participation

Given these signals, the market has seen an overall positive growth and deepening of participation from a much broader corporate range. Their involvement is helping to boost the market even further by increasing the competition. In 2006, non-Kyoto and non-EU ETS CER buyers deepened their involvement in the carbon market. The United States and Australian governments said that they were considering recognizing CERs for use in meeting their internal reduction targets.[1] This has influenced the current market and strengthened the belief in a post-2012 carbon market, evidenced by the increasing transaction amount of forward carbon trades that include both CERs and VERs (Verified/Voluntary Emission Reductions). The growth

[1] "An Australian Cost Curve for Greenhouse Gas Reduction", McKinsey & Company.

of the emissions trading systems is increasing at an exponential pace due to the success of the system so far as a popular form of emissions control. The continual development of the market will ensure the participation of more entities, both in range and depth. As such, there is a fundamental benefit in the development and linking of the numerous trading schemes available globally.

A key benefit of linking schemes is that a globally linked-up carbon market would allow businesses to look globally for the cheapest emissions permits, and so cut the cost of the emissions reductions. According to the International Emissions Trading Association (IETA), lower emissions abatement costs could arise from linking these carbon markets, because of increased efficiency, liquidity and competition. In order to strengthen the developing regional carbon markets of countries such as the United States, Canada and Australia, the support and robustness of the EU ETS has to come into play, as they have a relative maturity compared to the rest of the market that has to be passed on. Part of the EU's success so far in carbon trading is the genuine support from both the government and private sector for the framework. Their ideology towards their cause has garnered high levels of participation, helping to improve the level of competition in and functionality of the EU ETS. The second key benefit would be the assistance in helping to insulate such markets from local political decisions, both of the Designated National Authorities (DNAs) in host countries as well as the governments of countries with any carbon trading scheme. The pressure put on these governments under a global trading scheme to cooperate or lose out on the benefits is helping to reduce the risk for investors wishing to take part in these markets and therefore, once again, increasing liquidity.

In addition, the overall deeper participation of the developing world has to be increased in order to fulfil the purpose of CDM. The carbon market has to ensure this by facilitating the effective transfer of the newest climate change technology to these developing economies, so as to allow the maximum potential emissions reductions to occur. These developing economies often are unable to allocate budgets to research and develop technologies in energy efficiency, renewable energy or such technology, even though they are, in fact, the very economies that are best able to implement these technologies due to their developmental nature. It is always easier to influence the widespread adoption of such new technology whilst these economies are still in the developmental stage than to make more significantly sized infrastructural changes later on. The resulting pressures from equalizing the level of competition between developing and developed economies will push more industries to select technologies that will allow them to streamline their operations in an environmentally-friendly manner, based on the norm of being driven mainly by maximized productivity.

Another perspective is to take a look at the deepening involvement of various industries. The range of industries with CDM projects has been on the rise.[2] Coming in first place is still the energy industry (both renewable and non-renewable sources), which accounts for more than half (51.95%) of the total registered projects. These other industries form the remainder: waste handling and disposal (21.18%), agriculture (8.36%), fugitive emissions from fuels (7.25%), manufacturing industries (6.35%), fugitive emission from production and consumption of halocarbons and sulphur hexafluoride (1.67%), energy demand (1.45%), chemical industries (1.11%), mining/mineral production (0.45%) and transport (0.11%).[3] There is still a great opportunity for the development of new projects within the industries that have fallen behind, thus a further deepening of involvement is to be expected.

In terms of host country involvement, some have taken a more direct approach. China, with its large number of CDM projects, has implemented a few on its own, instead of the usual practice of bilateral cooperation. Such projects conducted unilaterally, will allow China to generate carbon credits for itself, such that if and when the second stage of the protocol necessitates China's own emissions reduction, it will have the ability to meet those targets. More developing countries will be expected to take this route, as governments

[2] "Efficient implementation of CDM in China", Fridtj of Nansen Institute.

[3] Source from the United Nations Framework Convention for Climate Change website: http://unfccc.int

eventually move climate change policy up the political agenda and give more attention and investment to environmental issues.

6. Towards a Commercial Market

Rather than a narrower appeal to sustainable-growth orientation entities, one measure of the Kyoto framework's success has been the significantly broader participation by pure investment funds, large industrial corporations, service providers, etc., into climate change as a viable commercial market in itself. The effect of climate change awareness has not been limited to the governmental level. It can also been seen on the corporate scene, as more companies begin to initiate efforts to contribute to the cause of alleviating climate change. This corporate social responsibility movement has already seen concrete measures being taken, whether for environmental purposes, for getting a head start in catching a ride on the green movement or for early-bird accruals of carbon credits in the case of companies with government-allocated reduction targets. As can be seen, the CDM is not only a mechanism for governments to apply carbon credits to their emissions reduction target, but it is also for the private sector, which propels the evolution of the current CDM towards an even more market-based structure.

Participation by other market players such as pure investment funds has also been seen. The main goal of carbon funds is to activate the market for project-based greenhouse gas emissions reduction within the Kyoto Protocol framework and to contribute to sustainable development. These funds provide cash investment to projects designated to reduce greenhouse gas emissions, which participants may be able to use in compliance with their expected greenhouse gas reduction obligations. Various governments and companies from multiple countries have contributed billions of dollars of investment through these funds. Large industrial corporations and services providers are also possible project developers. They strengthen the feasibility of the proposed projects by using their technical expertise and enhancing the relations between potential CER buyers and sellers, resulting in more projects implemented for increased emissions reductions.

As a requirement of the projects registered, sustainable development has to be explored with regard to the local environment. These investment funds are also able to contribute to this effort in sustainable development by providing tangible and important benefits to improve the standard of living for the developing economies, through a better environment and infrastructure.

The importance of these entities lies in their financial strength. As the lack of investment plagues various projects with potential, their ability and willingness to invest is crucial. By covering the gaps between the ideas and the execution of CDM, these corporate players have become an integral part of the market. They contribute years of corporate experience with their competent operating protocols to the efficiency of project execution, in addition to the geographic diversification that complement the international nature of CDM projects and allow developers to take advantage of their own local market awareness. In addition, they lend certain credibility to the system in providing proper financial evaluation and accounting methods, to ensure that the market operates with as much organization as possible. So far, the number of growth companies in the carbon market has indicated the relative well being of the market.

7. Potential Bottlenecks?

Such enthusiastic participation by the market has helped to identify the potential bottlenecks in both practical implementation and cracks in the underlying policy frameworks. Despite the apparent success and recognition of CDM so far, a number of teething problems have cropped up. Due to the developmental stage that CDM is still in, there are numerous items to address in its design. At the moment, the factors that limit investment in CDM projects have not yet been solved.

One of these problems is the investment cost barrier that is difficult to cross. The types of projects that really require CDM implementation are those that come across as difficult to execute even in the initial stages. If not, companies who found such projects as feasible would have already done so, to improve their profit margins.

The high transactions costs are what prevent an even greater number of projects from being completed. As this also affects the range of activities possible, there are still multiple areas in climate positive technology which should be tapped. As technology improves, hopefully this will lead to the net cost savings of CDM projects to be even more attractive. There are also critics who point out the possibility of a misalignment in priorities when CDM occurs, meaning that developers might be more keen on generating carbon credits than making actual contributions to the sustainable development of the project sites. However, despite this criticism, any project that generates carbon credits is ultimately helpful to the climate and should be implemented for that reason alone.

Another rising problem is that of disorganization in the market. As the number of CERs issued surpasses 50 million, the confusion and transparency over who owns what, coupled with the trading of European Union Allowances (EUAs) under the EU ETS, is proving to push this system into disorder. Hence, developments on the upcoming International Transaction Log (ITL) are closely watched. When the ITL was first being designed, it was scheduled to launch on April 2007, but its original due date has already come and gone. New estimates place the ITL operation for the end of this year, as many of the Kyoto Protocol countries are currently in the testing stage of linking the ITL to their own national registries. The ITL aims to provide a system to record and track the ownership of carbon credits being circulated on the market. Trading of carbon credits has been stifled due to the lack of a method in telling whether or not the carbon credits being traded had any actual emissions reduction tied to it, hence ITL should improve the overall credibility of the market, and help to ensure integrity over Kyoto accounting, as originally intended by the CDM Executive Board (EB). Monitoring of the system should be able to improve this bottleneck, in time for managing the CERs for projects in the CDM pipeline. The only worry is whether or not the ITL can be pushed through, as anticipation over its facilitation of carbon trading is already high.

Additionally, the process of registering projects and then verifying emission reductions is extremely slow and bureaucratic, with significant uncertainty throughout the process of the likely outcome. The constant changes in methodology and regulations are also hard to keep up with, complicating the assessment of the cost-effectiveness of projects. As it is impossible to know exactly when such changes will occur, many are having problems in determining the pace at which to plan their projects. Being a step too late could mean a great loss in terms of CERs. Since the financial incentive of these projects lie mostly in CER profits, the reduction in CERs being generated could greatly marginalise profits and render the project financially impractical, even though the net reduction in greenhouse gas emissions is still a good rationale for implementation by any means. For example, in 2006, changes in methodology[4] caused a slashing of CERs for manure management projects that brought the policy risks into the light. Now such risks are becoming more obvious as new versions of regulations keep changing, causing a greater financial threat to project developers.

Additionally, the lack of a stable CDM system is also proving to be a dampener on the full potential of this mechanism. Many delays are being caused by the lack of UNFCCC staff to process all the registrations in a timely manner. Thus projects with merit and potential are not being pushed through the registration process quickly enough. The waiting process can also be uninformative, as it is difficult for such a limited staff to communicate effectively to the developers on any problems with the project registration.

The risks are not limited to these problems. The necessary approvals of different bodies at multiple stages of the project development are also an obstacle. In the case of host country approval, projects have to be in line with the DNA aims or risk rejection. Without DNA approval, the project cannot be registered with the CDM for CERs. As mentioned before, this kind of host country risk can affect even the projects that provide real benefit to the global environment. Regardless of CERs, such projects should be executed, but in reality, there is usually no one willing or able to execute them without the facilitation of CDM.

[4] Approved Baseline and Monitoring Methodologies in UNFCCC website. AM0006 and AM0016 were replaced by ACM0010 and withdrawn from Sep 2006.

During the construction of the regime, it should be always kept in mind that the ultimate purpose is to reduce the global GHG emissions levels, and that the market based mechanism is just one feasible approach to involve more entities with mutual benefits for each. Misunderstandings about where the market should lead the participants can bring the structure into malformation, such as a market filled with CER speculators instead of CDM project developers aiming to reduce emissions. An example would be the European carbon market crash in May, 2006, when the carbon credit price fell more than 50% from €30 per CER in one week to about €13 per CER the following week. The crash was mainly caused by the public release of data showing the accumulation of CERs by many countries, thus implying the possible over-supply of CERs. This illustrates a necessity in guiding the carbon market, with possible policy adjustments to make the market more feasible and in order to avoid such scenarios from occurring. The risk involved in the failure of the market to predict the expected price of CERs on delivery is creating a much more daunting scenario to the participants, as carbon credits are not a traditional commodity whereby its supply and demand can be predicted with great accuracy.

8. For a Continual Market Development

With these imperfections to the current regime, there needs to be a look into what the corporate entities require for their continual and productive participation in the market. Each type of entity requires something different from the market. Namely, financial funds are looking for greater market liquidity and predictive certainty, which is critical to its pricing and transactional needs, whilst the large industrial corporations require the development of a continuous volume of sophisticated technical methodologies necessary to keep up with evolving industrial practices. As for service providers, greater mechanisms of input and adjustments and stricter timelines are necessary for the system to not become bureaucratically sclerotic.

The development of a secondary market in CDM shows a sufficient maturity in the market based mechanism, such that this sort of liquidity can be provided. The secondary market in CDM has grown from US$221 million to US$444 million,[5] with the amount of CERs transacted increasing from 10 to 25 million tonnes of CO_2 equivalent. With the help of a secondary market, developers and buyers will be more likely to invest in an even larger number of CDM projects. Previously, uncertainty about the long term potential of these projects made investors think twice about locking in their investments into a project that would require many years to terminate. However, they can now buy and sell these projects, once again re-evaluating the value of the projects and affirming the significance of CDM. This development allows the growth of funds, as they require a mature secondary market to provide liquidity to their operations. These funds can operate on the basis of trading projects in the most profitable manner, thus they are less inclined to hold on to projects to full term if the projects do not perform up to a sufficiently high level of expectations. This development also allows for the growth of an industry of service providers to facilitate these transactions. These companies would specialize in areas such as providing due diligence, verification services, technical consultations and others.

Industrial corporations should be involved and active enough in the development of technical methodologies so as to survive in the face of the evolving industrial practices. CDM methodologies are basic approaches and are also, to some extent, rules that CDM projects should abide by. Given a certain methodology, the scope of CDM projects is limited as the methodology itself has specified applications. As project developers pick and choose on the projects that are both easier to implement and will provide the highest return on CERs, the number of these projects will dwindle. By then, the industry giants should be ready to head into the next promising sectors, in order to avoid a shortage of favourable projects. A rise in investment costs might be possible, and this further highlights the calls for more innovation and new methodologies to conform to the developing market needs.

A number of problems in the process of CDM can be analyzed to provide the market with maximum functionality. Despite the enthusiasm

---

[5] 2007 World Bank study on the carbon market.

of some in developing CDM projects, the bureaucratic nature of an international organization like the UN can sometimes have a negative effect on these efforts to mitigate climate change. Project developers and some governments have expressed considerable frustration at the extended delays as the Executive Board and the Methodological Panel work through reviews and recommendations on the large number of submissions. There have been wide calls for the EB to improve the feedback loop into the process for proponents, as several months can pass before proponents receive information from the review of their projects, and there is minimal opportunity to request clarification.

Other complaints have centred round what some have viewed as overly stringent interpretations of additionality and the recent introduction of consolidated methodologies, which are argued to reduce potential project reductions in some cases by as much as 40 percent. If anything, the expensive and time consuming experiences of project developers to date have illustrated the difficult, if necessary, role of early actors in a brand new system.

It is also difficult to make the gas emission predictions with certainty. Emission baselines can take one of two forms: they can either be set at the beginning for the project lifetime to serve as a fixed reference system ("static"), or they may be revised during the project operation ("dynamic"). Static baselines are predictable and can increase certainty regarding the volume of credit generation throughout the project lifetime. Dynamic baselines are designed to be re-evaluated at points during execution of a project, in order to allow closer estimates of reductions. Rather than a set number for each year, the baseline fluctuates to account for factors that influence emissions such as economic variables, and other influencing projects and policies. As such, they may reflect more accurately what would have happened in the absence of the project and, therefore, in doing so, may ensure the continuing environmental additionality of a project more consistently than static baselines. The static baseline tends to be preferred as it helps project developers give a more solid outlook on emissions reductions, which is something that the market requires in estimating risk factors.

9. Post-Kyoto Mobilization

Evidently, projections of future climate change and climate impacts have remained uncertain. As the policy focus is now shifting towards the possible post-Kyoto landscapes, mobilizing such an enthusiastic subscription by the broader market is essential. Based on the global consensus that temperatures will continue to rise as a result of greenhouse gas emissions, the key concern now is the speed and magnitude of climate change, and its inevitable impacts on human society. Generally speaking, climate policy is also about managing risks. Governments assess the potential impacts of climate change, judge how likely it is that various impacts will occur, and determine what policy choices should be made and how these policy adjustments will affect those risks. Accordingly, the market based framework should revolve around a similar principle of risk, which is best managed by robust participation from the corporate entities.

While the demand for carbon credits in relation to the Kyoto Protocol looks substantial, there is a significant risk in this industry that there might be an absence of or at least a delay in reaching an agreement on the post-2012 regime. The value of carbon credits prior to 2012 is hotly debated but there is general agreement amongst industry commentators that without a regulated system of some sort these credits will lose their value and the companies established over the past few years to supply them will run into difficulties.

Previously, the advantages for linking trading schemes were discussed, with respect to corporate entities. The benefits are not just limited to the private sector and those who seek to profit from carbon trading. Essentially, linking the project-based mechanisms to the allowance-based EU ETS will provide gains on more than one end. First of all, a benefit is that of corporate involvement, which is already evident in the EU system. Those who have been given emissions allowances from their respective governments are more inclined to participate actively to develop the market based mechanisms according to their own vested interests. Secondly, it encourages the on-going transmission of technology to developing economies.

With additional linkages, this sharing of technology would accelerate from its current pace. The transfusion of new technology to developing countries tends to be slow, as it is only competitively permissible when this particular level of technology has already been surpassed in the developed country. Lastly, the creation of a globally linked market ensures a strong signal to all countries that have not yet ratified the Kyoto Protocol. A significant party would be the United States, whose historical emissions level more than justifies its ratification of the Protocol. However, due to the clause in the Protocol excluding developing countries from mandatory emissions reductions, the United States is reluctant to tie itself down to what is seen as an economic limitation. The presence of regional markets in the United States will be greatly influenced by a stronger market focused around the EU ETS, such that the attractiveness of the EU carbon market would boost the level of participation from the United States even further.

The increased possibility of parity between the different carbon credits is also an incentive to market participants. To encourage even wider participation in the carbon markets, there has to be some positive outlook on the prices of carbon credits. Currently, the prices of CERs and VERs have fallen behind those of EU Allowances (EUAs), due to uncertainty over the approval of non-EUA credits. The lack of benchmark pricing also makes it hard to predict which direction the CERs and VERs will head in over the next few years. Some of the benchmarks discussed have included pegging the price as a fraction of EUA forward rates, or using China's fixed CER prices as a pricing floor. The goal behind a linked market would be to see an equalization of prices, which will hopefully help to remove some of the uncertainty over the future prices of carbon credits post-2012. The closer a carbon credit is to being a guaranteed delivery, the more it is worth. The clear benefit of lower compliance costs for EU countries will drive them to tap on the possibility of lower overall prices on the various carbon credits. As these prices become more affordable, there will also be a greater demand for more CDM projects in order to fill the gap in EUA credits, as there will be a higher demand for cheaper credits. The broader market will boost new opportunities for corporations and investment funds in supporting the design of CDM projects, and promote research and development in cutting edge technologies. It also creates new prospects for host countries to achieve their sustainable development goals through technology diffusion and training of skilled personnel.

Most importantly, there is a need for confirmations of the post-2012 protocol. In the upcoming Bali conference talks at the end of 2007, the stage for developing the protocol further will be developed. The best way to ensure that market participants do not shy away from the carbon market is to prove the evidence of such developments for the market. Furthermore, with the verbal agreements of the "+5" large developing countries, China, India, Brazil, South Africa and Mexico, the way is being paved for an even larger market of carbon credit buyers. Although these developing countries are not under any mandatory reduction targets yet, and with a natural inclination to remain in this economically-beneficial stage in order to avoid the costs of cutting emissions, these countries have surprisingly pledged themselves to the Bali talks. This sparks hope that the second stage of the protocol might bring even more success in global cooperation.

10. A Broader Accommodation

As there is little doubt today about the effects of climate change, as proven by the erratic changes in weather, the possible policy changes now address more of the viability of the structure of the market cap-and-trade system than the arguments about the science of climate change. In order not to become a niche system of limited impact, the new regime must accommodate the broader societal participation. Thus, the next step to take is to ensure that whatever effort is put in to minimize the effects of climate change are successfully implemented. Although it is important to always remember the underlying goals of the Kyoto Protocol, there must be an acknowledgment that most of the change that has occurred is being driven by the market. Therefore, this is the direction that should be taken as the next steps are being planned.

The use of carbon trading has already been proven, to a certain extent, to be the preferred policy instrument to achieve the targeted reductions. The potential for growth in the carbon market has yet to be met; as such, the amount of activity seen in trading is bound to increase. However, due to the numerous obstacles discussed, there has to be a number of policy changes pushed through to ensure the smooth continuation of market activity, with the expected increased of participants. In view of this, there also has to be a stabilization of the CDM framework. With the constant revisions being made to policy and methodologies, the efficiency of the market cap-and-trade system is being compromised.

The cap-and-trade system was first proven to work under the direction of the United States to decrease the amount of sulphur dioxide in the country. The immense success of their policy in averting an increase in the amount of acid rain was touted as a way forward into designing the right policy in reducing greenhouse gas emissions. The structure of the EU ETS has been largely based on the US sulphur dioxide cap-and-trade system. The flexibility of the system has allowed the increased support of those involved, as opposed to mandatory reductions with monetary penalizations on the failure to meet targets. This flexibility, in return, enhances the viability of the system. Obligated corporations would rather spend relatively small amount of capital than make costly technical adjustment or adopt new procedures in order to meet their obligation. Meanwhile, active carbon funds provide sufficient allowances to meet the demand, together with the buy side companies, making the system sustainable.

Although companies rushing to be involved in the market have been spending a large amount on projects, the system has not yet yielded substantial environmental benefits. Quite a few organizations are buying worthless carbon credits that have not been proven with real emissions reductions, while some industrial companies are earning profits from the generation of carbon credits in projects from which they have already gained the monetary benefits of energy efficiency. The balancing point within the system is still to be found. However, it is clear that there should be more inclination to "cap" than "trade", meaning that carbon trading should be seen as the tool to allow the Annex I countries in maintaining their "cap", so as to ensure that a significant enough net overall reduction does occur, before it is too late for changes to be made.

There have also been previous discussions[6] on emissions taxing, which penalizes the polluters who exceed their limit, but eventually offers no solution for those who cannot find a sufficiently cost-effective way to reduce their emissions. Depending on the amount of taxes imposed, taxation might not be as effective as the market-based framework. For the private sector, resorting to paying emissions taxes might oftentimes be the quick and easy way to solve their problems if companies find it too much trouble or simply too expensive to justify such massive changes to their operations. Trading, on the other hand, if lucrative enough, might allow for the added economic incentive for companies to engage in emissions reduction voluntarily whilst recouping their losses for the project implementation cost through carbon trading, all in addition to meeting the overall reduction target. CDM has already provided for a generally climate positive rationale in the number of projects already registered and operating, hence the market system will actually serve to benefit the environment more than simple taxation.

11. A Volume Approach?

The market based framework has to change from the narrow orientation of growth tied to sustainability, in order to harness the maximum potential for the market. A decoupling from that orientation will lead to a shift for more volume based "climate-positive" and more "energy efficiency-positive" growth that is commercially sustainable. So far, the effort for climate change has revolved around the two pillars of emissions reduction and sustainable development, whilst driven by a CER hungry market. The outcome of this is a strict assessment on the "additionality" of the projects proposed. The fact that developers have to prove that the emissions reductions they make are beyond and above that of what would have occurred regardless of CDM, is acting as a barrier to the full proliferation of CDM.

[6] Catalyst Vol. 4, 2005, Cap-and-Trade Systems.

For example, if a project activity is conceived due to financial incentive in the form of VERs or CERs, then it should receive these benefits as long as it has a net beneficial effect on the global environment. It should be recognized that any emissions reduction is better than no reduction at all. Therefore, the framework should be remodelled to reflect this sentiment. Some projects are abandoned at the conceptual stage due to an inability of the developers to design the project such that it meets all of the CDM criteria. However, a common-sense look at the potential project suggests that it is generally "climate positive" and should be implemented in spite of not fully meeting the requirements.

Of the registered CDM projects in the UNFCCC, about half (485 out of 1032) are small-scale projects.[7] The current situation is that a number of participants have stayed focused mainly on the economic incentives of CDM projects and ignored the environmental impacts. It would make sense to invest in small-scale projects to decrease exposure to some of risks mentioned above, such as registration risk and host country risk. There needs to be a new direction whereby considerations of policy adjustment or incentives should reinforce any inclinations towards large-scale projects. This is not to say, however, that small scale projects are not worth the effort, although some limitations on small-scale CDM projects are needed to decrease the economic attractiveness in the coming decades, whilst host country governments need to increase their support for large-scale projects. This proposed new inclination is important for establishing public confidence for the market based mechanism in the future. The policy framework in favor of large-scale projects should also be developed with a great sense of urgency on a commercial basis, since this will be the main motivation for changes on the part of the market.

As mentioned before, the process of setting up the post-2012 regime is critical. It has to tie in the support of both host countries and developed countries in providing the proper framework for projects, which will help to improve confidence in the commitment of

[7] UNFCCC Website http://cdm.unfccc.int/Statistics/Registration/RegisteredProjByScalePieChart.html

host countries and lead to less uncertainty about host country risk for developers. The need for reassurance will coax more participation out of the various market players. Overall, the outlook of the future operation of an enduring carbon market is strong. Recent discussions have been encouraging as discussions over the upcoming Bali conference talks promise to be fruitful.

As the need for an increasingly market based framework is emerging, it is evident that the only way to continue the developments in the fight against climate change is to help facilitate an even broader participation. This does not just include the governments and corporations, but also the involvement of everyone in general. To do so, the new framework has to be agreeable to a large party of people, which can be done by proving the flexibility of the system. If the CDM regulations keep to its rigidity, it might end up losing its supporters on the industrial and financing sides, as real projects with merits are increasingly turned away over minute details.

CHAPTER 7

MAKING MARKETS WORK — A REVIEW OF CDM PERFORMANCE AND THE NEED FOR REFORM

Charlotte Streck*

Climate Focus, Minervahuis III, Rodezand 34, 3011 AN Rotterdam, NL, c.streck@climatefocus.com

Jolene Lin

Faculty of Law, The University of Hong Kong. jolene@hku.hk

The Kyoto Protocol's Clean Development Mechanism (CDM) is the first global market mechanism in international environmental law. It has been much lauded for its success. However, doubts as to whether the CDM governance structure is robust enough to meet the challenges of regulating an international market mechanism in the long term are emerging. The Executive Board (EB)'s decision-making practice is often not predicable and many of its decisions have come as a surprise to project participants and technical project experts. Members of the EB often have multiple responsibilities which result in a complicated situation of conflicting interests. Finally, private sector participants in the CDM, who have been aversely affected by EB decisions, have no right of recourse and essentially little if any due process rights. This paper argues that incorporating mechanisms to promote procedural fairness and creating an appeals process for aggrieved CDM participants will promote transparency and accountability in the

* This paper was jointly presented by the authors at the conference, *Crucial Issues in Climate Change and the Kyoto Protocol: Asia and the World*, organized by the Asia-Pacific Centre for Environmental Law (APCEL) of the Faculty of Law, National University of Singapore, 30–31 August 2007.

CDM decision-making processes. This is essential for the sound operation of the CDM regulatory regime which will have a direct positive effect on the international carbon market. After conducting a comparative analysis of other regimes in which international bodies take decisions that directly affect individuals, most notably the system of targeted sanctions of the UN Security Council and the Anti-Doping Regime, as well as examining the World Bank Inspection Panel and the European Ombudsman as models of international review mechanisms, the authors set out proposals for reform of the CDM, including professionalizing the EB and the panels, securing better and more consistent funding, the elimination of political interference and the introduction of administrative law-like processes.

Introduction

The Kyoto Protocol's Clean Development Mechanism (CDM) is the first global market mechanism in international environmental law. The success of the CDM thus far can only be described as breathtaking. Its performance has been beyond the imagination of its creators at the Kyoto Protocol negotiations who did not foresee that the CDM would spawn a market in a regulatory commodity that is worth billions of dollars,[1] not including the value of the ancillary service industries that this new market has engendered.[2]

There are few mechanisms that lend themselves better to pioneering the emerging legal discipline of international administrative law than the CDM.[3] The authors are of the view that the climate change

[1] Capoor & Ambrosi, 'State and Trends of the Carbon Market', World Bank, 2007, http://carbonfinance.org/docs/Carbon_Trends_2007-_FINAL_-_May_2.pdf (accessed 15 June 2007).

[2] The creation of markets to reduce greenhouse gas emissions has itself led to the development of 'carbon finance' which has been described as being broader than the trading of carbon credits because new financial instruments are being developed to facilitate the transfer to a carbon constraint economy.

[3] Kingsbury, Krisch, & Stewart, 'The Emergence of Global Administrative Law', 68:3–4 *Law & Contemporary Problems* (2005) 15, at 17. The authors have benefited from the research output of the Global Administrative Law Research Project at New York University School of Law and would like to register an acknowledgement of thanks.

regime, particularly the CDM regulatory framework, is a good case study of the emergence of global administrative law. The CDM is unique in regulating a market dominated by private players that depend, in the creation of the market's underlying asset, on a United Nations committee that approves calculation methods and projects.

Containing many commendable design features, the CDM serves as a useful model for other emission trading and off-set schemes. The conceptual underpinnings of the CDM are strong and it is likely that the idea of the CDM, or the mechanism itself, will survive in a post-Kyoto climate regime. However, despite its success and model character, serious doubts as to whether the CDM governance structure is sufficiently robust to meet the long-term challenges of regulating an international market mechanism are emerging. As more projects move from design to implementation, and more funds are not only promised but actually disbursed, the limitations of the CDM regulatory structure are becoming increasingly obvious. While Executive Board (EB)'s decisions have direct effect on private entity rights, the Board's decision-making practice is often not predicable and many of its decisions have come as a surprise to project participants and technical project experts. While the EB is effectively a regulatory agency whose decisions have significant legal and financial consequences for private sector participants in the CDM, the EB is not subject to the usual political and legal controls to which a domestic regulatory agency would be subject. For example, there is no independent tribunal within the CDM regulatory framework to which aggrieved entities may appeal for a review of an EB decision. This gives aggrieved entities, who may have suffered damage from EB decisions, no right of recourse and essentially little if any due process rights. That the EB received twelve threats of legal proceedings from project developers in 2007 alone, despite the lack of access to a review mechanism, is evidence of the growing discontent among private project participants.[4]

[4] GTZ Climate Protection Programme, Anja Wucke & Axel Michaelowa, *CDM Highlights 54*, (November 2007, COP edition), online: www.gtz.de/climate.

The type of governance undertaken by the EB can be understood and analyzed as administrative action: rule-making, administrative adjudication between competing interests and other forms of regulatory decision-making and management. To the extent that such administrative actions bear direct and significant impact on the rights of an individual, they should be subject to a body of rules and principles that ensure that when public officials take such action, it is in accordance with the rule of law.[5] The rule of law is an amalgam of standards, expectations and aspirations that encompasses ideas about individual liberty and natural justice, and more generally, ideas about the requirements of justice and fairness in the relations between the government and the governed.[6] Containing the values of legality, certainty, formal equality, accountability, due process and access to justice, the rule of law establishes the background for a number of due process principles which form the backbone of constitutional and administrative law.[7] Such principles include the right to be heard before a decision is made,[8] the right to have the decision made in an

[5] The right to procedural justice and due process of law are guiding constitutional and administrative principles in common as well as civil law systems. The essence of these principles is that the government must respect a person's legal rights when the government deprives a person of life, liberty, or property. The requirement to adhere to due process has been introduced into the US constitution with the 5th and the 14th amendments. It also forms part of the set of constitutional rights of citizens of the countries of the European Union. References in community law can be found in Art. 6I (fair process) and 13 (right to appeal) of the European Convention on Human Rights, as well as The Charter of Fundamental Rights of the European Union, which establishes in Art. 41 the Right to a Good Administration and in Art. 47 the Right to an Effective Remedy and Fair Trial.

[6] T.R.S. Allan, *Law, Liberty, and Justice: The Legal Foundations of British Constitutionalism* (1993) at 21.

[7] Harlow, 'Global Administrative Law: The Quest for Principles and Values', 17 *Eur. J. Int. Law*, (2006) 187, 190; Jowell, 'The Rule of Law and its Underlying Values' in J. Jowell & D. Oliver, *The Changing Constitution* (2007).

[8] The right to be heard emanates from the fundamental principle of procedural fairness. This includes the right to be notified of charges and given opportunity to be heard. (e.g. Art. 41 II of The Charter of Fundamental Rights of the European Union confirms this principle as one of the basic principles governing the interaction between EU institutions and citizens of the Union).

unbiased and impartial manner,[9] and the right to know the basis of the decision so that it can be contested.[10] These procedural rights may be said to fulfill the central aspiration of the rule of law — "the subjection of public power to controls that ensure it is exercised in the interests of those affected by it".[11] This paper argues that incorporating mechanisms to promote procedural fairness and creating an appeals process for aggrieved CDM participants will promote transparency and accountability in the CDM decision-making processes. This is essential for the sound operation of the CDM regulatory regime which will have a direct positive effect on the international carbon market.

If the CDM is to form part of the international carbon market beyond the Kyoto Protocol's first commitment period which ends in 2012, the mechanism (which is part of the UN climate change intergovernmental process) has to be brought in line with due process requirements and procedural justice guiding administrative processes in the majority of national legal systems. It is essential that the CDM is governed by rules and procedures that promote the rule of law in

[9] The rule against bias (*nemo judex in re sua*) is one of long standing in the common as well as civil law. The prerequisite of any decision-making system that seeks to lay claim to any degree of fairness is the provision of unbiased tribunals and decision-makers. It can be said that no amount of procedural safeguards (such as the right to legal representation or to cross-examine witnesses) is likely to deliver fairness if the deciding tribunal is, in the first place, biased in the sense of being inherently predisposed against (or for) the individual who is the subject of the decision.

[10] For the EU, Art. 253 of the Treaty establishing the European Community formulates the requirement to state the reasons for any act, decision, directive, or regulation adopted by the Community or Community organs. It is also embodied in Art. 41 II of The Charter of Fundamental Rights of the European Union. Most EU countries possess administrative procedure acts, which contain the obligation of public authorities to give reasons for their acts (e.g. §39 of the German Verwaltungsverfahrensgesetz). See also Subchapter II of the US Federal Administrative Procedure Act.

[11] Dyzenhaus, 'The Rule of (Administrative) Law in International Law' 68:3–4 *Law & Contemporary Problems* (2005) 126 at 130; See also Le Sueur, 'Courts. Tribunals, Ombudsmen, ADR: Administrative Justice. Constitutionalism and Informality' in J. Jowell & D. Oliver, *supra* note 5.

order to secure public legitimacy and to bolster market confidence. In this regard, the authors propose the introduction of additional administrative processes and rules into CDM decision-making.

A sound regulatory regime will go far in ensuring the successful creation of the first truly global environmental market-based mechanism. The call for sound regulation could not come at a more opportune time as recent controversial findings about the performance of off-set projects has not left unaffected the international carbon markets. Rising criticism about the legitimacy of many carbon off-sets brought to the market has cast a shadow on the integrity of the CDM market.[12]

Section 1 of this paper explains the conceptual origins of the CDM. The CDM regulatory framework is described in Sec. 2. Section 3 analyzes the role of the CDM EB as the regulator of the international carbon market. The areas of concern that have given rise to doubt about the EB's ability to regulate the market in the long-term will be explored. In Sec. 4, we conduct a comparative analysis of other regimes in which international bodies take decisions that directly affect individuals, most notably the system of targeted sanctions of the UN Security Council and the Anti-Doping Regime. Section 5 looks the World Bank Inspection Panel and the European Ombudsman as models for international review mechanisms. Section 6 draws lessons for a reform of the CDM. Amongst the proposed reforms are: professionalizing the EB and the panels, securing better

---

[12] See Michaelowa & Purohit, 'Additionality Determination of Indian Projects, Can Indian CDM Project Developers Outwit the CDM Executive Board?', 2007, available at http://medias.lemonde.fr/mmpub/edt/doc/20070608/920594_additionality_determination_of_indian_cdm_projects.pdf. Download press articles from the *Guardian* (UK) at http://www.guardian.co.uk/frontpage/story/0,,2093837,00.html and http://environment.guardian.co.uk/climatechange/story/0,,2093815,00.html, *Le Monde* (French) at www.lemonde.fr/web/article/0,1-0@2-3244,36-920043,0.html, *Tagesanzeiger* (German) at http://tages-anzeiger.ch/dyn/news/print/ausland/758571.html; and the *Financial Times In-Depth report on Carbon Trading*, online: http://www.ft.com/indepth/carbontrading. Also see Press Release by UNFCCC Executive Secretary, 'Confusion threatens to limit potential of key Kyoto Protocol mechanism', online: www.unfccc.it (accessed on 25 July 2007).

and more consistent funding, the elimination of political interference and the introduction of administrative law-like processes. Section 7 concludes the paper.

1. Setting the Context

1.1. *The Kyoto Protocol mechanisms*

The Kyoto Protocol establishes three market mechanisms to help industrialized countries (so-called Annex I countries)[13] meet their emission reduction commitments in a cost-effective manner. Joint Implementation, the Clean Development Mechanism and Emissions Trading are established under Articles 6, 12 and 17 of the Kyoto Protocol respectively. The Clean Development Mechanism or the CDM, which is the focus of this paper, establishes a mechanism under which an Annex I Party may receive carbon credits (Certified Emission Reductions) for an investment in an emission reducing project in a developing country. The aim of the CDM is not only to help Annex I Parties meet their emission targets in a cost-effective way, but also to promote sustainable development in the developing countries.[14] The CDM thus establishes a scheme of joint implementation between industrialized and developing countries and provides an important tool for involving developing countries in the Kyoto Protocol processes. One of the most innovative features of the CDM is the direct involvement of private entities in the compliance framework of the Kyoto Protocol. The CDM allows countries to authorize private sector entities to sell and acquire emission reductions from projects in developing countries. Parties have made and are making use of this right. Private entities that have come to dominate the CDM market are thus directly involved in the implementation of the flexible mechanism and indirectly in treaty compliance.

---

[13] Annex I refers to the Annex to the United Nations Framework Convention on Climate Change and the Kyoto Protocol which lists the countries which agreed to assume binding emission limitation and reduction targets. Such targets are set in Annex B of the Kyoto Protocol.

[14] See Article 12(2) of the Kyoto Protocol.

1.2. *The CDM's conceptual origins*

The CDM emerged as "the Kyoto surprise" in the early hours of the final day of the negotiations at the 3rd session of the UNFCCC COP.[15] Few, if any, of the negotiators in Kyoto could have foreseen the far-reaching impact of the mechanism that was being created on the basis of international emissions trading. Most of the negotiators were from their countries' environmental ministries and their experience laid in negotiating multilateral environmental agreements (MEAs), not in international trade and markets. As the negotiations were seen to pertain to a predominantly environmental problem, it was not deemed necessary to involve officials from the trade, industry, or energy ministries.

In designing the CDM, the negotiators drew on examples of financial mechanisms in other MEAs rather than examined how global markets operate. Many of the features of the CDM Executive Board were borrowed from the Executive Committee which manages the Multilateral Fund for the Implementation of the Montreal Protocol (MLF).[16] Beyond the organizational structures, the "additionality" requirement of both Joint Implementation and the CDM is a concept that is closely related to the incremental cost principle of the MLF and the Global Environment Facility.[17]

These analogies are not surprising as the creation of a market to deal with an environmental externality is unprecedented in international

---

[15] Werksman, 'Unwrapping the Kyoto Surprise' 7 *RECIEL* (1998) 147, at 151.

[16] The MLF is the financial mechanism created in 1990 by the London Amendment to assist developing (Article 5) countries meet the agreed incremental cost of fulfilling the control measures imposed by the Montreal Protocol.

[17] The additionality requirement may be explained as such: CDM project activities must result in reducing or absorbing (sequestering) GHGs that are 'real and measurable and would not have occurred in the absence of the proposed project activity'; (UNFCCC, Report of the Conference of the Parties on its seventh session, Marrakesh, 29 October–10 November 2001). Addendum part two: Action taken by the Conference of the Parties. Vol. II. FCCC/CP/2001/13/Add. 2, at pp. 20). In other words, to qualify for credits, a project activity must demonstrate that GHG emissions were reduced against the 'baseline scenario', a representation of GHG emissions under normal circumstances.

environmental law. Yet, the CDM is not merely a burden-sharing mechanism whereby the industrialized nations cover the compliance costs of the developing world and offers financial assistance and technology transfer. The CDM instead provides entities from Annex I countries an intrinsic incentive to invest in carbon abatement projects in non-Annex I countries where the costs of abatement are lesser than in the developed countries (which most Annex I countries are).

2. The CDM Regulatory Framework

2.1. *The CDM project cycle*

An explanation of the CDM project cycle at this stage will help one understand the regulatory framework governing the CDM. The basic operational principle of the CDM is the rewarding of greenhouse gas (GHG) emission reductions generated by a project activity implemented in a developing country Party to the Kyoto Protocol by tradeable Certified Emission Reductions (CERs) that can be sold on international carbon markets. Examples for CDM projects include renewable energy, energy efficiency or reforestation projects. The CDM project cycle consists primarily of eight stages:

(1) *Project Development, Evaluating Feasibility and Obtaining Permits.* There are no CERs without a project which realizes the reduction of GHG emissions compared to baseline emissions. Each issued CER is labelled with a project identifier which links it with the project that generated the corresponding emission reduction.

(2) *Methodology Approval.* If the project uses a new methodology for calculating baselines and monitoring of emission reductions, it must be submitted for approval to the EB before the project can be validated.

(3) *Host Country (and Annex I) Approval.* Host country approval is a condition for the validation of the project design by an internationally accredited auditor. If the project owner wishes to add Annex I project participants to the project (e.g. the buyers of the

CERs), those entities have to present evidence of authorization and approval from an Annex I country.

(4) *Validation.* An independent environmental auditor that has been accredited for that purpose (a Designated Operational Entity or DOE) validates the project design as described in a Project Design Document (PDD).

(5) *Registration.* Once the project has been validated, the DOE submits the project to the CDM Executive Board for registration. Provided that there are no objections regarding the registration, the project will be registered and thus formally approved as a CDM project.

(6) *Generation of CERs.* An operating CDM project has to calculate and monitor the emission reductions it generates.

(7) *Verification.* At periodic intervals, the project developer will contract an accredited independent auditor (a DOE different from the one that validated the PDD) to verify the emission reductions of the project.

(8) *Issuance of CERs.* The verification and certification reports of the verifying DOE constitute the basis on which the Executive Board issues CERs. These CERs are issued into a pending account in the CDM registry from where they are distributed to the accounts of the project participants. The project developer does not necessarily hold a registry account and often CERs are directly transferred to the accounts of the CER purchasers. CER accounts can be held in registries of Annex I countries.[18]

The majority of CDM projects are designed by project owners in developing countries. They prepare the relevant studies and documents, including the PDD, either alone or in cooperation with an entity from an Annex I country interested in acquiring the CERs.

---

[18] Non Annex I (developing country) Parties have no obligation to establish and manage registries. This means that, unless they maintain an account in an Annex I registry, non-Annex I country entities do not hold accounts in national registries. CERs of non-Annex I countries are held in the international CDM registry, managed by the EB and administered by the UNFCCC secretariat.

The acquisition of CERs takes, in most cases, the form of a forward purchase of CERs.

By the time of writing this article, an average of 150 projects enter the validation stage each month. The CDM pipeline now contains 2783 projects (excluding the 46 rejected and the 9 withdrawn projects) that have entered the process of validation. 859 of the projects are now registered and a further 149 are in the registration process.[19] The majority of the proposed projects (62%) generate electricity from renewable energy sources. Renewable energy sources, such as wind or hydropower, reduce GHG emissions against electricity generation from fossil fuels. The high number of projects does not correspond, however, to a similarly high percentage in expected CER yield. Only 29% of the projected CERs stem from renewable energy projects. More than 50% of the expected CERs come from the abatement of non-CO_2 gases, such as methane or Nitrous Oxide.

2.2. *Entities involved in the CDM*

The following section sets out to describe each of the entities involved in the CDM regulatory regime.

2.2.1. *The COP/MOP*

The CDM forms part of the compliance framework of the Kyoto Protocol, which is a protocol to the 1992 United Nations Framework Convention on Climate Change (UNFCCC). The COP/MOP is established under the Kyoto Protocol and refers to the "Conference of the Parties serving as the Meeting of the Parties" to the Kyoto Protocol.[20] It is an assembly of all the Parties to the Protocol which convenes annually and is the governing body of the Kyoto Protocol. The mandate of the COP/MOP is broadly drafted. Article 13(4) states that the COP/MOP "... shall keep under regular review the

[19] UNEP Risø CDM Pipeline, online: http://www.cdmpipeline.org/ (accessed 29 December 2007).

[20] See, for example, Article 2(1)(b) of the Kyoto Protocol.

implementation of this Protocol and shall make, within its mandate, the decisions necessary to promote its effective implementation". While engaging in rule making and thus exercising legislative functions, neither the UNFCCC COP nor the COP/MOP of the Kyoto Protocol has legislative powers *per se*. The decisions of the COP/MOP are not legally binding on any Party without its consent. The acceptance of COP/MOP decisions is founded on a Party's consent rather than through the legislative authority of the COP/MOP.[21]

The CDM was established in Article 12 of the Kyoto Protocol. However, the limited provisions of the Protocol do not provide sufficient guidance for the creation of an operational mechanism as complex as the CDM. The COP/MOP has thus adopted, with the so-called Marrakesh Accords, the "Modalities and procedures for a clean development mechanism".[22] The Marrakesh Accords formulate a set of decisions that guide the implementation of the CDM. Even though the CDM modalities and procedures do not have any formally binding effect on the Parties participating in the mechanisms, there is general agreement amongst the Parties that the COP/MOP decisions determine their position under the mechanisms.

2.2.2. *The CDM executive board*

International bodies normally do not have the authority to take administrative or legal decisions that directly affect non-state actors. The Executive Board of the CDM (EB) is an exception. The EB is composed of ten members and ten alternate members from Parties to the Kyoto Protocol who are elected by the COP/MOP. While the

---

[21] For a detailed review of law-making by COPs, see J. Brunnée, 'COPing with Consent: Law-Making Under Multilateral Environmental Agreements', 15 *Leiden Journal for International Law* (2002) 1, 32.

[22] The Marrakesh Accords were adopted by the 7th session of the UNFCCC COP held in Marrakesh, Marrocco, in December 2001 and confirmed by the 1st session of the COP/MOP in Montreal in December 2005 FCCC/KP/2008/8/Add.1 Decision 3/CMP.1 (Modalities and procedures for a clean development mechanism as defined in Art. 12 of the Kyoto Protocol).

COP/MOP is the ultimate authority of the CDM, the day-to-day supervisory work is undertaken by the EB.[23] Pursuant to Article 12(4) of the Kyoto Protocol, the EB supervises the implementation of the CDM "under the authority and guidance of the COP/MOP". The EB shall be "fully accountable" to the COP/MOP.[24] Such language suggests a delegation of authority by the COP/MOP to the EB to act as the regulatory authority. In a domestic setting, such delegation of power is usually from the legislature to a regulatory agency that is recognised to possess the technical expertise to implement the regulatory framework, including the making of subsidiary legislation to complement the primary legislation, adjudication of license applications, etc. Also, such delegation is common because the legislature usually does not have the requisite time and resources to undertake day-to-day supervision of the many areas in which the modern state now plays a role, for example, healthcare, monetary policy, welfare, environmental protection. This characterisation of the EB as akin to a regulatory agency is not far-fetched.[25] In its daily operations, the EB clarifies and interprets the decisions of the COP/MOP. The Board takes decisions on methodologies and projects, mandates reviews and revisions to project applications. It is assisted in this task by various expert panels. However, the recommendations of these panels are not binding and increasingly, the Board's decisions do diverge from the recommendations of these

---

[23] Article 12(4) of KP: 'The clean development mechanism shall be subject to the authority and guidance of the Conference of the Parties serving as the meeting of the Parties to this Protocol and be supervised by an executive board of the clean development mechanism.'; The scope of its role and its powers is set out in Part C of Annex "Modalities and procedures for a clean development mechanism", Decision 3/CMP.1, para 2–4.

[24] Decision 3/CMP.1, Annex, para. 5.

[25] R. B. Stewart, 'US Administrative Law: A Model for Global Administrative Law?', 68 *Law & Contemp. Probs.* (2005) 63, at 91, fn. 11; E. Meijer, 'The International Institutions of the Clean Development Mechanism Brought Before National Courts: Limiting Jurisdictional Immunity to Achieve Access to Justice' 39(4) *NYU Journal of International Law and Politics* (2007) 877, at 886.

panels.[26] The EB is mandated to approve baseline and monitoring methodologies, review provisions with regard to simplified procedures for small scale projects, and accredit DOEs. The Board also interprets the decisions of the COP/MOP and prepares technical and decision papers for review and adoption by the COP/MOP. In the process of interpreting the Kyoto Protocol and the COP/MOP decisions, the EB is effectively engaging in subsidiary law-making and adjudication. While the decisions of the EB are not legally binding in a formal sense, they have been accepted as *de facto* binding by entities that participate in the CDM, including both Parties to the Protocol as well as public and private sector legal entities. In this respect, the EB acts like a market regulator that is responsible for approving applications for licenses to carry out any activity within its regulatory purview.

The EB also prepares annual reports for the COP/MOP.[27] The COP/MOP responds to requests of the EB and provides guidance to the EB on its operations.[28] The COP/MOP therefore decides on the broader policy issues and on the strategic development of the CDM. It is the EB, however, which implements these policy directives at the project level.

2.2.3. *Panels/working groups/teams*

To assist it in the performance of its functions, the EB is entitled to establish panels, working groups or teams. In this regard, it "... shall

[26] The EB's divergence from the Methodologies Panel's (MethPanel) recommendations in approving methodologies is an example of such divergence in interpretations between the EB and its panels. A concrete example is the pending issue of addressing the double counting of emission reductions generated by the use of biofuels. The MethPanel has developed several proposals to address this issue, the last one being rejected at the 30th session of the EB from 21–23 March 2007. This renewed rejection of potential solutions to the double counting problem is likely to lead to a further indefinite delay in approval of such methodologies. Online: http://cdm.unfccc.int/EB/030/eb30rep.pdf, agenda item 3(b). (Accessed 15 June 2007).

[27] Pursuant to Decision 3/CMP.1, Annex, para. 4(a). The EB reports are available on the CDM website http://cdm.unfccc.int. The latest report covers the period from end-November 2005 to 21 July 2006 (accessed on 29 May 2007).

[28] Pursuant to Decision 3/CMP.1, Annex, para. 2–4.

draw on the expertise necessary to perform its functions, including from the UNFCCC roster of experts...[and] take fully into account the consideration of regional balance."[29] The expert panels do not take decisions. But they undertake the technical assessments upon which decisions of the EB will necessarily be based. Thus far, two panels, two working groups and one team have been set up.[30]

2.2.4. *The UNFCCC secretariat*

The UNFCCC secretariat provides organizational support to the various actors in the Kyoto Protocol institutional framework, including the COP/MOP and the EB. It establishes the link between the various actors participating in the CDM.[31] The Secretariat is responsible for, *inter alia*, preparing the minutes of meetings, drafting decisions and guidelines, arranging the meetings of the COP and the various Kyoto bodies, and monitoring the implementation of the UNFCCC through the collection and analysis of information provided by the Parties.[32] The UNFCCC secretariat does not take any decision. However, it serves to provide a collective institutional memory of the

---

[29] Decision 3/CMP.1, Annex, para. 18.

[30] See 'General Guidelines for Panels/Working Groups (version 02)', online: http://cdm.unfccc.int/Reference/Procedures/pnlguide.pdf for more information (accessed on 27 July 2007). The panels include the Accreditation Panel and the Methodologies Panel; the Working Groups assist the EB in matters relating to Afforestation and Small Scale Projects respectively. The CDM Registration and Issuance Team (RIT) assists the EB to consider the requests for registration of project activities and the issuance of CERS respectively that are submitted by DOEs.

[31] Netto & Schmidt, 'CDM Project Cycle and the Role of the UNFCCC Secretariat', in D. Freestone and C. Streck (eds), *Legal Aspects of Implementing the Kyoto Protocol Mechanisms* (2005) 175.

[32] More information about the UNFCCC Secretariat can be found at http://unfccc.int/secretariat/items/1629.php (accessed on 27 July 2007). It has been suggested that the in-depth review procedure of national communications under the UNFCCC (Articles 5, 7 and 8) gives the Secretariat a dominant position in evaluating and analyzing Parties' national climate policies which strengthens the position of the Secretariat; see S. Oberthür and H. E. Ott, *The Kyoto Protocol: International Climate Policy for the 21st Century* (1999) at 249.

negotiation process and the evolution of the CDM. While the members of the EB and the technical panels change, the secretariat supplies the climate change regulatory regime with long-term career staff. Taking into account the increasing complexity of the various subjects, the secretariat staff understands the context of the constellations of issues often better than the members serving on the Board and panels. Through the preparation of decisions and interpretation of EB rulings, it communicates its positions and influences the process in the direction of its interpretation of the decision text. While the UNFCCC secretariat operates in the background, its role is nonetheless a powerful one.

2.2.5. *Designated operational entities (DOEs)*

The responsibility of a DOE is to validate a CDM project by independently evaluating the Project Design Document against the CDM requirements, including substantive review of the baseline and monitoring methodology, and assuring that the CDM project has an adequate monitoring plan to safeguard against the overstatement of emissions reductions.[33] The DOE is responsible for validating that the project activity will result in a reduction of anthropogenic GHG emissions that are additional to any that would have occurred in the absence of the proposed project. During the implementation of a CDM project activity, the project participants prepare a monitoring report based on the registered monitoring plan and forward it to another DOE for verification/certification. After the second DOE verifies and certifies the amount of emissions reductions, it submits a report to the EB which will then issue the CERs.[34]

---

[33] The DOE's scope of work is set out in Section E of "Modalities and procedures for a clean development mechanism", Decision 3/CMP.1. The criteria that a Project Design Document must meet are set out in Appendix B, Decision 3/CMP.1.

[34] Information on the issuance of CERs is found in Section J of the Marrakesh Accords. A list of all accredited DOEs can be found on the CDM website: http://cdm.unfccc.int/DOE/index.html.

2.2.6. *Designated national authorities*

Every country that wishes to participate in the CDM is required to set up a Designated National Authority (DNA). Project participants are required to obtain the written approval from the host country DNA before its project can be registered by the EB. The DNA has to certify that a proposed CDM project contributes to its sustainable development.[35] Additional project participants, in most cases the purchasers of CERs, are required to have evidence of authorization and approval from an Annex I Party in order to participate in a CDM project.

2.2.7. *Project participants*

Project participants may be either private or public entities. These entities are the ones that develop CDM project activities or are interested in acquiring the resulting CERs. In order to obtain the requisite approval by the EB, a project participant will have to prepare the relevant documents and channel them through the describe CDM project cycle.

3. CDM Market Regulation

3.1. *A snapshot: The CDM and the carbon market*

The CDM market was worth more than five billion Euros in 2006 alone.[36] The asset creation takes place when registered CDM projects

---

[35] To obtain the host country approval for CDM projects, most DNAs require the official Project Design Document (PDD), as well as compliance with local regulation. For Chile, Peru, Argentina, Ecuador, Honduras, Panama, and Korea, meeting these conditions is sufficient to demonstrate that the project contributes to sustainable development in the host country. Other countries have established additional sustainable development criteria that a CDM project has to comply with. Countries which require compliance with sustainable development criteria in addition to compliance with local environmental laws include Cambodia, Indonesia, the Philippines, Thailand and Vietnam; Ministry of Foreign Affairs of Japan, FEALAC (Forum for East Asia-Latin America Cooperation), *Analysis of the Present Situation and Future Prospects of the Clean Development Mechanism (CDM) in the FEALAC Member Countries*, Study of EALAC for the 4th Economy and Society Working Group (June 7–8, 2006), online: http://www.mofa.go.jp/region/latin/fealac/index.html at 32.

[36] Capoor & Ambrosi, *supra* note 1.

generate emission reductions, which can be issued as CERs. The "primary market" is defined by the transactional relationship between the project and the purchaser of the emission right. Beyond these primary market transactions, a vibrant secondary and derivative market is emerging. The secondary market refers to transfers of issued CERs, while derivative trading involves the trading in options, futures, and carbon bonds.

The first CDM transactions were concluded between private project developers and sovereign carbon buyers (Annex I governments and international organizations). Today, an overwhelming majority of the entities trading in the CDM market are private entities.[37] They participate in the market either through investments in funds (for speculative purposes or compliance), through intermediaries, or through direct purchases. As of April 2007, there were (at least) 58 carbon funds in the market.

The Kyoto Protocol and the CDM have thus given rise to the first international market in an environmental commodity: certified emission reductions. The right that is created through the Kyoto Protocol is of a regulatory nature, and the market is a "permit" market.[38] In the case of the CDM, CERs are treaty-based rights which exist only in the context of the Kyoto Protocol.[39]

---

[37] The market segmentation in 2006 is as follows: Governments — 8%, Funds — 34%, Private buyers — 58% (Source: Point Carbon (2007)); K. Røine and H. Hasselknippe (eds), *Carbon 2007- A new climate for carbon trading*, at 17, Fig. 3.17). According to this same report, private buyers are dominating the market as more companies see the value of project credits. Most of the governmental buyers are European countries (at pp. 19).

[38] See B. Yandle, 'From Local to Global Commons: Private Property, Common Property, and Hybrid Property Regimes: Grasping for the Heavens: 3D Property Rights and the Global Commons', 10 *Duke Env L & Pol'y F* (1999); Streck & Zhang, 'Implementation of the Clean Development Mechanism in China: Sustainable Development, Benefit Sharing, and Ownership of Certified Emission Reductions', 16 *Yearbook of International Environmental Law* (2005) 259, at 264.

[39] Through the Linking Directive (Directive 2004/101/EC, OJ 2004 L 338/18), CERs would be eligible for use in Phase One of the EU ETS (2005–2007) while both CDM and JI credits would be eligible for Phase Two (2008–2012).

A functioning carbon market will direct money to the most cost-efficient ways to create CERs. This will in turn foster innovation in efforts to address climate change and create opportunities for market participants to take advantage of economies of scale. It is therefore important to create and maintain conditions that will encourage the deepening and increasing liquidity of the carbon market. Compared to traditional commodity markets, the success of a market in carbon rights (among others, in CERs) is more dependent on investor confidence in the robustness of the market and the regulatory framework simply because the creation, authenticity and consequent value of the commodity in question is entirely dependent on the regulatory framework. The existence of a transparent governance structure, which ensures market oversight and fair market access to all market participants, is a precondition for such confidence. Consequently, the long-term robustness of the CDM goes along with the existence of a capable market regulator charged to promote market integrity through the administration, interpretation and enforcement of rulings that apply to all regulated persons in the same manner.

3.2. *The EB as regulator*

As an instrument of international law, the provisions of the Kyoto Protocol apply only to States that have ratified the treaty.[40] However, from the moment of its creation, the negotiators of the Kyoto Protocol foresaw private participation in the flexible mechanisms established under the Protocol. Article 12.9 of the Kyoto Protocol explicitly states that "Participation under the CDM [.....] may involve private and/or public entities, and is subject to whatever guidance may be provided by the executive board of the clean development mechanism."

The governance structure of the CDM is evidence of the increased delegation of authority under treaties to various subordinated panels

[40] The general rule as stated in Article 34 of the Vienna Convention on the Law of Treaties (United Nations, Treaty Series, vol. 1155, p. 331) is 'A treaty does not create either obligations or rights for a third State without its consent'.

and constituted bodies. Such delegation is not undisputed as the increased distance between the legitimised law makers (the national governments) and the executive bodies to which significant powers have been delegated poses the risk of deficits in control and accountability.[41] Administrative effectiveness may increase through the delegation of executive tasks to specialized bodies such as the CDM EB. To the same extent, however, democratic accountability arguably diminishes through the increasing distance from legislative bodies and ratification procedures.[42] Further, while delegated decision-making in the domestic context exists within a broader system of checks and balances, the accountability regime in the international realm is much thinner. For example, there is no judiciary to check the legality of executive decision-making.[43] The Executive Board's decision or its omissions to decide have important consequences for any project participant, regardless of whether it is a Party or a private legal entity participating in a particular CDM project.[44] If actions similar to those of the Executive Board were performed by national agencies, there

---

[41] The delegation of 'legislative' or rule-making power from the COP/MOP to the EB poses the thorny question of abdication of responsibility on the part of the COP/MOP which has been given such powers by the Convention and the Protocol. The delegation of decision-making power to constituted bodies and subordinated panels would raises questions with respect to general principles of good governance such as accountability. While EB has not delegated any decision-making powers further. It is the role and authority of the EB itself and its accountability towards the COP/MOP or any legitimised law makers (i.e. the national governments) which is questionable.

[42] When governments and legislators that bind sovereign states through the process of signing and ratifying a legal instrument are put in place through democratic elections and legitimatised through transparent and democratic processes, treaty law also provides for democratic legitimacy on the international level.

[43] D. Esty, 'Good Governance at the Supranational Scale: Globalizing Administrative Law' 115 *Yale L.J.* 1490, at 1504.

[44] The conception that non-state actors are the regulated entities presents an evolution of international regulation away from a state-centric mode and towards a conception of global regulation of market actors, with states serving an intermediate position; R.B. Stewart, *supra* note 25, at 96 fn. 11.Other examples can be found in the context of the WTO and the Montreal Protocol.

would be little doubt of their administrative and regulatory character. But the CDM does not provide entities affected by decisions of the Executive Board with procedural rights comparable to those of domestic administrative law regimes.

The EB, under the guidance of the COP/MOP, acts as the market regulator for the CDM and is thus responsible for establishing and maintaining standards for the development of a fair, orderly, and efficient market. However, its role is not limited to pure market access regulation. The EB's responsibility includes the process of asset creation itself: The existence of any circulating CER depends on the EB's approval at various stages of the creation process.

The administrative rules that guide the operations of the EB foresee a number of safeguards. Individuals serving on the EB are required to have defined expertise, obliged to preserve confidentiality, undertake to have no interest in any project or operational entity and are required to take an oath of office.[45] However, there are no assessments or interviewing procedures to ensure that these requirements are met before an individual is appointed to the EB. Following UN tradition, the regional negotiations groups nominate representatives which so far have been approved by the COP/MOP without further questioning.[46]

3.3. *The EB in operation: An analysis*

Any evaluation of the performance of the CDM will have to strike a balance between the environmental integrity of CERs and the efficiency in their supply. It has not only been the CDM's environmental performance that has been the subject of debate and criticism, but its

[45] Decision 9/CMP.1, Annex, Guidelines for the implementation of Article 6 of the Kyoto Protocol (hereinafter "Decision 9/CMP.1"), para. 10; Decision 3/CMP.1, Annex, Modalities and procedures for a clean development mechanism as defined in Article 12 of the Kyoto Protocol (hereinafter "Decision 3/CMP.1"), para. 8(e). The Executive Board has adopted rules of procedures that regulate their operations.

[46] The procedure is based on CDM Modalities and Procedures, Annex, paras. 7 and 8. So far no objection to the candidates proposed by the regional groups has been recorded.

institutional performance as well.[47] The CDM can only be successful in creating CERs if participants in CDM projects trust that the procedure that leads to the eventual transfer of CERs to their registry accounts is administered in a transparent and fair manner. Once project participants are not confident that their investments will eventually be rewarded with CERs, they will shy away from investing in emission-reducing projects.

The continuous stream of project proposals that reaches the EB demonstrate that participants in CDM projects still evaluate the risks associated with the administrative procedures as bearable. The risks associated with the complicated and obscure CDM rules and the EB's performance are off-set by the opportunities of generating cost-efficient CERs. However, such confidence will not necessarily be perpetuated and investors will continue to evaluate CDM risks each time they make an investment decision. Once project participants perceive that they may not be treated with fairness or enjoy certain minimal due process rights and that there is uncertainty in the decision-making processes, the risks will be considered too high to justify participation in the CDM. It is therefore crucial that the CDM applies commonly accepted principles of due process to guarantee fundamental fairness, justice, and respect for property rights.

A look behind the scenes reveals, however, significant unhappiness with a number of the EB's rulings and its way of conducting business. While market actors generally appreciate the hard work by the individuals serving on the Board, there are mounting complaints about the continued lack of transparency in the Board's decision-making

---

[47] C. Streck, 'The Governance of the Clean Development Mechanism — the case for strength and stability', in D. Freestone and C. Streck (eds), 15(2) *Special Issue, Environmental Liability Journal: The Kyoto Protocol — Current Legal Status of Carbon Finance and the Flexible Mechanisms*, (2007) 91. See also IETA, '2006 State of the CDM, IETA Position on the CDM for COP/MOP2', available at http://www.ieta.org/ieta/www/pages/getfile.php?docID=1931 (accessed 25 July 2007). For discussion of the environmental integrity of the CDM, see E. Meijer & J. Werksman 'CDM — concepts, requirements and project cycle', D. Freestone & C. Streck (eds), *ibid.*, at 81.

and the lack of predictability.[48] The ingredients of a sound process are, at the same time, the conditions for fair and predictable decisions. In the following, the EB's performance will be measured against the procedural requirement of independency, transparency, efficiency, predictability, as well as the right of project participants to be heard and to appeal EB decisions.[49]

3.3.1. *Independency*

The EB and the DOEs are mandated to be neutral and independent. Any potential conflicts of interests are to be avoided, whether these interests are of a financial, political, or personal nature.

EB members act in their personal capacity and do not represent any country or constituency. They have to be properly qualified and are not supposed to have an interest in any project. To ensure that all decisions taken are unbiased, at the beginning of each meeting, the members of the Board confirm that they do not have any interest in any project or operational entity.[50] Affected entities can raise objections regarding such confirmations.

In practice, the interpretation of what constitutes "conflicting interests" is *de facto* reduced to the question of whether a financial conflict of interest exists. However, while it can be assumed that no EB member has a pecuniary interest in the CDM projects he/she

---

[48] IETA, 'Strengthening the CDM, Position Paper for COP 11 and COP/MoP 1', Position paper to COP12 COP/MOP 2, online: http://www.ieta.org/ieta/www/pages/getfile.php?docID=1132; IETA letter dated 6 October 2006 to the Chair of the Executive Board regarding the communication with the Board and its panels; World Bank, Reforming the Clean Development Mechanism, Background Paper for the Steering Committee Meeting, World Bank Carbon Finance Host Country Committee, http://carbonfinance.org/Router.cfm?Page=DocLib&CatalogID=5668 (accessed 10 August 2007). Center for Clean Air Policy, Washington DC, 'Summary of an informal workshop on streamlining the CDM, COP/MOP1', http://www.ccap.org/international/Summary%20of%20Day%201%20&%202%20Discussions.pdf (accessed 10 August 2007).

[49] The following list borrows heavily from Streck, *supra* note 47.

[50] Decision 3/CMP.1, Annex, para. 8(f).

reviews, political conflicts of interest are common and uncontested. In some instances, those conflicts appear to dominate the Board's decision-making practices. It is not seldom that EB members play multiple roles at the same time, including those of being UNFCCC/Kyoto Protocol negotiators for their country, representing their country's DNA for the CDM, or managers of large government CDM purchasing programs.[51] Although there may be no financial conflict of interest, many EB members are entangled in a complicated web of different interests which is tolerated by the COP/MOP and the members of the Board.

3.3.2. *Transparency*

The EB is a body which exercises regulatory functions but is not directly subject to the control of national governments or national legal systems. Transparency and public participation should provide the necessary counterbalance to the missing legal controls and is supposed to play a central role in the decision-making processes of the CDM.[52] The CDM project cycle provides the possibility of public participation at the project level and the CDM rules require that 19 separate types of CDM-related information be made 'publicly available'.[53] While there is a clear recognition of the importance of participation and transparency in the CDM project cycle, it is questionable whether the same degree of transparency applies to the work of the EB. In order to ensure transparency regarding some 'aspects' of its work, the EB decided, at the tenth

---

[51] A comparison of the list of EB members and the representatives of country DNA's (both listed on the UNFCCC website, http://cdm.unfccc.int/index.html (accessed 10 August 2007) reveal the overlaps in responsibility. Reviewing the portfolios of the individuals within their national Ministries, it also becomes clear that their responsibilities often cover all aspects of the CDM, including the purchase and sale of CERs on behalf of their respective countries.

[52] E. Meijer & J. Werksman, *supra* note 48.

[53] Eddy, 'Public Participation in CDM and JI Projects', in D. Freestone & C. Streck (eds), *Legal Aspects of Implementing the Kyoto Protocol Mechanisms*, (2005) 71, at 79.

meeting of the CDM Executive Board in July 2003, to take the following measures:

> *enhancement of the UNFCCC CDM web site; provision of CD-ROMs, selected documents/forms in all UN languages; organization of meetings with Parties and accredited observers; and the review of modalities for attendance by observers to meetings of the Executive Board.*[54]

To enable interested parties to follow the deliberations of the EB, the Board's meetings 'shall be open to attendance, as observers, by all Parties and by all UNFCCC accredited observers and stakeholders, except where otherwise decided by the Executive Board.'[55] The objective of transparency is to be weighted against the need for efficiency and the need to keep certain information confidential. This is expressed in rule 14 of the EB rules and procedures,[56] which authorizes the Chair of the Board to decide over the opening and closing of EB meetings without giving reasons for his/her decision. In addition, the EB may decide 'in the interest of economy and efficiency, to limit attendance at its meetings to members, alternate members and secretariat support staff.'[57] The public is neither allowed to attend the meetings of the expert Panels nor the meetings of the Working Groups.[58]

---

[54] EB10 Annex 6/Annex 6/Second Annual Report of the Executive Board to the Clean Development Mechanism to the Conference of the Parties (2002–2003)/ section IV, number 54–65.

[55] Paragraph 16 of the Annex to Decision 3/CMP.1 on modalities and procedures for a CDM, document FCCC/KP/CMP/2005/8/Add.1.

[56] The Board has further complied with its mandated and has proposed its Rules of Procedures which have been adopted by the 8th session of the COP (2002) and the 1st session of the COP/MOP (2005). Rule 26 of the Rules and Procedures emphasizes that the rule of transparency applies to all the work of the EB, http://cdm.unfccc.int/Reference/COPMOP/08a01.pdf#page=31 (accessed 20 June 2007).

[57] *Ibid*, Rule 27.

[58] The newly created 'Review and Issuance Team' has been created without any clear legal basis and, although it is highly influential, there is no international process of selecting and appraising of its members, nor are the meetings of the team open to the public.

In practice, there is a clear tendency to limit attendance to EB meetings and to conduct closed sessions. Recently, it has become more frequent that half of each EB meeting takes place behind closed doors. As a result, even those very entities that are directly affected by the decisions of the EB do not have access to the Board meetings. In additional, there is little opportunity for direct interaction between stakeholders and the EB and its bodies, as appropriate, e.g. between the Methodologies Panel and project participants, that could improve the transparency of the process for those directly affected. Participants in CDM projects are left on their own to try to understand the underlying rationale and the reasoning for a particular decision. Yet, one of the purposes of granting a fair hearing, apart from considerations of the efficient administration of law or policy, is to enable the affected party to play a role in the process of decision, thereby acknowledging his autonomy as a rational agent.[59] Further, closed door meetings do not promote adherence to the fundamental tenet that '... justice should not only be done, but should manifestly and undoubtedly be seen to be done.'[60] The appearance of bias offends the rule of law as much as its actuality.

3.3.3. *Efficiency*

Consistent with the aim of ensuring environmental and social integrity, the CDM process should minimize: (1) the duration of the review; (2) the transaction costs for project developers; and (3) the administrative costs of the international process. The review procedures should be as streamlined as possible.

Reality does not live up to this aim. The CDM process is lengthy and cumbersome. The approval of new methodologies can take between six months and two years. Different interpretations by the Methodologies Panel and the EB lead to delays and even the

[59] *Supra* note 6 at 29.
[60] Lord Hewart CJ in R. v Sussex Justices, ex parte McCarthy [1924] 1 KB 256.

mandatory 90-day period for the registration of a submitted project is set back in practice.[61]

It is true that the EB is confronted with an increasing workload because of the scope of guidance sought by project participants and the sheer number of projects presented for registration. It is also true that the EB consist of extraordinary, committed and hard working individuals. However, the individuals serve on the Board in addition to their full time jobs and part-time volunteers find it almost impossible to keep up with the work load of the EB.

In addition, the EB is confronted with an increasing number of technical issues which lie beyond the expertise of its members. While the EB members would be qualified to exercise an oversight function, they are overwhelmed by technical details of various project classes. The requirements of daily micro-management of project- and methodology-related issues lead to delays in the approval of projects and the review of methodologies.

3.3.4. *Predictability and certainty*

Decisions of the Executive Board (and DOEs) should be consistent and predictable.[62] The various rules and decisions have to be applied in a consistent manner to ensure that all project participants are treated fairly. Given the importance of stability to encourage the development and maturity of the emerging carbon market and the significant role played by the market regulator in creating and maintaining such stability, there is all the more need for the EB's decision-making to be consistent, fair and predictable.

In reality, the decisions and interpretations of the EB are often unpredictable.[63] The lack of predictability may, to a certain extent, be the logical consequence of ambiguous COP/MOP decisions which

---

[61] The UNFCCC secretariat has introduced the practice of counting the 90 period not from the moment of submission of the documents but from the moment the secretariat and the RIT consider the documents to be complete. Misspellings or other minor issues can thus lead to significant delays in the registration process.

[62] IETA, *supra* note 48.

[63] See e.g., *ibid.*

lie outside the control of the EB. However, there are a number of issues that increase the risk of inconsistent decision-making practices: (1) The lack of institutional memory and the rotation of the EB members;[64] (2) a process driven by politics rather than technocratic application of rules, and (3) insufficient technical expertise. The issues presented to the EB are of increasing technical complexity which gobeyond the training and expertise of its members. In addition, political interests and horse-trading may also exercise undue influence on the decisions of the EB.

3.3.5. *Review*

Where rights of private entities are affected by EB decisions, these entities should be granted the right to be heard and the right to have a decision that they wish to contest reviewed by an independent body.

It is a necessary condition of a fair administrative procedure that entities that are affected by the decisions of a regulatory body have access to a full and fair review of the decision in question. The COP/MOP decisions foresee a review procedure of some contested decisions when a decision improperly affects a Party's interest. The review is conducted by the enforcement branch of the Protocol's Compliance Committee.[65] These procedures however do not extend to non-Party participants in CDM projects. Under the existing UNFCCC and Kyoto Protocol guidelines, procedures and rules, the procedural rights of private parties are very limited. Under the current CDM regime, affected project participants are afforded no opportunity for the review of Board decisions. It is unacceptable that private entities that see themselves directly affected by administrative decisions of an administrative body have no access to a court or independent tribunal that reviews these decisions.[66]

---

[64] This point is partly being address by an increasing number of professional UNFCCC secretariat staff. However, currently the UNFCCC staff is still new and does not possess the required institutional memory.

[65] Decision 27/CMP.1, Annex, Procedures and mechanisms relating to compliance under the Kyoto Protocol, sections IX and X.

[66] E.E. Meijer, *supra* note 25 at 925.

The described short-comings of the CDM process increase the likelihood of disputes between private (and public) entities and international bodies such as the COP/MOP or the EB. In its daily operations, the Executive Board clarifies and interprets the application of the decision of the COP/MOP. The Executive Board serves as administrative body under the Kyoto Protocol and its decision or its omissions to decide have important consequences for any project participant, regardless of whether it is a Party or a private legal entity participating in a particular CDM project. Potential areas of dispute could be based on the following claims:

- Determinations of the Executive Board are ultra vires the Board's delegated authority
- Members of the Executive Board are not qualified and that the Board has taken decisions based on factually incorrect technical and scientific conclusions
- Members of the Board are conflicted and that conflicting interests make impartial decisions impossible
- Breach of confidentiality
- Non-conformity with the Executive Board's operational procedures that result in a violation of procedural rights under the CDM. Such rights include the right to open Executive Board meetings, access to information and hearings according to the CDM procedures and modalities.[67]

The CDM is dominated by private sector interests. Project participants take investment decisions based on the promise to receive CERs that reward achieved emission reductions. Any decision that impedes or reduces the likelihood to receive CERs will potentially cause project developers financial damage. Wherever the damage can be linked to a particular determination of the Executive Board, the affected party may seek compensation from the Board and/or its individual members for the damage suffered.

[67] E.g. Decision 3/CMP.1, Annex, para. 16 or 23.

International organizations must respect the rule of law. Accordingly, there must be processes which allow entities that claim to be harmed by such organizations to be heard. These processes may be through dedicated international bodies or, in the absence of such bodies, national courts. Without being granted procedural rights and access to appeal and review of EB decisions, private sector participants may turn to domestic courts as a forum of redress when their rights are perceived to have suffered infringement.[68] The EB and its Members are not granted any immunity under the UNFCCC or the Kyoto Protocol. And even if these treaties included provisions that extend immunity on constituted bodies of the Kyoto Protocol, immunity would not lead to impunity, and in the absence of an international dispute settlement mechanism, national courts may hear the claims put forward by private entities. Whether such suits would eventually be dismissed would depend on the substantive law of the State Party concerned.

While some authors argue that national courts would be a suitable forum for addressing the lack of accountability at the international level, it is our view that a review of CDM procedures by national courts would seriously put at risk the coherence of the mechanism that is unlikely to survive as a global mechanism if it were subject to litigation in various Member States and, consequently, differing judicial interpretations of the rights and obligations under the Kyoto Protocol.[69]

4. Comparative Analysis

The call for greater protection of individual rights in the interaction with international bodies is not limited to the carbon market. Relevant discussions conducted in other contexts may help us draw

[68] *Supra* note 3.

[69] E.E. Meijer, *supra* note 25. Meijer argues that the review of administrative CDM decisions by national courts should be possible. Assuming that the EB would be protected by judicial immunity, she proposes a lifting of such immunity for any decisions of administrative nature.

some conclusions for a proposed reform of the CDM. In this section, we will examine two examples of international bodies affecting rights of individuals with the objective of drawing lessons for a reform of the CDM. We will analyze the extent to which targeted sanctions applied by the UN Security Council and their effect on individuals holds comparative lessons for the CDM. Another case study is provided by the global anti-doping regime.

4.1. *The UN Security Council and its targeted sanctions*

Since the 1990s, the UN Security Council has applied 'targeted sanctions' towards individuals or entities which are considered by the Council to be a threat to global security and peace.[70] 'Targeted sanctions' are intended to have their impact focused on leaders, political elites and segments of society believed responsible for objectionable behaviour, while reducing collateral damage to the general population and third countries.[71] The concept of targeted sanctions as an alternative to comprehensive embargoes and sanctions is relatively new and has been driven by the intent to avoid the humanitarian impact of traditional sanctions on broader populations. However, shortcomings in the process of applying and reviewing target sanctions have triggered criticism and led to challenges in national courts. Problems with the failure to notify listed individuals and entities, as well as the lack of information regarding the basis for subjection to targeted sanctions, contribute to perceptions of unfairness.[72] Criticisms persist about procedures related to the designation or listing of individuals,

---

[70] The authors thank Prof. Michael Bothe for making the link between the CDM and the targeted sanctions of the UN.

[71] Hufbauer & Oegg, 'Targeted Sanctions: A Policy Alternative?', Paper presented at symposium on "Sanctions Reform? Evaluating the Economic Weapon in Asia and the World" February 23, 2000, http://www.petersoninstitute.org/publications/papers/paper.cfm?ResearchID=371 (accessed 27 July 2007).

[72] T. Biersteker and S. Eckert (eds), 'Strengthening Targeted Sanctions Through Fair and Clear Procedures', White Paper prepared by the *Watson Institute Targeted Sanctions Project*, Brown University, 30 March 2006, Executive Summary, p. 3.

or the process for the removal of individuals and entities from the list. Targeted sanctions clearly affect the fundamental rights of those affected; however, they are applied outside any administrative process and the entities 'targeted' are therefore without means for appeal or review. This means that while targeted sanctions have reduced impact on the broader public, the move towards targeted measures has created new issues, in particular, with regard to the rights of individuals and legal entities that have been listed wrongly. Concerns regarding the legality of such sanctions have been expressed by academics,[73] NGOs[74] and courts.[75] The Council of the European Union has also addressed the issue.[76]

Similar to the application of targeted sanctions, albeit with less severe impact on an individual or legal entity, the decisions of the CDM Executive Board affect the rights of parties that do not have access to any mechanism of review or remedy. Depending on the

---

[73] Frank, 'UN-Sanktionen gegen Terrorismus und europäische Menschenrechtskonvention', in L. Calfisch *et al.* (eds), *Liber Amicorum Luzius Wildhaber. Human Rights — Strasbourg Views/Droits de l'homme — Regards de Strasbourg* (2007) 237; C. Warbrick, 'The European Response to Terrorism in an Age of Human Rights' 15 *EJIL* (2004) 989; Cameron, 'UN Targeted Sanctions, Legal Safeguards and the ECHR', 72 *Nordic Journal of International Law* (2003) 1; Fitzgerald, 'Managing Smart Sanctions Against Terrorism Wisely', (2002) 36 *New England LR*; St. Schmahl, 'Effektiver Rechtsschutz gegen targeted sanctions des UN-Sicherheitsrats?' (2006) *Europarecht* 566; H. Aust/N. Naske, Rechtsschutz gegen den UN-Sicherheitsrat durch europäische Gerichte?, 61 *ZÖR* (2006), 587.

[74] Human Rights Watch, 'Sanctions Rules Must Protect Due Process', March 2002, online: http://hrw.org/english/docs/2002/03/04/global5839.htm.

[75] *Ahmed Ali Yusuf & Al Barakaat International Foundation v. Council and Commission, Court of First Instance*, Case 306/01, relating to S/RES/1267 (1999); *Organisation des Modjahedines du people d'Iran v. Council of the European Union*, Court of First Instance, Case T-228/02, Judgement of 12 December 2006, relating to S/RES/1373 (2001).

[76] Council of the European Union, 'Basic Principles on the Use of Restrictive Measures (Sanctions)', 7 June 2004, Doc. No. 10198/1/04 PESC 450 REV 1; cf. also Council of the European Union, 'Guidelines on implementation and evaluation of restrictive measures (sanctions) in the framework of the EU Common Foreign and Security Policy', 2nd December 2005, Doc. No. 15114/05 PESC 1084 Fin 475.

impact of a certain decision on participants in a CDM project or DOEs, courts may review such a decision, like UN sanctions, in light of the requirements of the right to an effective remedy. Improving the procedures for the application of sanctions — or decision that affect individual parties under the CDM — ensuring that they are fair and clear in their application, would therefore reduce the risk of judicial decisions.[77]

The discussions on a review mechanism of UN Security Council decisions on targeted sanctions can therefore be of value when deciding on the design of a similar mechanism in the context of the CDM. The rights to a fair trial and an effective remedy lie at the heart of the debate. Such rights are generally included in the Universal Declaration of Human Rights (UDHR), such as Article 8 which enshrines the right to an effective remedy. The right to review and appeal a decision is also enshrined in Articles 2(3) and 14 of the International Covenant on Civil and Political Rights (ICCPR)[78] or Article 6 of the European Convention on Human Rights. There is certainly room for discussion as to whether any of these provisions apply directly to cases such as the targeted sanctions or CDM related decisions.[79] The need for due process has however been acknowledged by the UN General Assembly which declared in its 2005 World Summit Outcome:[80]

> *109. We also call upon the Security Council, with the support of the Secretary-General, to ensure that fair and clear procedures exist for placing individual and entities on sanctions lists and for removing them, as well as for granting humanitarian exemptions ...*

---

[77] For UN targeted sanctions: *supra* note 72.

[78] International Covenant on Civil and Political Rights Adopted and opened for signature, ratification and accession by General Assembly resolution 2200A (XXI) of 16 December 1966 entry into force 23 March 1976.

[79] For UN targeted sanctions: L. van den Herik, N. Schrijver, 'Human Rights Concerns in Current Targeted Sanctions Regimes from the Perspective of International and European Law' in *supra* note 72, Executive Summary, p. 8.

[80] A/RES/60/1/.

This is reflected in a statement issued on 22 June 2006 by the President of the Security Council:[81]

> *The Council is committed to ensuring that fair and clear procedures exist for placing individuals and entities on sanctions lists and for removing them, as well as for granting humanitarian exemptions. The Council reiterates its request to the 1267 Committee to continue its work on the Committee's guidelines, including on listing and de-listing procedures.*

In addition, the Security Council has adopted a number of resolutions providing for a review of listing decisions and for a de-listing procedure.[82] Implementing these resolutions, the Sanctions Committees have elaborated guidelines for the review of listing decisions and for de-listing.[83] In addition to those guidelines, a number of procedural improvements as well as a revision and appeal mechanism have been proposed, such as the establishment of a focal point, ombudsman, expert teams or an arbitration panel reviewing the delisting of entities.[84] Similar approaches could be envisaged in relation to the CDM.

4.2. *The global anti-doping regime*

Another area wherein individuals are directly affected by decisions of international bodies is that of international sports, in particular, the anti-doping regime. Unlike the case of the UN Security Council, international sports have been regulated for the longest time by private institutions.

---

[81] S/PRST/2006/28.

[82] Resolutions 1617 (2005); 1730 (2006); 1735 (2006).

[83] Guidelines of the Security Council Committee Established Pursuant to Resolution 1267 (1999) Concerning Al-Qaida and the Taliban, adopted on 7 November 2002, as amended; Guidelines of the Committee Established Pursuant to Resolution 1636 (2005).

[84] *Supra* note 72, Executive Summary, p. 38.

In 1999, the International Olympic Committee (IOC) set up the World Anti-Doping Agency (WADA), legally a non-governmental body with a mixed representation of Olympic bodies and Governments.[85] WADA developed the World Anti-Doping Code (WADC) which, in itself, is a private text. Over 100 national governments supported the Code through the adoption of the so-called Copenhagen Declaration.[86] The mix of private and public elements in the governance system of the anti-doping regime makes for a useful comparison with the CDM. Just as anti-doping measures affect an athlete's fundamental rights, the decisions of the EB can affect the rights of CDM project participants. In both professional sports and the CDM, the participation of the actors is voluntary. However, in choosing to participate in sports or the CDM, the athlete or the CDM project participant respectively is forced to subscribe to the rules and procedures of the game. Taking into account the impact that decisions of the anti-doping regime may have on an athelete's professional life, most national courts would not let the private nature of the relationship between the athlete and the governing sport institutions stand in the way of an administrative law review of the process.[87] Similarly, in the CDM regime, it is likely that some legal systems may enable recourse to the courts in certain circumstances, in particular, if the court is convinced that the international system is not giving affected legal entities sufficient protection.[88] In both cases, courts may respect self-regulation up to the point at which they consider that the fundamental rights of the claimant have been affected, warranting judicial intervention under the pretext of public policy.[89]

---

[85] Official website: www.wada-ama.org (accessed 10 August 2007).

[86] A. Van Vaerenbergh, 'Regulatory Features and Administrative Law Dimensions of the Olympic Movement's Anti-doping Regime', New York University School of Law, IILJ Working Paper 2005/11 Global Administrative Law Series, p. 4.

[87] *Ibid.* at 17.

[88] Note by the Secretariat, 'Privileges and immunities for individuals serving on constituted bodies established under the Kyoto Protocol', FCCC/KP/COP/MOP/2005/6, paragraph 19 (Executive Board of the clean development mechanism), para. 27.

[89] *Supra* note 86 at 18.

While the international law character of the CDM may convince national courts not to interfere with the EB's jurisdiction, the fact that the CDM governance does not provide for any alternative appeal or dispute settlement mechanism may lead courts to the conclusion that due process rights need to be granted to allow the private entity to seek recourse when its rights have been violated. The calls for good governance in the private regime of anti-doping can therefore be compared to, and are as convincing as those in the area of the CDM.

The Court of Arbitration for Sport (CAS) is the independent international arbitral body established by the International Olympic Committee to handle disputes touching the world of sports, including matters involving athlete eligibility and commercial disputes relating to sports. Parties need to submit themselves voluntarily to the process of the CAS. WADA has designed the CAS as the exclusive arbitrational mechanism to resolve doping-related matters involving international athletes or occurring at international events. Similar to the CAS's specialized regime for arbitration in sport disputes, the Permanent Court for Arbitration (PCA) has reacted to the emerging market in environmental commodities. The PCA has adopted the PCA Optional Rules for Arbitration of Disputes Relating to Natural Resources and/or the Environment and has established a roaster of experts with specific climate change law expertise. The objective of the optional rules is to create a 'unified forum' to which States, intergovernmental organizations, NGOs, multinational corporations, and private parties could have recourse when they had agreed to seek resolution of disputes relating to the environment and/or natural resources.[90] While the CAS and the PCA optional rules create a specialized forum to address disputes in their respective areas, they are unlikely to provide affected parties the administrative remedy needed. They are designed to create space for dispute resolution between private parties or parties to a treaty, but they do not establish a standing appeal mechanism open to those that feel that they have been treated unfairly by a sports body or the Executive Board of the CDM.

[90] D. Ratliff, 'The PCA Environmental Arbitration and Conciliation Rules', 1(1) *Transnational Dispute Management* (February 2004).

5. Models for International Review Mechanisms

The establishment of a Review Mechanism is a crucial aspect of meeting due process requirements. A note prepared by the UNFCCC Secretariat on "Privileges and immunities for individuals serving on constituted bodies established under the Kyoto Protocol" confirms this conclusion for it considers that the absence of formal procedures for private or public legal entities to bring their concerns when affected by decisions of a constituted body of the Kyoto Protocol regime, such as the Executive Board, increases the risk that such entities will seek redress through litigation in the domestic courts.[91] An appeals mechanism is necessary to meet the frustrated expectations of participants in the carbon market, which makes this discussion of the institutional design of such a mechanism timely. Before we develop a proposal for a CDM review mechanism, we will look at potential models and precedents. We found such models in the World Bank's Inspection Panel and the European Ombudsman and, in this section, will discuss the extent to which these institutions can serve as models for a CDM review mechanism.

5.1. *World Bank Inspection Panel*

In 1993, the World Bank created the Inspection Panel in response to widespread criticism by civil society and some government agencies for its failure to abide by its own policies in the course of its involvement in certain highly controversial infrastructure projects. A key reason for the formation of the Panel was to check on the Bank's adherence to its own policies and procedures as well as to make its decision-making processes more transparent and accountable. The mandate of the Panel is to allow qualifying non-state actors to hold the Bank accountable for actions that cause or threaten to cause serious harm to the complainants and

[91] FCCC/SBI/2006/21, *Privileges and Immunities for Individuals Serving on Constituted Bodies Established Under the Kyoto Protocol.*

are inconsistent with the Bank's own operational policies and procedures.[92]

The Panel consists of three members who are nominated by the President of the Bank, after consultation with the Board of Executive Directors (Board), and are appointed by the Board.[93] Panel members are 'selected on the basis of their ability to deal thoroughly and fairly with the requests brought to them, their integrity and their independence from the Bank's Management, and their exposure to developmental issues and to living conditions in developing countries'.[94] The Resolution establishing this Panel included a number of requirements designed to ensure the independence of the members of the Panel. First, the Resolution stipulates that members cannot have worked in any capacity, including those of consultant and local consultant, for the Bank for the two years prior to their appointment.[95] Second, a panel member is disqualified from taking part in the hearing and investigation of any request related to a matter in which he/she has a personal interest or had significant involvement in any capacity.[96] Third, Panel members may only be removed by a decision of the Board and 'for cause'.[97] Finally, following the expiry of their term on the Panel, members are not eligible for employment by the Bank Group in any capacity.[98] The Panel is assisted by a Secretariat which is functionally independent of the World Bank Management and answers only to the Panel.[99]

---

[92] IBRD Resolution 93-10 and IDA Resolution 93-6, 'The World Bank Inspection Panel,' was adopted by the Bank's Board of Executive Directors on September 22, 1993. Also see the "About Us" section on the Inspection Panel website at www.worldbank.org/inspectionpanel (accessed on 10 August 2007).

[93] *Ibid.*, para. 2.

[94] *Ibid.*, para. 4.

[95] *Ibid.*, para. 5.

[96] *Ibid.*, para. 6.

[97] *Ibid.*, para. 8.

[98] *Ibid.*, para. 10.

[99] See "Panel Secretariat" section on the Inspection Panel website at www.world bank.org/inspectionpanel (accessed on 10 August 2007).

An investigation by the Panel can be launched in three ways. The most important trigger for an investigation is a complaint by local people who are or might be affected by a Bank-supported project.[100] In the request for inspection, the affected party must demonstrate that its rights have been affected by an action of the Bank as a result of a failure of the Bank to follow its operational policies and that such failure has had, or threatens to have, a material adverse effect.[101]

The Panel is a fact-finding body and does not make recommendations for the correction or remediation of any failures which are uncovered during the investigation process. Instead, when the Bank management responds to the Panel's Investigation Report, it will usually propose a course of remedial action to the Board. The Board decides whether to approve the management's recommendations. The request for information, management's response to the request, investigation report, management's response to the investigation report and its recommendations are all made available to the public on the Inspection Panel's website.

While the Panel has proved to be an important accountability mechanism, a significant shortcoming hampers its efficacy. The Panel does not play any role in monitoring the implementation of the Board's final decision regarding the remedial course of action proposed by management and the Panel's findings.[102] This effectively means that there is no checking mechanism to ensure implementation of proposed remedial action and that grievances that gave rise to the request for inspection in the first place are effectively addressed. In order to close this gap in the accountability process, the Board should monitor the implementation of any remedial action and a feed-back loop to the Panel should be introduced.

---

[100] Defined in the Resolution to be 'an affected party in the territory of the borrower which is not a single individual (i.e. a community of persons such as an organization, association, society or other grouping of individuals)'.

[101] *Ibid.*, para. 12.

[102] Bradlow, 'Private Complainants and International Organizations: A Comparative Study of the Independent Inspection Mechanisms in International Financial Institutions', 36 *Geo. J. Int'l L.* (2005) 403 at 419.

5.2. *The European Ombudsman*

On 9 March 1994, the European Parliament adopted the European Ombudsman Statute, laying the legal foundations for the creation of an ombudsman office to 'help uncover maladministration in the activities of the Community institutions and bodies, and make recommendations with a view to putting an end to it'.[103] Apart from the fact that there is no standardized definition of 'maladministration', the examples of 'maladministration' provided are potentially confusing as they fail to provide a clear delineation of the Ombudsman's mandate. However, the definition may be intentionally broad so that the Ombudsman enjoys wide latitude to correct whatever administrative irregularities it sees fit.

The procedures of the European Ombudsman are relatively straightforward. A complaint must be made within two years of the date when the complainant became aware of the facts on which its complaint is based.[104] A complainant need not be individually affected by the maladministration. However, the complainant must have contacted the institution or body concerned before it contacts the Ombudsman.[105] The Ombudsman registers the complaint and has to determine whether the complaint is within its mandate and if so, whether it is admissible. It may request to be furnished with more information and documents before reaching a decision.[106] In addition, the Ombudsman decides if there are sufficient grounds to justify making an inquiry into an admissible complaint.[107] If an inquiry is warranted, the Ombudsman informs the complainant and the institution involved and invites the institution to submit an opinion within a stipulated period of time.[108] Upon receiving the institution's opinion,

---

[103] Article 2 of the European Ombudsman Statute, adopted by Parliament on 9 March 1994 (OJ 1994 L 113) and amended by its decision of 14 March 2002 deleting Articles 12 and 16 (OJ 2002 L 92).

[104] Article 2.4, European Ombudsman Statute.

[105] 'How to Complain', European Ombudsman Website <http://www.ombudsman.europa.eu> (accessed on 27 June 2007).

[106] Article 3.1, European Ombudsman Statute.

[107] Article 4.1, *Ibid.*

[108] Article 4.3, *Ibid.*

the Ombudsman sends a copy to the complainant and invites it to make observations on the institution's opinion, which should be submitted within a month.[109] The Ombudsman's powers of investigation include the ability to require Community institutions and bodies and the authorities of Member States to supply information or documents for the purposes of an inquiry, and to inspect the file of the Community institution concerned in order to verify the accuracy and completeness of its replies.[110]

In accordance with Article 6 of the Implementing Provisions, in the event that the Ombudsman uncovers maladministration in the course of its investigation, it is obliged to as far as possible co-operate with the institution concerned in seeking a 'friendly solution' to eliminate it and to satisfy the complainant. In the event that the Ombudsman considers that a friendly solution is not possible or that the search for one has failed, it either closes the case with a reasoned decision that may include a critical remark or makes a report with draft recommendations.[111] This detailed opinion could consist of acceptance of the Ombudsman's decision and a description of the measures taken to implement the draft recommendations. If the detailed opinion still does not meet the Ombudsman's satisfaction, it may draw up a special report to the European Parliament in relation to the instance of maladministration.[112] It should be noted that the Ombudsman may undertake inquiries on its own initiative and enjoys the same powers of investigation when doing so as when inquiries are instituted following a complaint.[113]

6. Lessons Learned for the CDM

The previous section demonstrates that there is a clear tendency to demand the application of due process requirements to the decisions

---

109 Article 4.4, *Ibid.*
110 Article 5.1, *Ibid.*
111 Article 6.3, *Ibid.*
112 Article 8.4, *Ibid.* Further, a copy of this report will be sent to the complainant and the institution concerned.
113 Articles 9.1 and 9.2, *Ibid.*

of international bodies, whether they are established in the context of international organizations and law or as part of public-private regulatory regimes. Courts and law makers are increasingly of the view that decisions by international bodies cannot go unchecked and are not averse to having such decisions reviewed by national courts in the absence of review processes in the relevant international regulatory regime. The need for reform of the CDM governance has recently been acknowledged by the third COP/MOP and will lead to a process of review in the context of the formulation of a post-Kyoto climate regime.[114] We will put forward in the following section some proposals for the reform of the CDM which include the introduction of administrative law-like processes, professionalizing the EB and the panels, securing better and more consistent funding, the elimination of political interference as well as the establishment of a review and appeal mechanism.

6.1. *Adoption of due process rules*

6.1.1. *Administrative rules and procedures*

We propose reforming the procedural rules of the CDM. Currently, there are only a few formalized provisions governing the interaction between project proponents, the EB and its panels. Insecurities regarding communications, hearings and time lines often make processes cumbersome and opaque. From the perspective of project participants, there is a perception of insufficient and circuitous communication. At the same time, communication becomes unsatisfying, redundant and ineffective, when new queries are brought up in each round of review of a project and it is not clear how many of such review cycles may take place. As a result, there is an undefined period of legal and planning insecurity during which project participants have (i) to retain resources to answer an undefined and unlimited number of new questions, and (ii) have no indication on whether they

[114] See the decisions of the 3 sessions of the COP/MOP, among others http://unfccc.int/files/meetings/cop_13/application/pdf/cmp_guid_cdm.pdf (accessed 29 December 2007).

can move ahead with developing the corresponding CDM project activities.

We therefore recommend the adoption of administrative due process rules governing communication amongst the various CDM actors. The adoption of due process requirements would apply to any activities related to (i) the accreditation and withdrawal of accreditation of DOEs, (ii) the approval and review of baseline and monitoring methodologies; (iii) the registration, or refusal to do so, of CDM projects; and (iv) the issuance, or refusal of issuance, of CERs. The objective of such rules would be that any person (DOE or project participant) with a direct and material interest in any of the abovementioned processes would have a right to participate by: (a) expressing an opinion and its reasons, (b) having that position considered, and (c) having the right to appeal (see below). The administrative requirements have to go beyond the existing guidelines governing the internal proceedings of the EB and establish rights for affected third parties, thereby promoting equity and fairness.

6.1.2. *Establishment of a focal point*

Communication would be made more efficient and reliable with the establishment of a focal point within the UNFCCC secretariat who would handle complaints from project participants and DOEs. Such a focal point would not have decision-making responsibilities, but serve an entirely administrative function of ensuring efficient and consistent interaction with the EB, its panels, and the review mechanism.

6.1.3. *Compiling CDM rules*

In addition to the adoption of procedural rules, we recommend making available the complete set of CDM rules in a comprehensive and easy accessible format. As it stands today, anyone who does not spend a significant amount of time trying to understand how the CDM functions (including many project participants) will inevitably be lost in the thicket of decisions and interpretations that govern today's CDM. To facilitate fair and transparent application of all CDM relevant rules, the UNFCCC secretariat should make available a compilation of all rules governing the CDM. The three sources of these rules are: (i) the Kyoto Protocol, (ii) COP/MOP decisions, and (iii) EB decisions that are currently spread over uncountable documents and their annexes. The rules should be thematically organized, referenced, and indexed.

Such an official compilation of CDM rules should be maintained and updated by the UNFCCC secretariat. The EB should review the rules periodically and present the updated compilation annually to the COP/MOP for endorsement. In the event that there is more than one version of a rule and a dispute arises, the latest effective version on file with the division should be deemed the authoritative or binding version.

6.2. *Reform of the executive board and its panels*

6.2.1. *Professionalizing the EB*

The current EB has been established as a United Nations committee, rather than as a professional regulatory authority overseeing the carbon market. This is not to be unexpected, considering the roots of the CDM in international environmental treaty law. It is revealing that the EB's role and powers as set out in Part C of Annex "Modalities and procedures for a clean development mechanism" does not include any regulatory objectives or principles.[115]

Nonetheless, whatever its role was originally intended to be, the CDM EB today is in the position of a *de facto* market regulator. In order for the Kyoto Protocol to succeed, the EB must rise to the occasion and fulfil the role of a market regulator. A first step in this direction is to professionalize the EB. Presently, the majority of its members have a background in international environmental negotiations, not in market regulatory work (for example, work experience in financial regulatory authorities). As a result, the considerations of the EB tend to be oriented towards agendas raised during international

[115] Decision 3/CMP.1, Annex C.

negotiations rather than to the sort of issues related to the creation and maintenance of an efficient international market.

The professionalizing of the EB would require the recruitment of full-time salaried individuals whose collective experience spans the entire range of sectors (including project finance, law, business management, science) and is grounded in practical, project-level experience and knowledge of the CDM. Technical expertise should therefore be the governing criterion for the selection of EB members. The right of the various geographical constituencies to nominate EB members need not be affected, but nominations should be backed by the technical expertise and experience that the nominee can bring to the EB. The selection of a new member to the EB should require the approval of the existing EB members and while the EB members are formally hired as employees of the UNFCCC, they should report to the COP/MOP directly in order to minimise the potential for political interference. Further, the creation of a direct reporting channel between the EB members and the COP/MOP will help foster greater accountability.

Staffing the EB with professional staff will also help avoid conflict of interests since individuals are no longer made to serve several agendas and interests in parallel. To avoid conflicting interests before and after the time an individual serves on the EB, eligibility to the EB should be limited to individuals who did not hold a position that involved decision making on CDM-related matters for a defined period before serving on the Board and should be excluded from such offices for a time after they cease to be active EB members.[116]

6.2.2. *Funding and hiring of sufficient support staff*

An overworked and understaffed EB can hardly be expected to deliver results. An adequate staff should be made available to support the EB in its work. At the 1st session of the COP/MOP, it was

[116] See below for similar rules applying to the members of the World Bank Inspection Panel.

decided that the share of proceeds to cover the administrative expenses of the CDM (SOP-Admin) to be paid by project developers would be USD 0.10 per CER issued for the first 15 000 tonnes of CO_2 equivalent and USD 0.20 per CER issued for any amount in excess of the first 15 000 tonnes of CO_2 equivalent.[117] Given the high number of CERs that the EB is expected to issue in the coming years till 2012, the SOP-admin levy is an adequate and predictable source of funding.[118]

6.3. *Establishment of a review mechanism*

6.3.1. *Design features of a CDM review and appeal mechanism*

(a) Mandate

The CDM appeal mechanism's mandate or jurisdiction should be clearly defined in its constitutive document (preferably, and likely to be, a COP/MOP decision). To avoid opening the floodgates to all and sundry, it is important that the terms of reference for the operation of the appeal mechanism be clearly set out in the constitutive document. The World Bank's Inspection Panel, for example, has a clear mandate which also serves as a filtering process against frivolous claims. The European Ombudsman, on the other hand, has too broadly defined a mandate such that 70% of the complaints it receives actually fall outside its mandate.[119] The availability of an accessible and effective remedial mechanism should be made known to all parties which may be adversely affected by a decision of any of the constituted bodies under the Kyoto Protocol, but the scope of its operation should also be emphasised to prevent creating unrealistic expectations about what the appeal mechanism can achieve.

---

[117] Annex 35 of the Report of the 23rd Meeting of the Executive Board.

[118] See CDM statistics, online: http://cdm.unfccc.int/Statistics/index.html (accessed on 19 July 2007).

[119] However, the broad mandate may also be seen as a low threshold so that the Ombudsman is highly accessible to the public. An 'over-inclusive' mandate also prevents the accidental exclusion of otherwise valid allegations of maladministration.

(b) External Authorisation

The European Ombudsman enjoys more independence than the Inspection Panel in the sense that the Ombudsman does not require any external authorisation to proceed with any investigation and may in fact launch an investigation on its own initiative. The Inspection Panel, however, requires the authorisation of the World Bank before an investigation may be launched and cannot trigger an investigation on its own. To the extent that "own initiative investigations" can help uncover potential or existing maladministration before it becomes the subject of a complaint, granting such powers upon an accountability mechanism will give it the ability to be pro-active in improving the institution's decision-making process. However, the authors are of the view that investigations by the CDM appeal mechanism should only be triggered upon the receipt of a complaint and after an initial review of the facts show that there is, *prima facie*, a case to answer. The right to submit a complaint should be governed by the administrative requirements adopted by the COP/MOP and governing the process. This is in line with the dispute settlement function of the CDM appeal mechanism, which sets it apart from the European Ombudsman which also has a preventive function of eliminating potential maladministration. The CDM appeal mechanism should also not be expected to require any external authorisation, e.g. from the COP/MOP or the Executive Board before it can commence investigations as there is no apparent value to this additional administrative hurdle. On the contrary, such authorisation would reduce the perception of independency and objectivity that the appeal mechanism should have. The UNFCCC secretariat should receive a mandate to support the appeal mechanism by conducting preliminary reviews of the eligibility of complaints received. Provided that the secretariat would assign staff to ensure an effective and consistent response on complaints, such initial review would help increase the efficiency of the process.

(c) Power to Issue Binding Decisions

Unlike the European Ombudsman and the Inspection Panel, a potential CDM appeal mechanism should be empowered to issue binding decisions. Crucial elements of an effective appeal mechanism are: (i) an independent and impartial authority, (ii) decision-making authority, and (iii) accessibility.[120] Aggrieved parties, such as private sector project developers whose investments are at stake, want an efficient, fair and effective settlement of any disputes. The inability to render binding decisions will hamstring the appeal mechanism's ability to conclusively resolve disputes and make remedial orders. The CDM appeal mechanism must be a satisfactory alternative to litigation in the national courts; being empowered to issue binding decisions just like any other administrative tribunal or national court is important in this regard.

In the case of the Inspection Panel, there appears to have been cases wherein the Board's decisions were not implemented and the complainants have not seen any improvement in their situation. Problems have arisen because the Inspection Panel does not have the power to issue binding decisions and does not have the power to monitor the implementation of the Board's decisions. These issues should not arise in relation to the CDM appeal mechanism if its decisions are made final and binding on all parties.

(d) Investigation and Administrative Process

The CDM accountability mechanism should be given powers of investigation, including the ability to call for hearings, view all relevant files and other documentation, to interview staff members and to require them to give evidence, and the ability to conduct special inquiries if required. The investigation process should not be allowed to take too long and therefore strict adherence to deadlines should be mandated. The EB should be consulted and the aggrieved parties given the opportunity to be heard or to make written submissions. The aggrieved party should be informed of all steps taken during the investigation process so that it does not feel that the process has been taken out of its hands and that, once again, it is the victim of opaque and exclusive decision-making.

[120] *Supra* note 72 at 3.

(e) Budget

Both the Inspection Panel and the European Ombudsman are sufficiently funded to carry out its purposes. This should be the case for the CDM accountability mechanism. An inadequately funded office will lack true independence as it will not be able to perform its functions properly. The accountability mechanism should not be answerable to the EB for its budget, but to the COP/MOP which is the quasi-legislative body which is also very unlikely to be the subject of complaints by project participants (who have very little, if any, direct contact with the COP/MOP).

(f) Independence and Integrity

Independence and integrity of the members that comprise the accountability mechanism is essential for the credibility of the mechanism. An accountability mechanism is futile if it is perceived to be working in the interests of the institution(s) against which it is supposed to be exercising a checking function. The criteria for the European Ombudsman and the Inspection Panel members emphasise the need for independence and integrity. The European Ombudsman, for example, has to give a solemn undertaking before the Court of Justice of the European Communities upon taking up his duties that he will perform his duties with complete independence and impartiality.[121] In order to preserve the independence of the CDM accountability mechanism, it is recommended that persons who serve on it should be experts in the appropriate fields, with qualifications elaborated by the COP/MOP. As with EB members, they should not be former or existing staff of the CDM regulatory regime and should not be allowed to take up employment therein for a period of time after the end of their term on the accountability mechanism.

7. Conclusion

The wide participation of private and public entities from developing and developed countries alike makes the CDM one of the most widely

[121] Article 9.2, European Ombudsman Statute.

supported elements of the Kyoto Protocol architecture. The mechanism has introduced the concept of market-based mechanisms to the realm of international law and creates a framework for private-public partnerships which support the objectives of the Kyoto Protocol. With the definition and creation of tradable emission rights, CERs, the mechanism has given rise to a growing carbon market.

The private sector's enthusiastic embrace of the CDM puts the mechanism to the test. Unlike other existing financial mechanisms under MEAs, the CDM has left the realm of intergovernmental cooperation and its operations and demands are driven by the rules and forces of international markets.

In our analysis, we have questioned whether the current CDM governance can meet the demands of the private, profit-driven market. We conclude that the CDM requires reform to ensure its effectiveness as well as efficiency as a robust element of a post-Kyoto climate regime. Leaving others to review concerns relating to design features that endanger the environmental effectiveness of the mechanism (such as concerns regarding the lack of additionality of projects), our analysis has focused on the procedural aspects of the CDM.

The CDM's Executive Board acts as supervisor and day-to-day regulator of the CDM and the Board's decisions have direct impact on the property interests of private entities participating in the mechanism. The Board, modeled in the UN tradition as a committee of Party appointees, has however not been equipped with clear procedural rules that would guide its dealings with those participating in the CDM. Rules are sketchy and often improvised; one would search the CDM modalities in vain for any due process requirements such as the right to be heard or to have a decision reviewed.

An analysis of other areas of international cooperation and law has shown that national courts may decide to fill the legal gap when the fundamental rights of individuals and legal entities are affected by the decisions of international bodies and no recourse for the vindication of rights is available. Having local courts review decisions of the EB will threaten the international infrastructure of the Kyoto Protocol in general and the CDM in particular. The CDM is designed to operate in a uniform fashion world-wide and any disputes concerning the operation or participation in these mechanisms should be resolved consistently and in accordance with CDM rules and procedures. Leaving the settlement of disputes to local courts which may supplement international rules and procedures with national law would result in diverging interpretations of the CDM and its rules. Such diverging interpretations would put into question the functioning of the mechanism as a whole. They may also alter the delicate web of rights and obligations that were carefully negotiated by the parties to the Kyoto Protocol.

The risk of dispute between participants in the CDM and the EB is real and imminent. At the time of writing this article, more than 850 CDM projects have been registered with the Executive Board. Most of them are still in the construction phase and are yet to produce CERs. The big tests of the robustness of the CER transfer agreements are still yet to come. Bearing in mind the monies at stake both in the delivery of CERs as well as in potential penalties, there is no doubt that project participants will not hesitate to enforce their rights rigorously. They will claim contractual rights and where damage can be attributed to perceived or real failures of the EB, they will try all means to recover their losses given what is at stake and the costs of inaction.

In the absence of an international review and dispute resolution mechanism, local courts may be the only forum at which claims have a chance to be heard. While it will be difficult to hold the Board itself responsible, Board members lack protection under the laws of international immunity, which makes them potentially vulnerable to claims of conflict of interest, fraud or incompetence. At the very least, the possibility that some local courts may exercise their jurisdiction to hear cases against the EB cannot be entirely dismissed.

Preempting any conflicts and court rulings, rule of law principles demand an adaptation of the CDM to due process requirements. We therefore recommend the adoption of clear and transparent administrative procedures, professionalizing of the EB and its members as well as the establishment of a review and appeal mechanism under the CDM.

The CDM is an ambitious project with the potential of laying the foundations for international cooperation beyond the climate regime and environmental law. The pricing of externalities allows developing country entities to benefit from financial transfers which help promote sustainable development during a time when foreign direct investment increasingly benefits only a few countries. By expanding its scope and participation, the CDM also has the potential to become a cornerstone of the post-Kyoto regime. However, in order to realise its full potential, it is important that the negotiators take note of and deal with a number of teething problems. Only a procedurally and environmentally robust CDM can fulfill the promise and potential it holds: the creation of the first global environmental market mechanism.

Acknowledgments

The authors would like to register a note of appreciation to Asst. Prof. Thomas Cheng for a very useful discussion on regulation of markets in general.

THEME III

EMISSIONS TRADING IN THE EUROPEAN UNION AND ASIA

CHAPTER 8

REGIONAL FRAMEWORK: THE EUROPEAN UNION EMISSIONS TRADING SCHEME — PAST, PRESENT AND FUTURE

Kurt Deketelaere

Catholic University of Leuven (Belgium), Honorary Professor of Climate Change Law at the University of Dundee, and Chief of Staff of the Flemish Minister of Public Works, Energy, Environment and Nature
kurt.deketelaere@vlaanderen.be

Marijke Schurmans

Institute for Environmental and Energy Law,
Faculty of Law, University of Leuven
marijke.schurmans@gmail.com

This paper analyzes the EU ETS, the European Union Emissions Trading Scheme. The scheme was adopted by Directive 2003/87/EC of 13 October 2003, which had to be implemented by the EU Member States by 31 December 2003. The Scheme became operational on 1 January 2005. Based on this Directive, all installations subject to the EU ETS need to have a "GHGEP", a greenhouse gas emissions permit, and all Member States must draw up, for each reference period, a "NAP", a National Allocation Plan, which indicates, among others, the amount of allowances each EU ETS installation would get allocated for the next reference period. After more than two years of application, the EU ETS Directive is now under revision. One of the central questions in that revision is whether more centralization and harmonization is needed to make this a more efficient and effective market-based policy instrument to tackle climate change. This paper will give a general survey of the present status and expected developments of the EU ETS. The paper states the position as at 15 November 2007.

Introduction

The timing of the conference to which this paper contributes is well chosen: The period of the drafting, the presentation and the approval of the "Integrated Energy and Climate Change Package", as presented by the European Commission on the 10th of January 2007 and as approved by the Spring European Council of the 8th and 9th of March 2007, is just behind us.

The importance of the Commission's package and the Council's conclusions must be emphasized strongly. They set forward, at EU level, legally binding targets regarding the reduction of greenhouse gas emissions, energy efficiency, renewable sources of energy, and bio fuels. The Energy Action Plan 2007–2009, included in the annex to the Council's conclusions, will be the leading document for the future energy policy of the EU. The package, the conclusions and the plan, confirm once again the far reaching ambitions of the EU in the field of climate change policy.

The most important instrument of the EU climate change policy remains the EU ETS, the European Union emissions trading scheme. After the adoption of the Emission Trading Directive and its modification by the Linking Directive, the European Commission is now, on the basis of its first experiences, preparing a modification of the Emissions Trading Directive. After several hearings, concluded by the summer of 2007, the Commission is expected to present a proposal of directive in the fall of 2007 (during the COP/MOP in Bali), aiming at, as soon as possible, approval and entry into force, and member state implementation by 2013. Next to more harmonisation, simplification and enforcement, enlargement of the scope of application of the Emission Trading Directive is high on the agenda. As is well known, air traffic is one of the hard debated sectors to be included. The revision of the Emission Trading Directive is of great importance: it most make emission trading more effective and efficient, and more credible. After a first round of weak national allocation plans (2005–2007), the Commission is well aware of the fact that it must take a tough position in the second round of national allocation plans (2008–2012), and it does. Much to the dissatisfaction of some Member states, which even want to challenge the allocation decision of the Commission before the European Court of Justice.

Although emission trading remains the central instrument of the EU's climate change policy, the Commission published earlier this year an interesting green paper on the use of market based instruments in the environmental and energy policy of the EU. In the paper, an interesting link is made between energy taxation and emission trading. Obviously, in a number of situations, energy taxation is a more efficient and effective instrument than emission trading, and should indeed, again, be considered as policy instrument in the battle against climate change. The unanimity problem in the Council obviously remains, but perhaps the track of enhanced cooperation could finally and definitely be explored.

Obviously, there are not only European developments in the climate change policy. Also at international level a lot of action and deliberation is undertaken, and this within different settings: UNFCC, IPCC, World Economic Forum, G8, Security Council, etc. Several important meetings have taken place and will take place in 2007. All of them are aiming at getting the post-Kyoto-track or the post-2012-track, launched. Between 2007 and 2009, a new international framework should be negotiated and approved, in order to continue international cooperation after 2012. Due to remaining differences between the EU, the US, but also big countries like India, China and Russia, and the large group of developing countries, the launching of the post-2012-debate remains problematic and it is far from certain that a new and strong legal framework will be developed in time. Involvement of all countries, but based on differentiated responsibilities and capabilities, should however and anyway, remain the leading principle in this debate.

Next to the European and international level, a lot is happening at country level. Almost on a daily basis, individual countries announce specific targets or actions in the field of climate change policy. Reference can be made e.g. to important initiatives at state level in the USA and

Australia. It is clear that in many countries, climate change has taken a high position on the political agenda, and forces politicians to take action, in order to respond to societal demands and concerns.

We would like to conclude this introduction with a note of caution: In several countries and organisations, we see certain climate change "hype", created and increased by movies and reports, correctly or incorrectly. As one must be careful with hyperboles, so also with the climate change hype. It is clear that there is a problem and that action is needed. However, one must avoid over-reaction, certainly when it is emotionally driven and done in a sphere of over-bidding between countries or international organizations.

1. Introduction to the Current European Union GHG Emission Trading Scheme

1.1. *Background: The UNFCCC and the Kyoto Protocol*

The European Community is one of the Parties to the United Nations Framework Convention on Climate Change (henceforth: 'UNFCCC'), which was adopted in New York on 9 May 1992, and is also a signatory of the Kyoto Protocol. The Kyoto Protocol was adopted in Kyoto, Japan on 11 December 1997 by the Third Conference of the Parties ("COP") to the UNFCCC. The Kyoto Protocol entered into force on 16 February 2005 and the European Community ratified it by Council Decision 2202/358/EC on 31 May 2002.

The Kyoto Protocol shares the UNFCCC's objective, principles and institutions, but significantly strengthens the UNFCCC by committing Annex I Parties to individual, legally-binding targets to limit or reduce their greenhouse gas emissions.[1] Only Parties to the UNFCCC that have also become Parties to the Protocol (i.e. by ratifying, accepting, approving, or acceding to it) will be bound by the

[1] The targets cover six main greenhouse gases: carbon dioxide (CO_2), methane (CH_4), nitrous oxide (N_2O), hydrofluorocarbons (HFCs), perfluorocarbons (PFCs), sulphur hexafluoride (SF_6).

Protocol's commitments. 175 Parties have ratified the Protocol to date. Of these, 36 countries and the European Community are required to reduce greenhouse gas emissions below levels specified for each of them in the Kyoto Protocol.[2] The individual targets for Annex I Parties are listed in the Kyoto Protocol's Annex B. These add up to a total cut in greenhouse gas emissions of at least 5% from 1990 levels in the commitment period 2008–2012.[3]

The European Community committed itself under the Kyoto Protocol to reduce its emissions of greenhouse gases by 8% during the period of 2005–2012 in comparison with their levels in 1990. It made its own agreement to meet the 8% target by distributing different rates to its Member States[4] (cf. The Burden Sharing Agreement in Annex II of the Council Decision 2002/358/EC).

To meet the targets as set forward, the Kyoto Protocol introduced three flexible mechanisms, as explained below: the international emission trading scheme (henceforth: 'IET'), and the two project mechanisms, Joint Implementation (henceforth: 'JI') and Clean Development Mechanism (henceforth: 'CDM'). The Marrakech Accords of 2001 established the technical rules and mechanisms for IET, JI and CDM.

1.1.1. *International Emission Trading (IET): Allowance-based transaction*

International Emissions Trading (henceforth: 'IET') (Article 17 of the Kyoto Protocol) is an allowance-based system that permits Annex B Parties to buy or sell amongst each other "any part of an assigned amount," which is also referred to as a collection of assigned amount units (AAUs). Thus IET is not a project-based mechanism unlike JI and CDM, which is project-based.

---

[2] See http://unfccc.int/kyoto_protocol/items/2830.php. Last visited 20 August 2007.

[3] *Ibid.*

[4] EU Council Decision 2002/358/EC of 25 April 2002 concerning the approval, on behalf of the European Community, of the Kyoto Protocol and defining emission reduction targets for each Member State.

The IET can be briefly summarized as follows:

Primary participants in a trading regime will be the Parties to the Kyoto Protocol listed in Annex B of the Kyoto Protocol. These are the northern members of the OECD as well as most Central and Eastern European countries and some successor states of the former Soviet Union. Parties without legally binding emission reduction and limitation objectives under the Protocol are thus precluded from participating in trading. This concerns first of all the developing countries, but furthermore those industrialized countries that do not ratify the Protocol or that are not included in Annex B. Article 17 therefore provides an incentive to take on legally binding obligations in order to participate in trading. Restriction of participation to those Parties with legally binding targets should ensure that the overall amount of units circulating in the system are stable. This "cap" on the overall amount of units circulating in the system is not only required by ecological considerations, but should furthermore provide stability and allow a dependable market to develop.

The basic mechanism for a trading regime has been defined in Article 3, paragraphs 10 and 11 of the Kyoto Protocol. Any emission reduction units, or any part of an assigned amount, which a Party acquires from another Party shall be added to the assigned amount for the acquiring Party and shall be subtracted from the assigned amount of the transferring Party.

The commodity that may be traded is called "any part of an assigned amount". This term refers to the "budget" (assigned amount) allocated to the Parties by Article 3.7 and Annex B of the Kyoto Protocol. The Protocol thus does not follow the principle of per-capita distribution, as would be required according to ideas of equity and demanded, for example, by India.[5] Instead, Article 3.7 stipulates that the "assigned amount" for the commitment period 2008–2012 for each Party is equal to the percentage inscribed for it

---

[5] Dr. Hermann E. Ott, "Emission trading in Kyoto Protocol — finished and unfinished business", 1998, Wuppertal Institute for Climate, Environment and Energy, Wuppertal, Germany, 3–5.

in Annex B of its base-year emissions of the "basket" of gases listed in Annex A, multiplied by five.[6]

The Kyoto Protocol determines that any trading shall be "supplemental" to domestic actions for the purpose of meeting the obligations. The insertion of this clause was an attempt by the European Union to limit the possibility for countries to "buy themselves out" of their obligations. Putting limits on trade, although leading to greater inefficiency of the market, could be a legitimate concern under any trading system because of an overriding political emphasis on domestic action. The main reason for a limit is, however, posed by the significant emission reductions in Russia and the Ukraine due to their economic decline after transformation to market economies, the so-called "hot air".[7]

As mentioned above, the regulated gases in the so-called "basket" are first of all the three main contributors to climate change: carbon dioxide (CO_2), methane (CH_4) and nitrous oxide (N_2O). The basket furthermore comprises sulphur hexafluoride (SF_6) and two groups of industrial gases, hydrofluorocarbons (HFCs) and perfluorocarbons (PFCs). The reduction targets must be achieved for the basket as a whole. Since the respective radiating force for each of those gases is different, the "Global Warming Potential" of each gas, as determined by the Intergovernmental Panel on Climate Change (henceforth: 'IPCC'), is used to calculate the overall obligation. The overall emissions of these gases have been restricted by the Kyoto Protocol for a first "commitment period" from 2008 to 2012.

---

[6] *Ibid.*

[7] *See more* Brandt, U.S. and G.T. Svendsen, 'Hot Air as Implicit Side Payment arrangement: Could Hot Air provision have saved the Kyoto-agreement', (2005), Climate Policy, 4, 303–318; Pratlong, F., Van Regemorter, D., and Zagame, P., 'Hot air and market power in international trading', (2004), Universite Paris I, 1-12; Brandt, U.S. and G.T. Svendsen, 'Hot Air in Kyoto, Cold Air in The Hague- the Failure of Global Climate Negotiations', (2002), Energy Policy, 30, 1191–1199; Dr. Hermann E. Ott, "Emission trading in Kyoto Protocol — finished and unfinished business", (1998), Wuppertal Institute for Climate, Environment and Energy, Wuppertal, Germany, 5.

A second commodity that might be traded according to Article 3 paragraphs 10 and 11 is called an "emission reduction unit". This term refers to Article 6, where "emission reduction units" may result from Joint Implementation projects and thus clarifies that the "credits" generated may not only be used by the acquiring Party to meet its own obligations, but that these credits are transferable to other Parties.[8]

With regard to compliance and enforcement, the Kyoto Protocol, in Articles 5,[9] 7,[10] and 8,[11] contains the basis for reporting, monitoring and verification procedures that need to be adapted to the needs of a trading regime. The expert review teams established in Article 8 might also be enabled to check the data submitted on trades, nationally as well as internationally. Additionally, certified auditors linked to the UNFCCC Secretariat could perform some of those functions of verification. Those articles have been carried out by the Marrakesh Accords of 2001: '*The Marrakesh Accords aim to be transparent so that all data, except those designated as confidential, are to be made publicly available. At the conclusion of COP 7 some sections of the guidelines under Articles 7 and 8 were not completed, mainly because of the need to ensure coherence with other decisions. In addition, the COP requested*

---

[8] Dr. Hermann E. Ott, "Emission trading in Kyoto Protocol — finished and unfinished business", 1998–2002, Wuppertal Institute for Climate, Environment and Energy, Wuppertal, Germany, 5.

[9] Article 5 commits Annex I Parties to having in place, no later than 2007, national systems for the estimation of greenhouse gas emissions by sources and removals by sinks (Article 5.1). It also states that, where agreed methodologies (that is, the revised 1996 IPCC Guidelines for National Greenhouse Gas Inventories) are not used to estimate emissions and removals, appropriate "adjustments" should be applied (Article 5.2).

[10] Article 7 requires Annex I Parties to submit annual greenhouse gas inventories, as well as national communications, at regular intervals, both including supplementary information to demonstrate compliance with the Protocol. In addition, Article 7 states that the Conference of the Parties serving as the meeting of the Parties to the Protocol (COP/MOP) shall decide upon modalities for the accounting of assigned amounts prior to the first commitment period.

[11] Article 8 establishes that expert review teams will review the inventories, and national communications submitted by Annex I Parties.

the Subsidiary Body for Scientific and Technological Advice (SBSTA) to develop technical standards for the purpose of ensuring the accurate, transparent and efficient exchange of data between national registries, the clean development mechanism registry and the transaction log.'[12]

During COP/MOP 8, 9, 10 and 11, COP/MOP 1 adopted various decisions with regard to the implementation of Articles 5, 7 and 8 of the Kyoto Protocol.[13]

---

[12] *See:* http://unfccc.int/national_reports/accounting_reporting_and_review_under_the_kyoto_protocol/items/1029.php. *See also for example (not limited to):* UNFCCC, Decision 16/CP.7*:* Guidelines for the Implementation of Article 6 of the Kyoto Protocol, *FCCC/CP/2001/13/Add.2*, Marrakesh, 10 November 2001, p. 12, § 21 (c); UNFCCC, Decision 17/CP.7: Modalities and procedures for a clean development mechanism as defined in Article 12 of the Kyoto Protocol, *FCCC/CP/2001/13/Add.2*, Marrakesh, 10 November 2001, p. 32, §31 (c); UNFCCC, Decision 18/CP.7: Modalities, rules and guidelines for emissions trading under Article 17 of the Kyoto Protocol, *FCCC/CP/2001/13/Add.2*, Marrakesh, 10 November 2001, p. 52, § 2 (c); UNFCCC, Decision 23/CP.7: Guidelines for review under Article 8 of the Kyoto Protocol, *FCCC/CP/2001/13/Add.3*, Marrakesh, 10 November 2001, pp. 30–63; UNFCCC, Decision 22/CP.7: Guidelines for the preparation of the information required under Article 7 of the Kyoto Protocol, *FCCC/CP/2001/13/Add.3*, Marrakesh, 10 November 2001, pp. 19 e.v.

[13] 13/CMP.1 Modalities for the accounting of assigned amounts under Article 7, paragraph 4 of the Kyoto Protocol;

14/CMP.1 Standard electronic format for reporting Kyoto Protocol units;

15/CMP.1Guidelines for the preparation of the information required under Article 7 of the Kyoto Protocol;

19/CMP.1Guidelines for national systems under Article 5, paragraph 1 of the Kyoto Protocol;

20/CMP.1Good practice guidance and adjustments under Article 5, paragraph 2 of the Kyoto Protocol;

21/CMP.1Issues relating to adjustments under Article 5, paragraph 2 of the Kyoto Protocol;

22/CMP.1Guidelines for review under Article 8 of the Kyoto Protocol;

23/CMP.1Terms of service for lead reviewers;

24/CMP.1Issues relating to the implementation of Article 8 of the Kyoto Protocol–1 (Training programme for members of expert review teams);

25/CMP.1Issues relating to the implementation of Article 8 of the Kyoto Protocol–2 (Confidential information).

It is up to each government to implement into the domestic trading system a system for monitoring entity emissions and ensuring that entities surrender the correct number of credits to demonstrate compliance.[14] National governments should keep track of the impact of entity and government transactions on the national assigned amount and periodically report this information to the international community.[15] National governments should also be responsible for ensuring that changes in ownership of credits can be verified.[16]

Non-compliance with the rules of the trading system is conceivable in a variety of ways. Participants may fail to fulfill the reporting or monitoring obligations, they may violate the trading rules or cheat and they may fall into non-compliance because at the end of the commitment period they have exceeded their assigned amount of emissions.

Article 18 of the Kyoto Protocol handles the non-compliance and requires the Parties to "*approve procedures and mechanisms to determine and address cases of non-compliance*". In accordance with Article 18 of the Kyoto Protocol a Compliance Committee has been installed.[17] The Compliance Committee is made up of two branches: a facilitative branch and an enforcement branch. The facilitative branch aims to provide advice and assistance to Parties in order to promote

---

[14] UNFCCC, Decision 20/CP.7: Guidelines for national systems under Article 5, paragraph 1, of the Kyoto Protocol, *FCCC/CP/2001/13/Add.3*, Marrakesh, 10 November 2001, p. 2.

[15] UNFCCC, Decision 3/CP.1: Preparation and submission of national communications from the Parties included in Annex I to the Convention, Report of the Conference of the Parties on its First Session, Held at Berlin from 28 March to 7 April 1995. Addendum Part Two: Action Taken by the Conference of the Parties, *FCCC/CP/1995/7/Add.1*, Berlijn, 7 April 1995, pp. 13–14.

[16] UNFCCC, Decision 23/CP.7: Guidelines for review under Article 8 of the Kyoto Protocol, *FCCC/CP/2001/13/Add.3*, Marrakesh, 10 November 2001, pp. 30–63.

[17] UNFCCC, Decision 24/CP.7: Procedures and mechanisms relating to compliance under the Kyoto Protocol, *FCCC/CP/2001/13/Add.3*, Marrakesh, 10 November 2001, p. 65, Annex, section I, p. 65. *See also:* COP-MOP 1 (2005) Decision 27/CMP.1 Procedures and mechanisms relating to compliance under the Kyoto Protocol and COP-MOP 2 (2006) Decision — /CMP.2 Compliance Committee relating to the rules of the procedure of the Compliance Committee.

compliance, whereas the enforcement branch has the responsibility to determine consequences for Parties not meeting their commitments. Both branches are composed of 10 members, including one representative from each of the five official UN regions (Africa, Asia, Latin America and the Caribbean, Central and Eastern Europe, and Western Europe and Others), one from the small island developing States, and two each from Annex I and non-Annex I Parties. The Committee also meets in a plenary composed of members of both branches, and a bureau, made up of the chairperson and vice-chairperson of each branch, supports its work. Decisions of the plenary and the facilitative branch may be taken by a three-quarters majority, while decisions of the enforcement branch require, in addition, a double majority of both Annex I and non-Annex I Parties.[18]

Nonetheless, questions still arise as to whether the decisions of the enforcement branch are legally binding. For example, '*Whereas the determination by the Enforcement Branch of a party's non-compliance should be considered binding, an important unresolved issue is the legal status of the enforcement consequences. Suspension of eligibility to use the flexibility mechanisms, as well as requirements to adopt a Compliance Action Plan and to submit progress reports, may be considered to be within the competence of the Enforcement Branch. It may, however, be argued that Article 18 of the Kyoto Protocol means that binding deductions of assigned amounts at a penalty rate require an amendment of the Protocol. This may be a challenge to the effectiveness of the enforcement regime. It may also be asked whether the chosen consequences will work as effective deterrents.*'[19] And '*The regime's ability to operate and impose consequences will likely depend at least as much on its perceived legitimacy as on its legal form. Indeed, its adoption in binding form may*

---

[18] http://unfccc.int/kyoto_protocol/compliance/introduction/items/3024.php; *See more on enforcement of emission trading:* J.K. Stranlund, C.A. Chavez & B.C. Field, *Enforcing emissions trading programs: theory, practice and performance*, Paper, 2nd CATEP (?) Please check on this? Workshop, London, 25–26 March 2002, p. 29; T. Tietenberg , *Emissions trading, an exercise in reforming pollution policy*, Resources for the Future, Washington D.C., 1985, p. 168.

[19] O. Schram Stokke, J. Hovi & G. Ulfstein, *Implementing the climate regime international compliance*, Earthscan, London, 2005, p. 59.

create a false sense of assurance, diverting attention from the need to nurture the persuasive power of the climate change regime. If it enters into force, the Kyoto Protocol will provide a testing ground for compliance theories. But disentangling the factors that promote compliance will require not just an examination of the compliance regime's facilitative and enforcement-oriented elements. It will also require close attention to the legal forms and processes that anchor these elements.'[20]

One might argue that Article 18 on non-compliance does not lend itself well to the establishment of a sanctions regime for emissions trading. Although it requires the Parties to "*approve procedures and mechanisms to determine and address cases of non-compliance*" with the Protocol, it contains a clause that makes "*binding consequences*" dependent on a formal amendment. Since an amendment to the Protocol requires ratification by at least three-fourths of the parties, its entry into force appears to be very unlikely or would at least take an unacceptably long time.

One of the most essential elements for ensuring compliance is the need to track the Assigned Amount Units (AAUs) that each country must hold to cover its emissions under the terms of the Kyoto Protocol. Therefore 'national registries' are proposed to track countries' holdings on AAUs. Next to these 'national registries' the UNFCCC Secretariat will keep a log of all transactions from one registry to another to ensure that AAUs are located in one registry and are not being used more than once.[21] However the international transaction log (henceforth: ITL) is only very recently (since 14 November 2007) operational.

The European Union Emission Trading Scheme (henceforth: 'EU ETS') is one of those 'domestic' emissions trading system, which is created in conjunction with the Kyoto Protocol. EU ETS is one of the largest multi-national, greenhouse gas emissions trading scheme in the world and has its own community transaction log. The paper will

---

[20] J. Brunnee , The Kyoto Protocol: testing ground for compliance theories?, *ZaöRV-Heidelberg Journal of International Law*, 63/2, 2003, p. 280.

[21] OECD and International Energy Agency, 'International Emission Trading: from concept to reality' (2001), France, 74–75.

focus on the EU ETS, what the current situation is and what we may expect from the reform.

Next to the allowance trading system, the EU ETS allows operators to use credits from the Kyoto Protocol project mechanisms JI and CDM, as described below, to meet their targets in place of emission cuts within the EU (created by the so-called 'Linking Directive').[22]

1.1.2. *Joint Implementation (JI): Project based-transaction between an Annex I-Party and another Annex I-Party*

JI is a project-based mechanism defined by Article 6 of the Kyoto Protocol. It provides for Annex I[23]-Parties to implement projects that reduce emissions in other Annex I-Parties, in return for Emission Reduction Units (henceforth: 'ERUs') which is equal to one metric ton CO_2 equivalent, calculated in accordance with the Kyoto rules. Those ERUs generated by JI projects can be used by Annex I-Parties to help meet their emission targets under the Kyoto Protocol.

JI projects must have the approval of all Parties involved, and must lead to emission reductions or removals that are additional to any that would have occurred without the project. The crediting period cannot start before 2008. Projects starting from the year 2000 that meet JI requirements may be listed as JI projects but ERUs may only be issued in relation to periods from 2008 onwards.[24]

At the Seventh Conference of the Parties (COP7) to the UNFCCC in 2001, the operational aspects of JI were clarified and two different procedures were adopted.

---

[22] Directive 2004/101/EC of the European Parliament and the Council of 27 October 2004 amending Directive 2003/87/EC establishing a scheme for greenhouse gas emission allowance trading within the Community, in respect of the Kyoto Protocol's project mechanisms, *Pb L* 338/18.

[23] So-called industrialized countries and countries with economies in transition. These countries are listed in Annex I of the UNFCCC.

[24] *See also:* Michaelowa, A., Incentive aspects of joint implementation of greenhouse gas reduction, *Mitigation and Adaption Strategies for Global Change*, Vol. 1, no 1, 1996, p. 95.

The first procedure (also referred to as 'JI Track One') may be applied when the Annex I Party hosting the project fully meets all the eligibility requirements to participate in the mechanisms. In this situation, the host Party may apply its own national rules and procedures to the selection of JI projects and the estimation of emission reductions from them. The host Party may also issue ERUs (through converting existing AAUs[25] and RMUs[26]) and transfer them to project participants.

The second procedure (also referred to as 'JI Track Two') must be applied if the host Party does not meet all eligibility requirements to participate in the mechanisms. In such cases, the project and the quantity of ERUs it generates must be verified under rules and procedures supervised by the Supervisory Committee, as mentioned in Article 6 of the Kyoto Protocol, namely the Joint Implementation Supervisory Committee ("JISC"[27]).

Depending on whether it is Track One or Two the public entity which is developing a JI project needs to go through various stages of a project cycle.[28] An updated information and further details on the project cycle can be found on the UNFCCC website 'http://unfccc.int/kyoto_mechanisms/ji/items/1674.php'.

---

[25] AAUs stands for Assigned Amount Unit which is a unit pursuant to the UNFCCC and the Kyoto Protocol, that can be transferred on the basis of Article 17 of the Kyoto Protocol.

[26] A 'removal unit' or 'RMU' is a unit issued pursuant to the relevant provisions in the annex to decision 13/CMP.1 and is equal to one metric tonne of carbon dioxide equivalent, calculated using global warming potentials defined by decision 2/CP.3 or as subsequently revised in accordance with Article 5 (see Decision 3 of CMP 1); *See also:* R. DE WITT WIJNEN, 'Emissions trading under article 17 of the Kyoto Protocol', in: D. Freestone & C. Streck, *Legal aspects of implementing the Kyoto Protocol*, Oxford Univ. Press, Oxford, 2005, p. 409.

[27] The JISC will hold its eight meeting on 18 and 19 October 2007 in Bonn (see http://ji.unfccc.int/Sup_Committee/Meetings/008/index.html).

[28] *See also* S. Starckx, 'Inzicht in en validatie van Joint Implementation (JI) en Clean Development Mechanism (CDM) projecten', in: J. De Mulder, *Kyoto: recent beleid en praktijk*, Uitg. VandenBroele, Brugge, 2005, p. 239.

1.1.3. *Clean Development Mechanism (CDM): Project based-transaction between an Annex I-Party and a non Annex I-Party*

CDM is the mechanism laid down in Article 12 of the Kyoto Protocol with the purposes (1) to assist non Annex I-Parties in achieving sustainable development and (2) to contribute to the ultimate objective of the UNFCCC and (3) to assist Parties included in Annex I in achieving compliance with their quantified emission limitation and reduction commitments as described in the Kyoto rules. A CDM project can generate Certified Emission Reductions (henceforth: 'CER').

Compared with JI, the CDM Executive Board was installed quite immediately with the first COP/MOP[29] and will hold its 33th meeting in September 2007. The CDM Executive Board is essentially a panel of UNFCCC appointed experts. The CDM Executive Board supervises the CDM, under the authority and guidance of the Conference of the Parties serving as the meeting of the Parties to the Kyoto Protocol (COP/MOP), and is fully accountable to the COP/MOP.

Every country that wishes to participate in the CDM is required to set up a Designated National Authority (DNA). DNAs are public bodies integrated in the environmental or economic administration of the Governments of Parties to the Kyoto Protocol. The project proponent is required to obtain the written approval from the host country DNA before its project can be registered by the Executive Board. The CDM project must also satisfy two broad criteria: (i) sustainable development[30] and (ii) additionality.[31]

---

[29] *See* Decision 3 CMP/1.

[30] CDM activities must contribute to the sustainable development process of the country hosting the project. Each host country may choose its own sustainable development criteria and assessment process.

[31] CDM projects must reduce emissions below those emissions that would have occurred in the absence of the CDM project activity. The additional GHG reductions are calculated by comparing project emissions with defined emissions baseline representing the business as usual scenario.

The project needs to go through various stages of a project cycle. The CDM Executive Board has set out Guidelines for completing a project design document ("PDD"); this is just one stage of the project cycle.

Project developers must also have an independent certifier, called a Designated Operational Entity (DOE) affirm that their project satisfies the requirements as set forward in the Marrakech accords of 2001. This is known as validation, so the DOE is commonly known as the validator. If the validator determines that the requirements have been met then they recommend to the CDM Executive Board that the project be registered, which constitutes final approval.

If the Board does not disagree with this recommendation within 8 weeks, the project is automatically registered and can begin monitoring and claiming credit for the reduction of emissions.

The reductions must be verified by another DOE — not the same one that did the validation — before the Executive Board can issue Certified Emission Reductions (CERs), i.e. carbon credits. Monitoring, verification and issuance of CERs will continue for the entire period during which the project claims credit for reducing emissions.

More detailed information on the CDM project cycle can be found on the website of the UNFCCC 'http://cdm.unfccc.int/Projects/pac/index.html'. We also refer to the other papers of this Conference on Crucial Issues in Climate Change and the Kyoto Protocol in Asia, that discuss the CDM project cycle more in detail.[32]

To date, the CDM Executive Board has registered 761 CDM project activities and refused registration for 26 projects activities.[33]

---

[32] See paper of Yang Xing and Wang Xi, *CDM in China*; paper of Jolene Lin, *An overview of the clean development mechanism in Southeast Asia;* paper of Charlotte Streck and Jolene Lin, *Making Markets work: A review of CDM performance and the need for reform;* Wiliam I.Y. Byun and Felix H.C. Chan, *Empirical Considerations in the Development of CDM Projects in Asia.*

[33] http://cdm.unfccc.int/Projects/index.html. Last visit on 20 August 2007.

1.2. *The EU greenhouse gas emission trading scheme: Current situation and legislation*

1.2.1. *The European Climate Change Programme (ECCP)*

The European Council of Environment Ministers requested the European Commission in October 1999 to put forward a list of priority actions and policy measures to achieve the Kyoto Target. The Commission responded by launching, in May 2000, the European Climate Change Programme (henceforth: 'ECCP'), whose goal was to identify and develop all the necessary elements of an EU strategy to implement the Kyoto Protocol.

(a) The first ECCP

During the first phase of the ECCP (2000–2001) various working groups were set up to consider and give recommendations on the options to reduce the greenhouse gases cost-effectively. The findings of the working groups have been collected in the EPPC Report of October 2001 of the European Commission.[34] The EPPC Report brought forward a package of three broad measures to tackle climate change:

- Firstly, an Action Plan for the EPPC. This Action Plan has been drafted in the form of a communication with an overview of the financial implications and measures that have to be taken to implement the international emissions trading scheme.[35]
- Secondly, the European Commission put forward a proposal to the European Council to ratify the Kyoto Protocol. The European Community ratified the Kyoto Protocol on 31 May 2002.[36]
- Thirdly, the European Commission proposed a directive on the emission trading scheme.

---

[34] *See* http://ec.europa.eu/environment/climat/eccpreport.htm

[35] COM (2001) 580.

[36] Council Decision 2002/358/EC of 25 April 2002 concerning the approval, on behalf of the European Community, of the Kyoto Protocol to the UNFCCC and the joint fulfillment of commitments thereunder, OJ L 130, 15 May 2002, p. 1.

During the second phase (2002–2003) the EPPC focused itself on the actual implementation of the priorities identified in the first phase. This meant, not only the implementation of the international emission trading scheme, including the flexible mechanisms, but also other measures that could reduce greenhouse gasses, such as a proposal on promotion of bio fuels, vehicle taxation, proposal on combined heat and power.

In its communication on post-Kyoto 2012 of February 2005 the European Commission launched the proposal to have a new phase of the EPPC in 2005 merely focused on post Kyoto. The European Commission launched the Second EPPC at a conference in Brussels on 24 October 2005.

(b) The second ECCP

The focus of the second ECCP concerned key areas of current and future European climate change policy, including ECCP I review, geological carbon capture and storage, adaptation, aviation, passenger road transport, energy efficiency, renewable energy and technology policy. Six working group meetings have taken place.[37] Most of these working group meeting took place in 2006 and 2007.

The first working group meetings discussed the ECCP I Review with 5 topical groups (energy supply and demand, transport, Non-CO_2 gases, agriculture and forestry). They decided to abandon the plans for a communication on the review of the ECCP. Progress on specific policies discussed is being made. The reports of the groups will provide input to the Post-2012 Green Paper which was planned for October 2006.

The second working group meetings took place with ten sectoral groups. The topic of the second working group meetings concerned the impacts and adaptation. The reports from the topical groups under this working group meeting were meant to provide input to a Green Paper on Adaptation to be launched at a Brussels conference in late November 2006.

[37] All meeting group reports and speeches are available online: http://ec.europa.eu/environment/climat/stake_conf.htm

The third working group meeting discussed the topic of carbon capture and geological storage. This working group looked specifically into the necessary regulatory framework for CCS technology. In November 2007 the European Commission launched a proposal on geological storage of carbon dioxide. The proposal ensures that CO_2 capture is regulated under Directive 96/61/EC concerning integrated pollution prevention and control for certain industrial activities, and that both CO_2 capture and pipeline transport are regulated under Directive 85/337/EEC on assessment of the environmental impacts of certain projects (as amended by Directive 97/11/EC). The main purpose of the proposal is the regulation of CO_2 storage and the removal of barriers in existing legislation to CO_2 storage.

The fourth working group meeting discussed design options for integrating aviation in the EU ETS. A legislative proposal has been adopted by the European Commission, namely a proposal for a Directive of the European Parliament and of the Council amending Directive 2003/87/EC so as to include aviation activities in the scheme for greenhouse gas emission allowance trading within the Community, and has been forwarded on 22 December 2006 to the European Council. The objective of the proposal is to provide a model for aviation trading that can be a point of reference in the EU's contacts with key international partners and to promote the development of similar systems worldwide. Since this proposal has a great impact on the tackling of aviation emissions, it will be discussed more in detail in this paper.

The fifth working group meeting handled the integrated approach to reduce CO_2 emissions from light-duty vehicles. On 30 May 2007 the European Council confirmed the adoption of the Euro 5 emissions standards for cars.[38] Euro 5 puts new caps on pollutant emissions from diesel and petrol cars, limiting, in particular, nitrogen oxides

[38] EC Regulation no 715/2007 of the European Parliament and of the Council of 20 June 2007 on type approval of motor vehicles with respect to emissions from light passenger and commercial vehicles (Euro 5 and 6) and on access to vehicle repair and maintenance information.

(NO_x) and particulate matter (PM) which pose the most serious health and environmental problems. The next generation of standards (known as Euro 5) for new models of cars will enter into force in September 2009, and in January 2011 for all new cars. Even more tighter emission standards (known as Euro 6[39]) will enter into force as of September 2014 for new models of cars and in September 2015 for all new cars. Heavy-duty trucks and buses, off-road diesel vehicles and motorcycles are subject to separate emission regulations.[40] The European Commission will also propose to amend the fuel standards.

Next to these standards, the EU tries to raise the awareness among consumers. The EU Directive[41] requires the display on each new car of a label showing its fuel consumption and CO_2 emissions, as well as publication of fuel efficiency information in other formats, including printed advertisements. Also the promotion of fuel-efficient cars through fiscal measures has been adopted in EU legislation.[42]

The sixth working group meetings took place very recently in order to discuss the review of the EU ETS. This topic will be discussed later in this paper.

1.2.2. *The EU Directive 2003/87/EC: The establishment of a Community-wide greenhouse gas emission allowance trading scheme*

Once the Kyoto Protocol had been ratified on 31 May 2002 by the European Community, the European Parliament and Council approved the Directive 2003/87/EC of 13 October 2003 establishing a scheme for greenhouse gas emission allowance trading within the Community and amending Council Directive 96/61/EC.[43]

---

[39] Euro VI will set significantly lower emission limits for NO_x emissions from diesel.

[40] The European Commission presented on 16 July 2007 a range of possible caps on emissions from heavy-duty vehicles that pose serious health and environmental problems, such as particulate matter (PM), nitrogen oxides (NO_x), hydrocarbons, carbon monoxide and ammonia.

[41] Directive 1999/94/EC.

[42] COM(2005)261 final.

[43] OJ L 275, 25 October 2003, pp. 32–46.

The ETS Directive is based on the results of the Green Paper from the European Commission on greenhouse gas emission trading within the European Union.[44]

The EU ETS commenced operation in January 2005 with all 25 (now 27) member states of the European Union participating. It contains the world's only mandatory carbon trading program. The program caps the amount of carbon dioxide that can be emitted from large installations, such as power plants and carbon intensive factories and covers almost half of the EU's carbon dioxide emissions.

As mentioned above, the adoption of the Kyoto Protocol by the Commission and its Member States in 2002 commits those to reduce their greenhouse gas emissions by 8% in relation to 1990 levels between 2008 and 2012. It made its own agreement to meet the 8% target by distributing different rates to its Member States[45] (cf. The Burden Sharing Agreement in Annex II of the Council Decision 2002/358/EC). The ETS Directive, by establishing a market in greenhouse gas emission allowances, will help the Community and its Member States to meet the commitments made in the framework of the Kyoto Protocol in an effective manner which respects economic development and employment.

Member States had to implement the ETS Directive by 31 December 2003 at the latest.[46] Some of them were too late, and were convicted by the European Court of Justice (e.g. Finland and Italy).

Whilst the first phase (2005–2007) has received much criticism due to oversupply of allowances and the distribution method of allowances (via grandfathering rather than auctioning), the European Commission have been tough on Member States' Plans for Phase II, dismissing many of them as being too loose again. In addition, the first phase has established a strong carbon market. Compliance has also been high in 2006, increasing confidence in the scheme.

---

[44] COM (2000) 87 final.

[45] EU Council Decision 2002/358/EC of 25 April 2002 concerning the approval, on behalf of the European Community, of the Kyoto Protocol and defining emission reduction targets for each Member State.

[46] Article 31 of the ETS Directive.

(a) Scope of the ETS Directive

- Sectors and greenhouse gas covered

While emission trading has the potential to involve many sectors of the economy and all six greenhouse gases controlled by the Kyoto Protocol, the scope of the EU ETS is limited during its initial phase. Consequently, during the first trading period, from 2005 to 2007, and the second trading period, from 2008 to 2012, the EU ETS covers only CO_2 emissions from large emitters in the power and heat generation industry and in selected energy-intensive industrial sectors: combustion plants, oil refineries, coke ovens, iron and steel plants and factories making cement, glass, lime, ceramics, pulp and paper.[47]

Even so Article 2 of the ETS Directive applies to emissions from activities listed in Annex I and greenhouse gases listed in Annex II which covers a list of the six greenhouse gases mentioned in the Kyoto Protocol. Also the definition of 'tonne of carbon dioxide equivalent', that is also used in the definition of 'allowance', refers to '*one metric tonne of carbon dioxide (CO_2) or an amount of any other greenhouse gas listed in Annex II with an equivalent global-warming potential*,[48] and thus allows the application of all six greenhouse gases. However, the limitation to CO_2 is to be found in the list of activities, as laid out in Annex I of the ETS Directive. To broaden the scope of the EU ETS to the other five greenhouse gases, Annex I has to be amended.

Despite this limited scope, more than 12,000 installations in the 27 Member States are covered, accounting for around 45% of the EU's total CO_2 emissions or about 30% of its overall greenhouse gas emissions.

The group Arcelor, producer of pig iron and steel in France, Spain, Germany and Belgium brought (on 15 January 2004) an action against the European Parliament and the Council of the European Union.[49] Acelor argued that the EU ETS infringes its fundamental

---

[47] COM (2001) 581 final, p. 10.

[48] Article 3, j) of the ETS Directive.

[49] T-16/04.

right to property and the pursuit of an economic activity, by requiring it to operate its plants under economic conditions that are unsustainable. Amongst other arguments, Arcelor invoked a breach of the principle of equality, alleging that other sectors in direct competition with the applicant and with comparable or even higher emissions of greenhouse gases, such as producers of non-ferrous metals and chemicals, were not subject to the ETS Directive. In the same context Arcelor further alleged that steel producers find themselves in a unique lock-in situation preventing them from passing on to their clients any monetary fines imposed for excessive emissions. The applicant also submitted that the contested provisions infringe the freedom of establishment within the European Union by affecting its right to transfer production freely from a less efficient plant in one Member State to a more efficient plant in another Member State. Arcelor further claimed that there was very little technological potential for steel producers to reduce their greenhouse gas emissions beyond the 18% reduction already achieved since 1990 and that, therefore it is an infringement of the principle of proportionality to subject such installations to the ETS Directive. Finally, Arcelor invoked a breach of the principle of legal certainty, claiming that the ETS Directive imposed obligations whose financial implications are unforeseeable. The European Court of First Instance has not yet ruled on this interesting case.

- Opt-in

Article 24 of the ETS Directive allows for the unilateral inclusion (known as 'opt-in') of other activities and gases as from 2008. This means that from 2008, Member States may apply emission allowance trading in accordance with the ETS Directive to activities, installations and greenhouse gases which are not listed in Annex I, provided that inclusion of such activities, installations and greenhouse gases is approved by the European Commission in accordance with the procedure referred to in Article 23, 2, taking into account all relevant criteria, in particular effects on the internal market, potential distortions of competition, the environmental integrity of the scheme and reliability of the planned monitoring and reporting system.

From 2005–7 member states may "opt-in" smaller installations[50] within the above sectors. Discussions are being held as to whether small installations would also fall mandatorily under the scope of the EU ETS. Some Member States and institutions are opposed to it, because including smaller installations or installations operating only on a seasonal basis or a stand-by or back-up capacity means that the costs of participation in the EU ETS would outweigh the benefits of being covered by the scheme.[51]

- Opt-out

From 2005–2007 member states may apply to "opt-out" specified installations, under the condition that it must be ensured that there will be no distortion of the internal market and it is subject to Commission approval. From 2008 onwards all eligible installations must be covered.

According to Article 27 of the ETS Directive the public can also make its comments on the temporary exclusion or 'opt-out'. Subsequently, the European Commission will decide whether the installations will:

- as a result of national policies, be required to limit their emissions as much as would be the case if they were subject to the provisions of the ETS Directive.
- be subject to monitoring, reporting and verification requirements which are the equivalent of those mentioned in the ETS Directive.
- be subject to penalties at least equivalent to those referred to in Article 16 in the case of non-fulfilment of national requirements.

According to some jurisprudence the costs for opting out are much higher than complying with the EU ETS monitoring, reporting and verification itself.[52]

---

[50] Meaning installations carrying out activities listed in Annex I below the capacity limits referred to in that Annex (Article 24, 1 of the ETS Directive).

[51] *See* more below (EU ETS under review).

[52] A. Hobley & P. Hawkes, The EU emission allowance trading scheme, *Environmental Finance*, May, 2003, p. 3.

According to the review of September 2006 of the European Commission, 8980 installations have been opted out.

- Pooling

Article 28 of the ETS Directive allows Member States to allow operators of installations carrying out one of the GHG activities to form a pool of installations from the same activity for the first and/or second trading period. Those operators have to inform the competent authority of the Member State of their pooling activities and they have to nominate a trustee to which the total amount of allowances of the pooling installations will be issued, and will be responsible for surrender of the allowances equal to the total emissions from the installations in the pool. The Member State shall inform the European Commission. The European Commission will decide within three months, but may reject the proposal of pooling.

During the first trading period none of the Member States have proposed a pooling of activities.

(b) Combustion installation

In accordance with Annex A of the Kyoto Protocol, the ETS Directive lists the activities covered by the ETS Directive in its Annex I. The activity of combustion installations with a rated thermal input greater than 20MW includes individual boilers or other combustion units in the same installation or on the same site which cumulatively reach that threshold (e.g. two 15MW combustion units on a site will be covered by the ETS Directive). The competent authority of a Member State has an important role in determining the boundaries of the installation and of the site. Where an installation comes within the scope of the ETS Directive because of combustion taking place, emissions from the installation other than from combustion are only covered where these are covered by another activity listed in Annex I to the ETS Directive.

Because Member States were troubled by the words 'combustion installation' as mentioned in Annex I, and 'operator' as mentioned and defined in Article 3(f)[53] in the ETS Directive, the European Commission drafted in September 2003 a non paper on the installation coverage of the EU emissions trading scheme and the interpretation of Annex I.[54]

The "Replies to some frequently asked questions on the EC emissions trading proposal"[55] state that "*a combustion installation with a rated thermal input exceeding 20 MW operating in connection with an installation of the chemicals industry would fall into the category of energy activities.*" In the non paper the European Commission clarified that it should be understood that installations in the chemical industry only come within the scope of the ETS Directive in respect of their on-site combustion capacity, while emissions of other greenhouse gases from other activities are not initially covered. This contrasts with the coverage of oil refineries, where emissions of carbon dioxide from all activities (i.e. including non-combustion activities) of an oil refinery are covered.

The term "*combustion*" is used in a wide range of EC legislation including the ETS Directive, the Integrated Pollution Prevention and Control Directive 1996/61/EC (henceforth: 'IPPC Directive'), the Large Combustion Plants Directive 2001/80/EC, and Council

---

[53] 'Operator' means a stationary technical unit where one or more activities listed in Annex I are carried out and any other directly associated activities which have a technical connection with the activities carried out on that site and which could have an effect on emissions and pollution'. In the non paper of September 2003 the European Commission stressed that The definition of "operator" in the ETS Directive is identical to the IPPC Directive (Article 2(12). There will always be an operator of an installation, and the identification of the operator is a task for the competent authority. In case an installation is co-owned by several legal entities, it is for the competent authority to determine one single operator based to a large extent on national legislation and administration. Article 6(1) of the ETS Directive makes clear that one operator can hold a single greenhouse gas emissions permit in respect of more than one installation on the same site. Thus, an installation cannot have more than one operator, and all installations under a single permit must have the same operator.

[54] Not published.

[55] http://europa.eu.int/comm/environment/climat/emissions_faq.pdf, 23 April 2002.

Directive 1999/32/EC of 26 April 1999 relating to a reduction in the sulphur content of certain liquid fuels. It is logical for the meaning of combustion in the context of ETS Directive to be construed by reference to other EC legislation where definitions are included. The Large Combustion Plants Directive states that "*combustion plant*" means "*any technical apparatus in which fuels are oxidised in order to use the heat thus generated*" (Article 1(7)). Further guidance can be drawn from looking at the list of combustion plants, which are specifically excluded from the Large Combustion Plants Directive in Article 1(7), where the ETS Directive does not provide for such exclusion. Certain activities that are specifically excluded by the Large Combustion Plants Directive are also excluded from the ETS Directive, such as "*(h) any technical apparatus used in the propulsion of a vehicle, ship or aircraft*" because the ET Directive only applies to stationary technical units (Article 3(e)). The ETS Directive therefore covers neither transportation in general nor greenhouse gas emissions arising from traffic on the site of an installation.

Article 1(7) of the Large Combustion Plants Directive states that it is to apply to:

> "*combustion plants designed for production of energy with the exception of those which make direct use of the products of combustion in manufacturing processes*", and that it "*shall not apply to the following combustion plants*":
>
> "*(a) plants in which the products of combustion are used for the direct heating, drying, or any other treatment of objects or materials e.g. reheating furnaces, furnaces for heat treatment;*
>
> *(b) post-combustion plants i.e. any technical apparatus designed to purify the waste gases by combustion which is not operated as an independent combustion plant;*
>
> *(c) facilities for the regeneration of catalytic cracking catalysts;*
>
> *(d) facilities for the conversion of hydrogen sulphide into sulphur;*
>
> *(e) reactors used in the chemical industry;*
>
> *(f) coke battery furnaces;*
>
> *(g) cowpers;*

> (h) *any technical apparatus used in the propulsion of a vehicle, ship or aircraft;*
> (i) *gas turbines used on offshore platforms;*
> (j) *gas turbines licensed before 27 November 2002 or which in the view of the competent authority are the subject of a full request for a licence before 27 November 2002 provided that the plant is put into operation no later than 27 November 2003 without prejudice to Article 7(1) and Annex VIII(A) and (B),*"

Plants powered by diesel, petrol and gas engines are also excluded from the scope of the Large Combustion Plants Directive by Article 1(7).

Given that the ETS Directive makes no similar specific exclusions, all of the above types of combustion installations are included within the scope of the ETS Directive where the threshold is met or exceeded.

It should also be noted that the same definition of combustion as in the Large Combustion Plants Directive is used in Council Directive 1999/32/EC of 26 April 1999 relating to a reduction in the sulphur content of certain liquid fuels (Article 2(5) — "*combustion plant means any technical apparatus in which fuels are oxidized in order to use the heat generated*").

Any installation(s) which include one or more pieces of stationary technical apparatus in which fuels are oxidized in order to use the heat thus generated, that together on the same site have a rated thermal input exceeding 20MW, is therefore subject to the ETS Directive. This includes apparatus where the heat is used in another piece of apparatus, through a medium such as electricity or steam, and apparatus where the heat resulting from combustion is used directly within that apparatus, for example, for melting other substances.

The Commission Communication on the Council's common position on the ETS Directive (SEC(2003)364) notes in its conclusion that "*The Common Position incorporates many of the amendments proposed by the European Parliament at its first reading. The scope of the emissions trading scheme includes energy, heat and steam production of installations above 20MW, while Member States can extend the coverage of the scheme as from 2005 to lower thresholds*".

Further guidance comes from Annex I to the ETS Directive itself, because this specifically excludes municipal and hazardous waste incineration facilities from the scope of the scheme. The combustion of e.g. hazardous waste is clearly an integrated part of the normal process undertaken by hazardous waste installations. If the Directive were to be interpreted as to not apply to installations where combustion takes place as an integrated part of the installation's processes, municipal and hazardous waste installations would not have been specifically excluded as they would in any case have fallen outside its scope. Their specific exclusion is further confirmation that it is the presence of combustion with a rated thermal input exceeding 20MW that determines the ETS Directive's coverage of stationary combustion installations.

In the light of above points, there is no clear justification to limit the meaning of "combustion installation" to exclude any stationary units where combustion takes place, for example units providing heat input to chemical reactors, where these satisfy the threshold set down in Annex I to the Directive. In particular, there is no clear basis for claiming that combustion does not include the burning of fuels where the products of combustion are used directly as an integrated part of the production process. Energy produced by combustion may be in the form of electricity, heat, hot water or steam, and the distance between the production of energy and its eventual use is not relevant for competent authorities to use to decide whether or not an installation is subject to the ETS Directive.

Furthermore, it is well-established that industries can fall under more than one heading of the IPPC Directive, for example, integrated steel works carry out several Annex I activities, and refineries include combustion installation *of more than* 50MW. Given the similarities between the IPPC and ETS Directives, there is no reason to take a different approach to its interpretation. In particular, a different approach cannot be justified by the separate listing of the steel and cement industries, given that both produce substantial emissions from (chemical) processes in addition to their emissions from combustion.

Nonetheless, in spite of the non paper of the European Commission, in preparation of the national allocation plans, the Member States interpreted the term 'combustion installation' differently, which led to complications. Some Member States used a broad definition[56] (e.g. the Netherlands, Belgium (Wallonia), Ireland); others used a medium broad definition.[57] None of them used a narrow definition.

The use of a harmonized definition within the EU ETS could bring an end to the considerable differences regarding emissions and installations covered in the various Member States. Direct process heating could be explicitly included via Annex I, albeit requiring a change of the ETS Directive. At the same time it can be argued that they are implicitly included via the broad interpretation.[58]

(c) GHG permit

With effect from 1 January 2005, all installations carrying out any of the activities listed in Annex I to the ETS Directive (activities in the energy sector, iron and steel production and processing, the mineral industry and the wood pulp, paper and card industry) and emitting the specific greenhouse gases associated with that activity must be in possession of an appropriate permit issued by the competent authorities. This is prescribed by Article 4 of the ETS Directive, stating that Member States shall ensure that, from 1 January 2005, no installation undertakes any activity listed in Annex I resulting in emissions specified in relation to that activity unless its operator holds a permit issued by a competent authority in accordance with Articles 5 and 6, or the

[56] 'All combustion installations that produce electricity, heat or steam, even if their main purpose is not energy production, but e.g. the production of ethylene or ammonia (e.g. naphtha crackers or ammonia plants)'.

[57] 'All combustion installations that produce electricity, heat or steam, with the purpose of energy production, including those that are process-integrated, e.g. a steam plant integrated in e.g. chemical industry is included, but process furnaces such as crackers in the petrochemical industry are excluded'.

[58] *See* report of McKinsey & Company and Ecofys: http://ec.europa.eu/environment/climat/emission/review_EN.htm

installation is temporarily excluded from the Community scheme pursuant to Article 27.

Applications for greenhouse gas emissions permits must describe[59]:

- the installation, its activities and the technology used;
- the materials used which could emit the greenhouse gases listed in Annex II;
- the sources of gas emissions;
- the measures planned to monitor and report emissions.

The authorities will issue a permit provided that they are satisfied that the operator of the installation is capable of monitoring and reporting the emissions. A permit may cover one or more installations on the same site operated by the same operator.[60] The permit will contain details of[61]:

- the name and address of the operator;
- the installation's activities and emissions;
- the monitoring methodology and frequency;
- the reporting requirements in respect of emissions;
- the obligation to surrender, during the first four months of each year, a quantity of allowances commensurate with the total emissions over the previous year.

Most installations covered by the ETS Directive are also the subject of IPPC permits. For administrative simplicity, Article 8 of the ETS Directive allows Member States to combine the permitting procedure for greenhouse gas emissions trading with that for the IPPC Directive. The Commission services would expect Member States to want to take advantage of this possibility. Nevertheless, Member States would not be obliged to combine these procedures. Where Member States choose not to combine the procedures, Article 8

[59] Article 5 of the ETS Directive.

[60] Article 6, 1 of the ETS Directive.

[61] Article 6, 2 of the ETS Directive.

requires them to co-ordinate the conditions of, and procedure for, the issue of an emissions trading permit with permitting under the IPPC Directive. In Member States with different competent authorities for the IPPC permit and the emissions trading permit, the emissions trading permit may be issued at a different time than the IPPC permit although the competent authority for the IPPC permit must be consulted. The information required for an IPPC permit will tend to include the information required for an emissions trading permit, and it would be useful for the relevant authorities to be able to check the consistency of the applications. Belgium is one of the few Member States which combined the GHG permit with the IPPC permit.

(d) National allocation plans

The elaboration of a National Allocation Plan (henceforth: 'NAP') by each Member State is one of the most important tasks to be accomplished prior to the commencement of the allowance issuing and trading. It is at the same time the subject of much debate.

The NAP and allocation of allowances is governed by Articles 9, 10 and 11 of the ETS Directive as well as Annex III. Allocations are to be made before the beginning of each period (the first period being 2005–2007, and the second period being 2008–2012).

Member States draw up a NAP which is a statement of how they intend to allocate allowances to individual operators.

The objective in drawing up a National Allocation Plan is to fix a cap on greenhouse gas emissions from installations participating in the EU ETS and to ensure a reasonably fair share-out of the task of emission reductions:

- between sectors participating in the trading scheme and the rest of the economy;
- among sectors participating in the trading scheme, and
- among installations in the participating sectors.

In this context the NAPs have to respect the Commission Decision 2006/944/EC of 14 December 2006 determining the respective emission levels allocated to the Community and each of its Member States under the Kyoto Protocol pursuant to Council Decision.

The NAP must contain an allowance methodology, which does not favour particular sectors or firms unless it can be justified under the Annex III criteria. The allocation should avoid unjustifiable differences between covered sectors and sectors which are not covered, and between and within covered sectors. The principle of equality and transparency is explicitly laid down in Article 9(1) of the ETS Directive ('*The plan shall be based on objective and transparent criteria*'). With regard to this principle of non-discrimination, Arcelor Atlantique and Lorraine posed the following question to the European Court of Justice: '*Is Directive 2003/87/EC of 13 October 2003*[1] *valid in the light of the principle of equal treatment, in so far as that Directive makes the greenhouse gas emission allowance trading scheme applicable to installations in the steel sector without including in its scope the aluminium and plastic industries?*' The Court has not ruled yet.

The NAP has to meet at least the (binding) criteria of Annex III, optionally also the non-binding criteria of Annex III.[62] The criteria of

[62] A Member State has an obligation to apply all elements of criteria (2) — assessments of emission developments), (5) — non-discrimination between companies and sectors), (9) — involvement of the public) and (10 — list of installations), and some elements of the criteria (1 — the Kyoto commitments), (3 — potential to reduce emissions) and (4 — consistency with other legislation). It can, therefore, choose whether it wants to take specific action with respect to some elements of criteria (1 — the Kyoto commitments), (3 — potential to reduce emissions) and (4 — consistency with other legislation), and the criteria (6 — new entrants), (7 — early action), (8 — clean technology) and (11). The Commission will not reject a plan if all mandatory criteria and mandatory elements of criteria are applied in a correct manner. The Commission will not reject a plan if optional criteria or optional elements of criteria are not applied. However, if these optional criteria or optional elements of criteria or additional transparent and objective criteria are applied, the Commission will assess their application. In all cases, the Commission does require information from a Member State with respect to criteria (7 — early action) and (8 — clean technology), even if this is only to state that a criterion has not been applied. In respect of criterion (6 — new entrants) a Member State must state the manner in which new entrants will be able to begin participating in the Community scheme in that Member State (See the Guidance of the European Commission of 7 January 2004).

Annex III allow Member States to take a variety of approaches to establish absolute quantities (including historical emissions or a "national benchmarking" approach). A benchmarking approach can be used to set in the NAP absolute quantities of allowances by multiplying input or output data with an emission factor.[63] They remain applicable also in context of future auctioning (in whole or in part). The Common Position does not lay down how Member States determine the quantities of allowances allocated to each operator.

As prescribed in Article 9(1) of the ETS Directive, the European Commission issued in January 2004 a guidance on the implementation of the Annex III criteria pursuant to Article 9(1) of the ETS Directive as well as on the circumstances under which force majeure is demonstrated.[64] This guidance applies to both the first and the upcoming second trading period. In the meantime, the European Commission adopted a second guidance in relation to the NAPs of the second trading period.[65]

Once the NAP has been adopted by the Member State, the plan is notified to the Commission.[66] Furthermore it needs to be published at the latest upon notification in order to allow the general public to express comments prior to a decision being taken on allocating of the allowances. The public consultation is needed to be in accordance with the Aarhus Convention. The public consultation is also explicitly foreseen in Article 17 of the ETS Directive ('*Decisions relating to the*

---

[63] The Common Position does not allow for determining absolute quantities of allowances *ex-post* by using actual production per installation in the period. If benchmarking should be based on output during the period 2005 to 2007, it can only be done using expected output (See Non-paper of the EU Commission of 1 April 2003 'How to develop a National Allocation Plan').

[64] Communication from the Commission, dated 7 January 2004, on guidance to assist Member States in the implementation of the criteria listed in Annex III to ETS Directive establishing a scheme for greenhouse gas emission allowance trading within the Community and amending Council Directive 96/61/EC, and on the circumstances under which *force majeure* is demonstrated, COM (2003) 830.

[65] The second guidance document can be found at: http://ec.europa.eu/environment/climat/pdf/nap_2_guidance_en.pdf

[66] Non-paper of the EU Commission of 1 April 2003 'How to develop a National Allocation Plan'.

allocation of allowances and the reports of emissions required under the greenhouse gas emissions permit and held by the competent authority shall be made available to the public by that authority subject to the restrictions laid down in Article 3, 3 and Article 4 of Directive 2003/4/EC').

The Member States also have to bear in mind that the National Allocation Plans will constitute state aid under Article 87 (1) EC and will therefore have to be notified to the Commission for assessment under state aid rules. Competition policy procedural rules will apply in this respect. The European Commission intends to take at the same time the two decisions legally required on the NAP as regards the assessment as required in the common position and the state aid assessment.[67]

Within three months of notification of the NAP by a Member State, the Commission can reject a plan, and ask for changes to be made.[68]

With regard to the first trading period, the European Commission rejected none of the NAPs outright, but certain aspects of the NAPs of Finland, France, Portugal, the Slovak Republik, Germany, Ireland, Austria and United Kingdom have been rejected (see the Communication of 20 October 2004 from the Commission to the Council and to the European Parliament on Commission Decisions of 20 October 2004 concerning national allocation plans for the allocation of greenhouse gas emission allowances of Belgium, Estonia, Finland, France, Latvia, Luxembourg, Portugal, and the Slovak Republic in accordance with ETS Directive[69] and Communication from the Commission to the Council and to the European Parliament on Commission Decisions of 7 July 2004 concerning national allocation plans for the allocation of greenhouse gas emission allowances of Austria, Denmark, Germany, Ireland, the Netherlands, Slovenia, Sweden and the United Kingdom in accordance with ETS Directive).[70]

---

[67] *Ibid*, p. 2.
[68] Article 9, 3 of the ETS Directive.
[69] COM (2004) 681.
[70] COM (2004) 500.

EnBW Energie Baden-Württemberg AG, Karlsruhe (in relation to the NAP of Germany),[71] United Kingdom[72] and the Slovak Republik[73] contested the decision of the European Commission before the European Court of First Instance with regard to the subsequent corrections made in and proposed amendments to the NAPs. Even private parties, such as AITEC[74] lodged an appeal against the decision of the European Commission on the compatibility of the NAP notified to it by Italy and US Steel Kosice lodged an appeal against the decision of the European Commission regarding the first NAP of the Slovak Republik.[75] With regard to the US Steel Kosice case the Court of First Instance declared on 1 October 2007 the request as inadmissible because the contested decision of the European Commission regarding the NAP does not in any event affect the applicant individually and directly.[76]

On 23 November 2005 (T-178/05) the European Court of First Instance ruled its first judgment over the corrections made in the NAP of the UK and it annulled the decision of the European Commission. In the contested decision, the Commission rejected as inadmissible the amendments of the NAP proposed by the U.K. because they would have resulted in the total quantity of allocated quotas exceeding the quantity authorised by the Commission in its Decision of 7 July 2004. Therefore, the Commission did not consider itself obliged to carry out any examination of the proposed amendments on their merits and on their compatibility with the criteria set out in Annex III to or with Article 10 of the ETS Directive. The Court argued that "*the NAPs developed under the ETS Directive must take due account of accurate data and information*

---

[71] T-374/04, OJ C 284, 20 September 2004, p. 25.

[72] T-143/05, OJ C 115, 14 May 2005, p. 39; T-178/05, OJ C22, 5 May 2005, p. 14.

[73] T-489/04, OJ C 82, 20 December 2004, p. 29.

[74] T-371/05, OJ C 296, 26 November 2005, p. 36.

[75] T-489/04.

[76] T-489/04, 1 October 2007.

relating to emission forecasts for the installations and sectors covered by the Directive. If a NAP was based in part on incorrect information or erroneous evaluations relating to the level of emissions in certain sectors or certain installations, the Member State in question would have to be entitled to propose amendments to the NAP, including increases to the total quantity of allowances to be allocated, in order to address those problems before the market began functioning. That notwithstanding, on order to ensure that the environmental objectives of the Directive are respected, the Commission must still assess whether the amendments proposed by the Member State are compatible with the criteria listed in Annex III to or with Article 10 of the ETS Directive".

Following this annulment, the European Commission adopted a new decision concluding that the proposed amendment was inadmissible. This decision was again being challenged by Drax Power and others.[77] On 25 June 2007 (T-130/06) the European Court of First Instance judged that the applicants may not be considered to be directly concerned by the contested decision in the sense of the fourth paragraph of Article 230 EC and, as a result, the action must be rejected as inadmissible without it being necessary to rule on the other grounds of inadmissibility raised by the Commission.

Another appeal was lodged by Cemex UK against the European Commission to annul the Commission decision of 29 November 2006,[78] concerning the national allocation plan for the allocation of greenhouse gas emission allowances notified by the United Kingdom in accordance with ETS Directive; insofar as the latter decision failed to object to/approved an allocation of allowances to the applicant in respect of its Rugby plant which was inadequate and unlawful to the extent of 343 838 tonnes; the latter decision failed to object to/approved an allocation to cement manufacturers in competition with the applicant which was excessive and unlawful to the extent of

---

[77] T-130/06, 15 July 2006.
[78] T-13/07, OJ C 56, 10 March 2007, p. 37.

the 343 838 tonnes comprising as it did the under-allocation to the applicant; the latter decision failed to object to/approved the allocation methodology laid down in paragraphs 3(7) and 3(8) of the UK national allocation plan, and paragraphs 28 and 30 of Appendix C to the UK national allocation plan insofar as the latter methodology treats a cement plant as commencing operations in a year in which the plant was undergoing commissioning, and treats this year as the first year of operation of such a plant, and calculates emission allowances on the basis of average emissions for the baseline period 2000–2003, excluding the lowest year's emissions, regardless of the actual length of the commissioning period of the plant in question. On 6 November 2007 The European Court of First Instance declared the Cemex UK case also inadmissible because of lack of direct and individual concern.[79]

On 16 August 2007 Cemex UK cement lodged another appeal in order to request the annulment of the Commissions' decision of 12 June 2007 rejecting the complaint made by the applicant concerning the NAP of UK.[80] Compared with the other cases, the appeal concerns a complaint and not the NAP as a whole. We have to wait and see whether this time the Court of First Instance will accept the admissibility of the appeal.

On the date of this paper, the European Court of First Instance has not yet pronounced itself over AITEC and the Slovak Republik cases. The case of EnBW Energie Baden-Württemberg AG, Karlsruhe (Germany) was pronounced inadmissible.[81]

Compared with the first trading period, the European Commission was much stricter in its ruling over the NAPs of the second trading period which will commence in 2008. The European Commission wants to avoid over-allocation.

Several Member States and private partners acted against the refusals of the European Commission. Most of the refusals concern either the reduction of allocation for the second trading period, or the

[79] Order T-13/07, 6 November 2007.
[80] T-313/07, OJ C 235, 6 October 2007.
[81] Order T-387/04, 30 April 2007.

non compatibility of the proposed NAP with the Annex III criteria.[82] The European Court of First Instance has already ruled upon the U.S. Steel Kosice case[83] and the Fels-Werke case,[84] and declared it as inadmissible because of lack of direct or individual concern.

(e) Allocation and issue of allowance

• The decision of allocation by the Member States

Once the NAP has been adopted and approved by the European Commission, the Member State has to decide upon the total quantity of allowances it will allocate for that period and the allocation of those allowances to the operator of each installation. For the first trading period, the decision has to be taken at least three months before the beginning of the period; for the second trading period this decision shall be taken at least twelve months before the beginning of the relevant period.[85] The decision to allocate must be based on its NAP, taking due account of comments from the public.

---

[82] T-32/07, OJ C69, 24 March 2007, p. 29 (the Slovak Republik); T-27/07, OJ C69, 24 March 2007, p. 25 (US Steel Kosice regarding the NAP of the Slovak Republik); T-183/07, OJ C155, 7 July 2007, p. 44 (Poland); T-208/07, 5 June 2007 (BOT Elektrownia Belchatow regarding the NAP of Poland); T-203/07, OJ C 170, 21 July 2007, p. 39 (Cemex Polska regarding the NAP of Poland); T-199/07, OJ C 170, 21 July 2007, p. 39 (Cementownia Odra regarding the NAP of Poland); T-198/07, OJ C 170, 21 July 2007, p. 39 (Cementownia Warta regarding the NAP of Poland); T-197/07, OJ C 170, 21 July 2007, p. 38 (Grupa Ozarow regarding the NAP of Poland); T-196/07, OJC 170, 21 July 2007, p. 38 (Dyckerhoff Polska regarding the NAP of Poland); T-195/07, OJ C 170, 21 July 2007, p. 37 (Lafarge Cement regarding the NAP of Poland), T-193/07, OJ C 170, 21 July 2001, p. 36 (Gorazdze Cement regarding the NAP of Poland); T-221/07, OJ C 199, 25 August 2007, p. 41 (Hungaria), T-194/07, OJ C 199, 25 August 2007, p. 38 (the Czech Republic), T-263/07, OJ C 223, 22 September 2007, p. 12 (Estonia), T-369/07, OJ C 269, 10 November 2007, p. 66 (Latvia).

According to the EU press release of August 2007 Lithuania lodged an appeal before the European Court of First Instance on 17 August 2007.

[83] T-27/07, 1 October 2007.

[84] Order, T-28/07, OJ C, 269, 10 November 2007.

[85] Article 11, 1 and 2 of the ETS Directive.

Note that allowances are valid for emissions during the period, first or second, for which they are issued.[86]

It is important to note that, in accordance with Article 11 of the ETS Directive, initial allocation of allowances can only be made to operators of installations covered by the NAP. Hence installations not covered by the NAP cannot be allocated any allowances, although they may purchase and hold allowances as any other person.

Only a proportion of allowances will be issued each year to an operator mentioned in the NAP, the total quantities to be allocated to each operator for the whole period will be known from the outset of the NAP. Further "adjustments" to an operator's holding have to be carried out through buying and selling with other participants in the scheme. Any allocation discussion *during an ongoing trading period*, can only concern the initial allocation for the *next* period. Any possibility to make revisions of the NAP during the trading period would create uncertainty for businesses. If revisions were possible, companies might rely on special pleading rather than the market to acquire additional allowances. Furthermore, companies that made investments to reduce their emissions on the basis of the quantities initially allocated would be disadvantaged by the subsequent issue of extra allowances (by increasing the supply, the price of allowances would be expected to fall). *Ex-post* adjustments are not allowed, except in two cases. The ETS Directive allows for two adjustments following the decision referred to in Article 11(2) of the Directive where such retroactive change does not occur or does not have a detrimental impact on the functioning of EU ETS: firstly, where an installation is closed during trading period, that member States determine that there is no longer an operator to whom allowances will be issued; and secondly, where allocation takes place to new entrants from the reserve, that Member States determine the exact allocation to each new entrant.[87]

---

[86] Article 13 of the ETS Directive.

[87] See e.g. decision of 16 January of European Commission regarding second NAP of Belgium.

Buzzi Unichem,[88] an Italian private association, brought an interesting case before the European Court of First Instance. The appeal concerns the annulment of the Commission Decision of 15 May 2007 concerning the NAP notified by Italy for infringement of the EC Treaty and the principles and rules of law adopted in its application to the extent that the NAP must be altered so as to render no longer permissible rationalization measures which envisage that the operator may maintain part of the allocated allowances in the event of 'closure due to processes of production rationalization' (Article 1(4) and Article 2(4) of the Decision). According to Buzzi Unichem the European Commission failed to explain the reasoning which led it to hold that the critized scheme was incompatible as "ex-post adjustment".

On 7 November 2007 the European Court of First Instance ruled a very interesting case with regard to the German NAP and measures for the ex-post adjustment of the amount of allowances allocated to installations.[89] The Court annulled the Commission's decision. Germany launched its case against the commission in 2004 when the EU executive rejected its plans for ex-post adjustments in the first trading period. Germany wanted to be able to withdraw allowances in cases where a company's production — and therefore emissions — turned out to be much lower than projected. The Commission said operators should be free to reduce their emissions by cutting production levels while keeping emission allowances. But in its ruling the court of justice's Court of First Instance declared that the Commission had "misconstrued" the ETS Directive and that ex-post adjustments did not contradict the essence of the trading scheme: "*The mere fact that the ex-post adjustments at issue are liable to deter operators from reducing their production volume and, therefore, their emission rates is not sufficient to call into question the adjustments' legality in light of the directive's objectives as a whole*".

The decision of allocation should also specify the proportion of allowances that the Member State will issue each year to the operator.

---

[88] T-241/07, OJ C 211, 8 September 2007, p. 38.

[89] Judgment, T-374/04, 7 November 2004.

The Common Position leaves Member States some freedom to allocate the quantities of allowances appropriate to their national circumstances. This reflects also that the burden sharing targets agreed by Member States vary from –28% (Luxembourg) to +27% (Portugal). However, Member States must comply with the *method* of allocation laid down in Article 10, and, as stipulated in Article 11 of the ETS Directive, their allocations "*shall be in conformity with the requirements of the Treaty, in particular Articles 87 and 88 thereof*". This is a very firm affirmation that allocations to individual operators or sectors must not constitute incompatible state aid (i.e. aid which would distort or threaten to distort competition to an extent contrary to the common interest).

At the moment, not a lot of case-law has been ruled yet. On 27 September 2004 EnBW Energie Baden-Württemberg AG, Karlsruhe (Germany) brought an action before the European Court of First Instance against the NAP of Germany.[90] In particular, the applicant complained of a rule in respect of transfers contained in the plan which allocates to a power station operator who decommissions an old installation and replaces it with a new one the quantity of allowances which it had for the decommissioned installation for a period of four years. In the applicant's view, that gave rise to an over-allocation of allowances which amounts to State aid within the meaning of Article 87(1) EC and cannot be justified. Further, in breach of Article 88(2) EC, the defendant failed to initiate the formal State aid procedure, although it must have had considerable doubt about the compatibility of that rule with the EC Treaty. In addition, the contested decision infringes Article 9(3) of Directive 2003/87/EC and criterion 5 of Annex III thereto, as the over-allocation of emission allowances unduly favors competitors of the applicant, which are strengthened because undertakings like the applicant, which have to decommission nuclear power stations in the near future owing to statutory rules, are unjustifiably placed at a disadvantage. By Order of 30 April 2007 the European Court of First Instance declared the application inadmissible without further motivation.

[90] T-387/04, OJ C 6, 8 January 2005, p. 38.

Also the application of AITEC in relation to the contest of the Italian NAP,[91] as already mentioned in this paper, and the application Fels-Werke GmbH in relation to the contest of the German NAP,[92] brought some arguments regarding the non-compliance with state aid rules. As mentioned above, the Court declared the appeal of Fels-werke GmbH inadmissible. The Court has not ruled upon the case of AITEC yet.

Worthwhile mentioning is the action brought by the Netherlands against the European Commission before the European Court of Justice concerning the state aid measure relating to tradable NO_x emission rights.[93] In support of its action, the Netherlands Government submits that there has been an infringement of Article 87 EC. The Netherlands scheme makes no use of State resources within the meaning of Article 87(1) EC. The measures relating to emission rights are financed entirely by private funds. Any profit made by undertakings on the sale of credits does not arise from the allocation of rights by the State but from the undertakings' own earnings made as a result of above-standard performance. The undertakings are not placed at any advantage. There is therefore no question of aid distorting competition or affecting trade. The Court has not ruled yet.

Another critical issue concerns the legal status of an "allowance". "Allowance" means the entitlement to emit a tonne of carbon dioxide or an amount of any other greenhouse gas with an equivalent global warming potential during a specified period. Allowance is distinguished from a GHG permit. However, the interpretation of the legal status of the allowance differs amongst most Member States. Is it a property right, a license to pollute, a financial effect, comparable with an intellectual property right? These questions are extremely important in order to know if allowances are revocable, securities can be taken, risks can be transferred, etc. The legal status of an allowance is also important to

---

[91] T-371/05, OJ C 296, 26 November 2005, p. 36.
[92] T-28/07, OJ C 69, 27 March 2007, p. 26.
[93] C-388/03, 5 September 2003.

know whether allowances fall within scope of Investment Services Directive 93/22/EEC[94] and Markets in Financial Instruments Directive 2004/39/EC.[95]

According to some jurisprudence the exact legal nature of the allowances depends very much on the Legal system under which the allowance is held and may include an authorization or permit to emit, an administrative or public right, a good or a commodity. In most cases the allowance will show features of a number of these categories and have a mixed nature.[96] Clarification on the legal status would certainly improve liquidity and "bankability".

- The allocation method: grandfathering, benchmarking and auctioning

In general, three different methods of allocation are distinguished: grandfathering, benchmarking and auctioning. Other methods of allocation also exist.

According to Article 10 of the ETS Directive, at least 95% of the allowances for the initial three-year period must be allocated to the installations free of charge. For the five-year period beginning 1 January 2008, Member States must allocate 90% of the allowances free of charge.[97] Furthermore, Member State may not auction more than 5% of the total quantity of allowances allocated during the first trading period and 10% in the second trading period.

---

[94] Generally, a company that provides investment services to third parties in relation to specified financial instruments will fall within scope of Investment Services Directive. There is significant divergence of opinion between Member States whether an allowance falls within this scope.

[95] Discussions are held as to whether options, futures and other derivatives contracts relating to allowance trading may fall within the scope of Markets in Financial Instruments Directive.

[96] M. Wemaere & C. Streck , "Legal ownership and nature of Kyoto units and EU allowances", in: D Freestone & C. Streck, *Legal aspects of implementing the Kyoto Protocol Mechanisms*, Oxford Univ. Press, Oxford, 2005, p. 42.

[97] *See also:* Brandt, U.S. and G.T. Svendsen (2004), 'Rent-seeking and Grandfathering: The Case of GHG Trade in the EU', Energy and Environment, 15, 69–80.

Member States will ensure the free circulation of allowances within the European Community. Each year, no later than 30 April, they will also make sure that the operators of the installations surrender the correct quantity of allowances commensurate with the total emissions over the previous year. The surrendered allowances are subsequently cancelled.

Grandfathering and benchmarking are the two methods that provide allowances free of charge. The issuance of allowances in the case of grandfathering is based on historic emissions, while benchmarking is based on a performance-based standard combined with a certain production or input indicator. Auctioning is different, in that operators buy their allowances according to their own estimated needs. In 2006 Ireland tried auctioning of allowances.

In the case of grandfathering and benchmarking, allowances can be allocated directly to individual installations or to sectors first, after which the sectoral cap is distributed over installations in that specific sector. The use of sectoral caps allows differentiation of allocation rules between individual sectors. However, determining sectoral caps adds complexity to the allocation process, because (for example) clear definition of sectors and sector boundaries is very important.

During the first trading period and looking at the NAPs of the second trading period, Member States have used very different approaches to many of the allocation methodology elements. This has led in Phase I to differences in the ambition levels reflected in the overall national caps, subsequently translated into installation-level allocations.[98] It has affected the overall environmental effectiveness of the scheme as a whole (the amount of emissions reductions achieved) and resulted in (perceived) differences in treatment of similar installations in different Member States. Harmonization of allocation methodologies across the EU post-2012 should prevent competition distortion effects between Member States and strengthen incentives for investments in clean technology.[99]

---

[98] Report of ECCP, October 2006, *Harmonisation of allocation methodologies*, p. 11.

[99] *Ibid.*

A lot of criticism has been spread on the use of grandfathering because of the allocation free of charge. Bohringer and Lange[100] summarized the issue of grandfathering as follows:

> *"(...) We have shown that it is generally impossible to retain efficiency when requiring free allocation of allowances and non-discrimination of firms across countries. Thus, the stated objectives of the European Directive — efficiency, harmonization ("rule neutrality" and "competition neutrality"), and free allocation of allowances — are incompatible. Along the policy-relevant example of lump-sum grandfathering, we illustrated the huge differences in (endogenous) assignment factors which give rise to major concerns regarding competitive distortions from a legal point of view. In our simple partial model the associated differing transfers do not affect firms' emission choice and overall economic efficiency. However, in a more complex setting that incorporates income effects, economic distortions may also be non-negligible.*
>
> *An efficient way out of the current regulation dilemma posed by the multiple objectives of the EU Directive could be to relax the requirement of free allocation. Then, the minimal fraction of allowances which must be allocated is given by the minimum of assignment factors across EU Member States. As a means to heading towards a more harmonized allocation of allowances, a continuous increase of the auctioned ratio of tradable allowances may prove a realistic policy option."*

- Force Majeure

During the first trading period, Member States could apply to the European Commission for certain installations to be issued with additional allowances in cases of force majeure.[101] Article 29 of the ETS

[100] Bohringer, C. and Lande, A., "Mission Impossible!? On the Harmonization of National Allocation Plans under the EU Emissions Trading Directive", Centre for European Economic Research (ZEW), Discussion Paper, No. 04–15 2004, p. 9 (to be found at: ftp://ftp.zew.de/pub/zew-docs/dp/dp0415.pdf).

[101] Article 29 of the ETS Directive.

Directive derogates from the general principle of the EU ETS, under which Member States allocate allowances before the beginning of the relevant trading period. Applications for force majeure allowances may therefore cause uncertainty in the allowance market, and, if granted, give an advantage to certain companies, which affects trade between Member States. Article 29 is therefore explicitly without prejudice to the Treaty, and the Commission will carefully consider the justification and potential effects of any application for such allowances.

In January 2004, the Commission issued guidance on the implementation of the Annex III criteria pursuant to Article 9(1) as well as on the circumstances under which force majeure is demonstrated.[102] The European Commission considers force majeure to be exceptional and unforeseeable circumstances, which cause a substantial increase in annual direct emissions of greenhouse gases covered by ETS Directive at an installation, which could not have been avoided even if all due care had been exercised. The circumstance must have been beyond the control of the operator of the installation concerned and of the Member State submitting an application to the Commission under Article 29 of the Directive with respect to the installation of that operator (e.g. natural disasters, war, threats of war, terrorist acts, revolution, riot, sabotage or acts of vandalism).

The presence of force majeure has to be demonstrated at installation level and on a case-by-case basis. An application under Article 29 of the ETS Directive should include, in respect of each installation, the Member State's best estimate of the increase in emissions resulting from the circumstance for which force majeure is pleaded and a substantiation of that estimate. A Member State should submit an application under Article 29 to the European Commission by 31 January the year following the year of the trading period during

[102] Communication from the Commission on guidance to assist Member States in the implementation of the criteria listed in Annex III to Directive 2003/87/EC establishing a scheme for greenhouse gas emission allowance trading within the Community and amending Council Directive 96/61/EC, and on the circumstances under which force majeure is demonstrated — COM/2003/0830 final.

which the circumstance occurred for which force majeure is pleaded. Based on this information, the European Commission shall decide whether force majeure is demonstrated.

(f) New entrants

According to Article 3(h) of the ETS Directive, a "new entrant" means any installation carrying out one or more of the activities indicated in Annex I, which has obtained a greenhouse gas emissions permit or an update of its greenhouse gas emissions permit because of a change in the nature or functioning or an extension of the installation, subsequent to the notification to the Commission of the NAP.

The definition of new entrants, in this Article puts new installations on equal footing with existing installations extending capacity. The definition in relation to an updated permit applies only to the extension of an installation, and not to the entire installation, nor to the increased capacity utilization at an existing installation.

It is required under Annex III that a NAP contains information on the manner in which new entrants will be able to begin participating in the emissions trading scheme. The way in which this is done is a matter for each Member State to decide. It is important to keep in mind that the new entrants issue is of a transitory character. What is a new entrant in one trading period will be an existing installation in the subsequent period and be treated in the same manner as all other existing installations.

In deciding how to treat new entrants a Member State needs to consider that the Treaty's provisions on the right of establishment in the internal market have to be respected. It is crucial that new entrants have access to allowances, as without this they could be prevented from establishing themselves in business and competing in the underlying product market of the sectors covered by the ETS Directive. Guaranteeing this is the essence of the second paragraph in Article 11(3) of the ETS Directive.

A broad and liquid EU-wide market in allowances should in itself provide access to new entrants. Moreover, competition law at national and European level would be applicable in the unlikely event that uncompetitive market practices with respect to allowances should be used to erect market entry barriers.

Member States may choose to let new entrants buy allowances on the market, as is done under many existing emissions trading schemes. Operators of existing installations may have made investments before they were aware of the need for them to contribute to greenhouse gas emission reductions. In contrast, operators of new installations will make investment decisions with certainty of the need for them to take their climate change impact into account. If a Member State chooses to let new entrants buy allowances on the market, this obligation should not be imposed during a disproportionate period. Moreover, if a Member State chooses this option, there is no need to reduce the total allocated quantities to provide for new entrants.

Alternatively, a Member State could choose to build a reserve of allowances for new entrants. In this case, the basis according to which new entrants will be allocated, and the total number that could potentially be allocated, should be specified clearly and objectively in the NAP. Up to the amount of allowances in the reserve, new entrants would be given a free allocation. If there are fewer new entrants than expected, the Member States needs to determine what it will do at the end of the period with the unused allowances. In case a Member State should decide to build a reserve, it needs to modify the allocations to individual installations defined in the previous step accordingly.

According to the report of McKinsey and Ecofys of July 2006 with regard to the review of EU ETS all Member States have opted for setting aside a number of allowances in a so-called New Entrants Reserve (NER). In all cases in the first phase, the allowances are provided to new entrants for free for at least a portion of the allowances — some countries will also be having auctions which will be open to new entrants and others.[103]

(g) Monitoring, reporting and verification

According to Article 14 of the ETS Directive the European Commission established with its Decision 2004/156/EC of

---

[103] http://ec.europa.eu/environment/climat/pdf/ecofys_new_entrant.pdf

29 January 2004 guidelines for the monitoring and reporting of greenhouse gas emissions pursuant to ETS Directive (henceforth: "Guidelines").[104] The 11 Annexes to this Decision contain guidelines for monitoring and reporting greenhouse gas emissions. Annex I contains general guidelines. Additional guidelines for specific activities are contained in Annexes II to XI. The guidelines are intended to ensure regular and accurate monitoring and reporting of greenhouse gas emissions in the Community and are legally binding.

Operators of installations under the EU ETS must hold a GHG emissions permit to operate from 1 January 2005. Member States must ensure that the GHG emissions permit sets out the monitoring methodology in accordance with the Guidelines. Member State are not allowed to give derogations to the application of the Guidelines unless explicitly provided for in the Guidelines.

The monitoring methodology (often also referred to as "monitoring plan" or "monitoring protocol") specifies how an operator of an installation will carry out the monitoring and reporting of CO_2-emissions for that specific installation. The approval of the monitoring methodology is thus part of the permit granting process. After approval by the competent authority, it becomes part of the permit (Article 6(2)(c) of the ETS Directive). The detailed contents of an installation's "monitoring methodology" are specified in Section 4.2 of Annex I of the Guidelines. Once approved, the installation has to implement and execute the monitoring of its CO_2 emissions in accordance with the "monitoring methodology". It provides the main reference for the verifier to assess whether an operator is meeting his monitoring obligations.

If the emissions of an installation are not monitored in full compliance with this monitoring methodology, the operator of that installation is exposed to compliance sanctions determined pursuant to Article 16(1) of the EU ETS Directive. In addition, the verifier must reject the installation's emissions report if he considers deviations to be material. The non or late submission of the emissions report in turn leads to the blocking of the transfer of allowances from

---

[104] OJ L 59 of 26 February 2004.

that installation and any further sanctions determined pursuant to Article 16(1) of the ETS Directive.[105]

Member States approach the monitoring process differently. Most Member States uses the installation-specific "monitoring methodology" between the competent authority and operator as part of the permit granting procedure for each installation. Other Member States use "general binding rules", setting out the monitoring and reporting obligations a company needs to apply, to define part of the monitoring methodology. Such "general binding rules" must however be anchored in the permit and the combination of the permit and the general binding rules must constitute the monitoring methodology containing the elements specified in Sec. 4.2 of Annex I of the Guidelines.[106]

Once the reports have been submitted by the operators, the Member States are obliged by Article 15 of the ETS Directive to verify these reports with the criteria set out in Annex V, and ensure that the competent authority is informed thereof, by 31 March each year.[107] In case the report has not been verified as satisfactory by 31 March each year for emissions during the preceding year, the operator cannot make further transfers of allowances until a report from that operator has been verified as satisfactory.[108]

(h) Banking and Borrowing

Banking allows participants with emissions below their allocated limits to save surplus allowances for use during a later compliance period. Borrowing is the opposite of this, permitting use of allowances from a future period (with the implicit commitment that repayment will be made in the form of equivalent reductions in a future period).[109]

---

[105] *See* Questions and Answers regarding monitoring: http://ec.europa.eu/environment/climat/emission/pdf/monitoring_report_faq.pdf

[106] *Ibid.*

[107] Article 15, par. 1 of the ETS Directive.

[108] Article 15, par. 2 of the ETS Directive.

[109] Perkhaus and Baumert, "Greenhouse gas emission trading", (2001), p. 22 (to be found at: http://wwwlib.murdoch.edu.au/adt/pubfiles/adt-MU20060615.132356/05CHAPTER4.pdf)

ETS Directive allows banking and borrowing within the same trading period, but not from one trading period to another.

Explanation thereof can be found in Article 12(3) of the ETS Directive which states that Member States shall ensure that, by 30 April each year at the latest, the operator of each installation surrenders a number of allowances equal to the total emissions from that installation during the preceding calendar years as verified, and that these are subsequently cancelled. Allowances are only valid during the trading period for which they are issued.[110] If allowances which are no longer valid and not have been surrended and cancelled four months after the beginning of the first or second trading period, the allowances shall be cancelled by the competent authority. For the first trading period, Member States may, and for the second trading period, Member States, shall issue allowances to persons for the current period to replace any allowances held by them which are cancelled.[111]

According to Mr. Ellerman of the Massachusetts Institute of Technology, the carry-over of unused allowances from the 2005–2007 period into the 2008–2012 period is prohibited and this provision has created a significant price disparity between 2007 and 2008 allowances that will lead to strange abatement behavior around the turn of the year as firms and traders arbitrage this price difference. Aside from this restriction, the EU ETS does allow complete intra-period banking and borrowing. This has resulted in very stable price relationships among years in each trading period and efficient abatement within these periods.[112]

However, it must be stressed that intra-period banking and borrowing may cause problems, e.g. in case of depreciation of carbon pricing of allowances banked from a previous trading period to the next one.

---

[110] Article 13 of the ETS Directive.

[111] Article, 13, 2 and 3 of the ETS Directive.

[112] Full Committee Roundtable Discussion on Carbon Dioxide Trading in the European Union, 26 March 2007, p. 16.

(i) Closures, mergers and acquisitions

Article 12 of the ETS Directive prescribes the possibilities of transfer, surrender and cancellation of allowances, but does not foresee specific rules with regard to closures, mergers and acquisitions of GHG installations.

The basic questions relating to closure are how to define ownership of allowances received to cover emissions after closure, and how to define a closure for such purposes. For instance, according to the Belgian Constitutional Court no. 92/2006, 7 June 2006 (regarding Wallonia) a GHG operator can only claim ownership over the allowances which have been issued to him in a specific trading year, but cannot claim ownership over the allowances for the following trading years (in one trading period) which have not been issued yet, nonetheless the allowances were allocated to him in the NAP for that specific trading period. If different Member States rule differently over e.g. ownership of allowances, difficulties in the application of EU ETS arise.

Transfer rules are a special type of closure rules that have been defined by several Member States in the first phase of the EU ETS. Transfer rules enable a closing plant to retain allowances where production is being transferred to a new installation or an extension of an installation that carries out the same process (within one Member State).

It is particularly important for NAPs to deal with closure as the overall cap, including the new entrant provisions, has to be set out in the NAP. If there is no provision for returning allowances to a new entrants reserve on closure, there could be a shortfall in allowances available to the overall pool of participants where an operator ceases to exist and therefore cannot sell the allowances intended for the years after the year of closure. Where an operator of a closing installation does sell the allowances, they would remain available to other participants, but at a cost, while the closed installation makes a profit. This profit could be interpreted as a windfall profit that is the direct result of a government grant received on the basis of expected emissions (i.e. based on need) yet used for a different purpose entirely, unless closure is considered a legitimate abatement option instead.

According to the report of McKinsey and Ecofys with regard to the review of EU ETS almost all Member States have also included statements about closure in their NAPs and relevant legislation, including definitions of closure that include partial and temporary closure in some cases. In most cases an installation is considered closed if it ceases operation completely. However, in some cases a threshold of emissions in the reference period is used as a proxy for closure. However, results of the report shows how different Member States have approached new entrant, closure and transfer rules differently. This range of approaches has two main implications. Firstly, there is a distortion of the market for operators or investors in different EU countries. Secondly, Member States are given the ability to promote the use or investment in certain clean technologies, in certain sectors as relevant to their national circumstances. Harmonization of approaches between Member States may solve some of the competition issues, but it may negate some of the clean technology benefits. These results do, of course, depend on the way in which the harmonized rules will be formulated in the reviewed EU ETS.

(j) Carbon trading

The legal framework of ETS Directive does not explicitly regulate how and where the market in allowances takes place. An installation that emits more CO_2 than it has allowances for would need to buy additional allowances in the market, while one that emits less has the possibility to sell its surplus allowances (AAUs) to persons within the European Union and to persons in third countries where such allowances are recognized.[113] Thus companies with commitments may trade allowances directly with each other, or they may buy or sell via a broker, bank or other allowance market intermediary (e.g. MCCF, NERCOF, European Climate Exchange).

To purchase or sell allowances, an Emission Reduction Purchase Agreement ("ERPA") is used. Three European standard ERPA agreements are made available (EFET Allowance Appendix, IETA

[113] Article 12 of the ETS Directive.

Emission Trading Master Agreement for the EU Scheme and ISDA Master Agreements); however these standard agreements slightly differ from each other (e.g. differences in events of default or termination events; differences in calculation of termination payments; differences in definition of force majeure, settlement disputes, penalties, confidentiality, etc.).

In principle, companies which carry out emission reduction projects outside the EU through JI or CDM can convert the credits they earn from those projects into allowances that can be used for compliance under the EU ETS. The Linking Directive[114] will therefore further lower the cost to EU industry by offering more options for complying with the requirements of the ETS. CDM is operational since 2004, so European companies running CDM projects have already used the earned credits (CERs) for compliance during the first trading period. ERUs of JI projects will kick off in 2008 with the beginning of the first commitment period under the Kyoto Protocol, which also marks the beginning of the second trading period under the EU ETS (2008–2012).

Derivative markets play an important role in the EU ETS. Derivatives involve the trading of obligations (futures) and rights (options) based on an underlying product, without necessarily directly transferring that underlying product. The most familiar derivative instruments are exchange-traded futures and options based on an underlying product. Approximately 95% of the total volume in the European carbon market are seen in derivative trades (forwards, futures and options) with the remaining in spot trades. This can partly be explained by the initial delay of national registries and final allocations in many of the Member States which prevented the execution of instant delivery for spot contracts. Another reason may be that in such a new and volatile market, derivative instruments are crucial tools to optimize the value of your emissions portfolio.[115]

Every Member State has an electronic registry system that keeps track of the ownership of emission allowances as they change hands in

[114] See more hereinafter.

[115] http://www.europeanclimateexchange.com/default_flash.asp

the market. This registry system is separate from trading activity — not all trades result in changes in ownership of allowances, but where a trade culminates in a change in ownership, there is a transfer of allowances between accounts in the registry system.

In this way, the registry system is similar to a banking system which keeps track of the ownership of money in accounts, but does not track the deals made in the goods and services markets, which are the cause of the money changing hands. So the registry system is not a marketplace; the way in which allowances are traded is a decision made by the participants in the market. Each company with a commitment and any person interested in buying or selling allowances needs to open an account.

The system consists of a national component in each Member State where the allowances are held and of a hub at European level, which will conduct automated checks on each transfer of allowances to ensure that the rules of the Directive are respected. Some of the data held in the registry will be released periodically.

(k) Carbon price

The carbon price is the price of an allowance to emit one metric ton of carbon dioxide — the only greenhouse gas covered by the EU ETS. Point Carbon, a carbon trade provider, provides daily carbon price forecasts and analysis of greenhouse gases.[116]

According to the European Climate Exchange there are three key factors that should be considered in assessing market prices in the EU ETS: policy and regulatory issues, market fundamentals (e.g. estimating and forecasting CO2 production, impact of weather, emission-to-cap, and role of fuel prices) and technical indicators.[117]

The EU ETS puts a price on carbon dioxide (CO2) emissions in a twofold approach to reducing emissions. In the short term, the aim is to encourage heavy emitting industries to take measures to lower emissions from their current technology. The longer term aim is to

---

[116] www.pointcarbon.com

[117] www.europeanclimateexchange.com

spur investment in new low- or zero-emission technologies. But, according to a report in the Financial Times, the European Commission's energy chief Andris Piebalgs says that at current levels carbon prices are not enough incentive for significant clean-tech investment and will need to rise. He forecasts the price of a carbon permit in the third phase of the EU ETS from 2012 will need to be €20 to €30. Each permit, an EUA, is a licence to emit one tonne of CO_2. Permits for phase two of the EU scheme have traded in the range of €12 to €20 over the past six months with the December 2012 forward contract currently trading at €16.45 on the European Climate Exchange. The third phase of the EU ETS is not yet agreed in the EU and depends somewhat on the emergence of a successor treaty to the Kyoto Protocol which expires in 2012.

(l) Tax

According to a guidance of the European Commission on the application of VAT to emission allowances, "the transfer of greenhouse gas emission allowances as described in Article 12 of Directive 2003/87/EC of the European Parliament and of the Council of 13 October 2003, when made for consideration by a taxable person, is a taxable supply of services falling within the scope of Article 9(2)(e) of Directive 77/388/EEC. None of the exemptions provided for in Article 13 of Directive 77/388/EEC can be applied to these transfers of allowances".[118]

According to a report of PwC[119] several inconsistencies and uncertainties remain regarding the tax treatment of emission trading in the different Member States. Among its findings, the report reveals that:

- Few Member States have issued binding guidelines for the tax treatment of carbon dioxide emissions trading, which has led to significant differences in their tax and accounting treatment.

---

118 http://ec.europa.eu/environment/climat/pdf/vat_guidelines.pdf

119 http://www.pwc.com/cz/eng/ins-sol/publ/Alerts/PwC_2006-06-26_TLA12_en.pdf

- The tax deductibility of penalties varies between the Member States.
- The rules regarding the VAT treatment of emission trading between companies in different Member States remain unclear, despite an earlier agreement of the EU VAT committee.
- Several issues are unclear regarding Joint Implementation (JI) and Clean Development Mechanism (CDM) projects, which are used to reduce greenhouse gases.
- The tax and accounting treatment of emissions rights varies between the Member States.

While the lack of uniformity may create certain difficulties for companies engaged in emissions trading, it can also create opportunities, particularly for companies operating in multiple jurisdictions.

(m) Compliance and enforcement

Installations must meet their targets by reducing emissions or by buying allowances which can be surrendered against their target. Any operator failing to surrender, by 30 April at the latest, the quantity of allowances commensurate with the emissions from his installation during the previous year will be required to pay an excess emissions penalty. The penalty will be EUR 100 for each tonne of carbon dioxide equivalent (EUR 40 during the first three-year period starting on 1 January 2005) and will not release the operator from the obligation to surrender an amount of allowances equal to the excess emissions. Each Member State was to determine its own sanctions regime covering infringements of this proposal and notify the Commission accordingly by 31 December 2003 at the latest.

DG Environment of the European Commission gave its interpretation on the use of next phase allowances under Article 16, 4 of the ETS Directive: "*In DG Environment's view, the wording of the second sentence of Article 16(4) ETS Directive ("Payment of the excess emission penalty shall not release the operator from the obligation to surrender an amount of allowances equal to those excess emissions when surrendering allowances in relation to the following calendar year.") does not specify the type of allowances (phase 1 or 2) to be used for making up under-surrendered emissions in addition to the penalty. Accordingly, it is legally possible to use phase 2 allowances to make up for under-surrendering after 1 May 2008 with respect to phase 1 emissions. There are no provisions in the Registry Regulation that could support a different view. This means in other words that if an operator is out of compliance on 1 May 2008 for the years 2005, 2006 or 2007 its entry in the compliance status table in the national registry for 2007 is negative and so its starting point in 2008 is also negative. It goes without saying that the operator will have to pay the excess emissions penalty for being in this situation. However, according to Article 16(4) ETD operators are permitted to surrender phase 2 allowances against the year 2008 in order to bring it back to 0, i.e. back to compliance, for 2008.*"[120]

1.2.3. *The EU Regulation of 21 December 2004: Community Independent Transaction Log (CITL)*

The Community Independent Transaction Log (CITL) records the issuance, transfer, cancellation, retirement and banking of allowances that take place in the registry. CITL is regulated by Commission Regulation (EC) No 2216/2004 of 21 December 2004 for a standardized and secured system of registries pursuant to Directive 2003/87/EC of the European Parliament and of the Council and Decision No 280/2004/EC of the European Parliament and of the Council,[121] amended by Commission Regulation (EC) No 916/2007 of 31 July 2007.[122]

The task of CITL is to connect the Member State registries and to maintain an independent record of the issuance, transfer and cancellation of the allowances. It is mandatory for each Member State to have a national registry. Each of the 12,000 GHG installations which hold a GHG permit will need to have an "operator holding account" in its national registry, into which its own allowances will

---

[120] http://ec.europa.eu/environment/climat/pdf/use_next_phase.pdf

[121] OJ L 386, 29 December 2004.

[122] OJ L 200, 1 Augustus 2007.

be issued. Any individual or organization wishing to participate in the market will be able to open up their own "person holding account" in any of the registries. The Community transaction log's web pages (http://ec.europa.eu/environment/ets/) shows when a national registry has gone live, when accounts have been opened, contact details for these accounts and when allowances have been allocated in accordance with the final "national allocation plan". Records of other types of transactions will be made available after a period of five years has passed. On 15 May of each year, these web pages also show the compliance position of each installation across the EU-27. CITL will ensure the accurate accounting of all units under the Kyoto Protocol plus the accurate accounting of allowances under EU ETS.

CITL will be linked to ITL. However, the go-live of the CITL with the ITL, which was planned for Mid-November 2007, has been postponed. At the moment Member States are technically testing and completing the initialization of their registry connections with ITL. The EU registries will join the ITL simultaneously together with the CITL. The linking of all the registry systems is to be expected beginning of 2008.

1.2.4. *The EU Linking Directive 2004/101/EC*

As a consequence of article 30(3) of Directive 2003/87/EC,[123] a first modification of Directive 2003/87/EC has already taken place

[123] "Linking the project-based mechanisms, including Joint Implementation (JI) and the Clean Development Mechanism (CDM), with the Community scheme, is desirable and important to achieve the goals of both reducing global greenhouse gas emissions and increasing the cost-effective functioning of the Community scheme. Therefore, the emission credits from the project-based mechanisms will be recognized for their use in this scheme subject to provisions adopted by the European Parliament and the Council on a proposal from the Commission, which should apply in parallel with the Community scheme in 2005. The use of the mechanisms shall be supplemental to domestic action, in accordance with the relevant provisions of the Kyoto Protocol and Marrakesh Accords."

in 2004, by Directive 2004/101/EC, the so-called "Linking Directive".[124]

Directive 2004/101/EC reinforces the link between the Union's emission allowance trading scheme and the Kyoto Protocol, by making the latter's "project-based" mechanisms (Joint Implementation and the Clean Development Mechanism) compatible with the scheme. This enables operators to use these two mechanisms in the allowance trading scheme to fulfil their obligations. The result will be lower compliance costs for installations in the scheme. It is estimated that annual compliance costs in the period 2008–2012 for all covered installations in the enlarged Union will be reduced by more than 20%.

Directive 2004/101/EC thus recognises ERUs (JI credits) and CERs (CDM credits) as equivalent to EU emission allowances (1 EUA = 1 CER = 1 ERU), except for those from land use, land use change, forestry and nuclear activities.

The degree of use must be supplemental to reductions achieved through domestic policy action, and needs to be fixed by each Member State in its NAP by specifying the maximum amount of such credits. Member States are free to choose whether to apply the limit individually in respect of each installation, or collectively to all installations. For greater flexibility, the Commission is recommending that Member States apply the limit for the entire trading period and collectively to all installations.

The limit on the use of credits does not imply that a company cannot generate and sell more of them. In fact, even if the limit should be reached in a Member State, the credits could be used by other companies in other Member States. And even if the limit were reached in all Member States, a company could sell credits to governments (both in Europe and beyond) or to other companies (both in Europe and beyond), or could keep them for the next ETS trading period.

---

[124] Directive 2004/101/EC of the European Parliament and of the Council of 27 October 2004 amending Directive 2003/87/EC establishing a scheme for greenhouse gas emission allowance trading within the Community, in respect of the Kyoto Protocol's project mechanisms, OJ, L 338, 13 November 2004.

Directive 2004/101/EC also takes steps to prevent ERUs and CERs being counted twice where they result from activities which also lead to a reduction in, or limitation of, emissions from installations covered by Directive 2003/87/EC.[125]

2. Review of the European Union GHG Emission Trading Scheme

2.1. *General*

After the first modification of Directive 2003/87/EC by Directive 2004/101/EC, several other modifications are presently coming up. On the one hand, there is the specific proposal to include aviation activities in the EU ETS, and on the other, there is the announced general proposal for a global revision of the EU ETS, as a consequence of article 30, 2 of Directive 2003/87/EC.

Both proposals also have to be seen in the light of the Presidency Conclusions of the EU Spring Council of 8/9 March 2007, which decided to go for an integrated climate and energy policy for the EU.

2.2. *EU Spring Council 8/9 March 2007*

At the Spring Council Meeting of 8/9 March 2007, the leaders leaders made a plea for an integrated climate and energy policy along the following lines[126,127]:

> "*27. The challenges of climate change need to be tackled effectively and urgently. Recent studies on this subject have contributed to a growing awareness and knowledge of the long-term consequences, including the*

[125] Commission Decision 2006/780/EC of 16 November 2006 on avoiding double counting of greenhouse gas emission reductions under the Community emissions trading scheme for project activities under the Kyoto Protocol pursuant to Directive 2003/87/EC of the European Parliament and of the Council (OJ, L 316, 16 November 2006).

[126] Italics and bold added.

[127] Council of the European Union, Brussels, 2 May 2007, 7224/1/07 REV 1.

consequences for global economic development, and have stressed the need for decisive and immediate action. The European Council underlines the vital importance of achieving ***the strategic objective of limiting the global average temperature increase to not more than 2°C above pre-industrial levels****.*

28. Given that energy production and use are the main sources for greenhouse gas emissions, ***an integrated approach to climate and energy policy is needed to realise this objective****. Integration should be achieved in a mutually supportive way. With this in mind, the Energy Policy for Europe (EPE) will pursue the following three objectives, fully respecting Member States' choice of energy mix and sovereignty over primary energy sources and underpinned by a spirit of solidarity amongst Member States:*

- *increasing security of supply;*
- *ensuring the competitiveness of European economies and the availability of affordable energy;*
- *promoting environmental sustainability and combating climate change."*

Regarding the problem of climate change, the following considerations were formulated and conclusions taken [128]:

"29. The European Council underlines the leading role of the EU in international climate protection. It stresses that international collective action will be critical in driving an effective, efficient and equitable response on the scale required to face climate change challenges. To this end negotiations on a global and comprehensive post-2012 agreement, which should build upon and broaden the Kyoto Protocol architecture and provide a fair and flexible framework for the widest possible participation, need to be launched at the UN international climate conference beginning at the end of 2007 and completed by 2009. In this connection, the European Council endorses the elements identified by the Council (Environment) of 20 February 2007

---

[128] Italics and bold added.

as essential parts of an effective and appropriate framework beyond 2012, which would include inter alia the development of a shared vision to reach the ultimate objective of the UN Framework Convention on Climate Change, the strengthening and extension of global carbon markets, the development, deployment and transfer of the necessary technology to reduce emissions, appropriate adaptation measures to deal with the effects of climate change, action on deforestation and addressing emissions from international aviation and maritime transportation. All countries should be invited to contribute to the efforts under this framework according to their differentiated responsibilities and respective capabilities.

30. The European Council reaffirms that absolute emission reduction commitments are the backbone of a global carbon market. ***Developed countries should continue to take the lead by committing to collectively reducing their emissions of greenhouse gases in the order of 30% by 2020 compared to 1990. They should do so also with a view to collectively reducing their emissions by 60 % to 80 % by 2050 compared to 1990.***

31. In this context, ***the European Council endorses an EU objective of a 30% reduction in greenhouse gas emissions by 2020 compared to 1990 as its contribution to a global and comprehensive agreement for the period beyond 2012, provided that*** *other developed countries commit themselves to comparable emission reductions and economically more advanced developing countries to contributing adequately according to their responsibilities and respective capabilities. It invites these countries to come forward with proposals for their contributions to the post-2012 agreement.*

32. ***The European Council emphasises that the EU is committed to transforming Europe into a highly energy-efficient and low greenhouse-gas-emitting economy and decides that, until a global and comprehensive post-2012 agreement is concluded, and without prejudice to its position in international negotiations, the EU makes a firm independent commitment to achieve at least a 20% reduction of greenhouse gas emissions by 2020 compared to 1990.***

33. The European Council decides that a differentiated approach to the contributions of the Member States is needed reflecting fairness and transparency as well as taking into account national circumstances and the relevant base years for the first commitment period of the Kyoto Protocol. It recognises that the implementation of these targets will be based on Community policies and on an agreed internal burden-sharing and invites the Commission, in close cooperation with the Member States, immediately to start a technical analysis of criteria, including socio-economic parameters and other relevant and comparable parameters, to form the basis for further in-depth discussion. Given the great importance of the energy intensive sector, the European Council emphasises that cost-efficient measures are needed to improve both the competitiveness and the environmental impact of such European industries.

34. The European Council notes the increasing share of greenhouse gas emissions from developing countries and the need for these countries to address the increase in these emissions by reducing the emission intensity of their economic development, in line with the general principle of common but differentiated responsibilities and respective capabilities. The European Council stands ready to continue and further strengthen its support for developing countries in lessening their vulnerability and adapting to climate change.

35. ***Given the central role of emission trading in the EU's long-term strategy for reducing greenhouse gas emissions, the European Council invites the Commission to review the EU Emissions Trading Scheme in good time with a view to increasing transparency and strengthening and broadening the scope of the scheme and to consider, as part of the EU ETS review, a possible extension of its scope to land use, land-use change and forestry and surface transport.*** *The European Council stresses the necessity of an efficient, safe and sustainable European transport policy. In this context, it is important to proceed with actions to increase the environmental performance of the European transport system. The European Council notes the European Commission's ongoing work regarding the assessment of external costs for transport and their internalisation."*

Obviously, these far reaching and important conclusions of the Council regarding climate change and emission reductions, will make it necessary that a new "Burden Sharing Agreement (BSA)"[129] is drafted, this time between the 27 Member States of the EU (and not as originally the EU 15), to realize a fair and transparent distribution of that figure of −20% over the 27 Member States. In this exercise, national circumstances obviously will have to be taken into account. Presently, technical analysis of criteria and parameters in view of further in-depth discussion is going on at Member State and Commission level, and a proposal is expected to be on the table, when the UNFCCC parties meet in Bali in early December 2007.

This whole exercise for a new BSA obviously will complicate the revisions of the EU ETS which will have to take place, also later this year. A "linkage" by the Member States of the BSA and the revisions is very likely to happen.

And talks obviously will be even further complicated and linked, by the other, energy-related, decisions the Council has taken at its Spring meeting[130]:

- increase of energy efficiency with 20% by 2020;
- increase of the use of renewables with 20% by 2020;
- increase of the use of biofuels with 10% by 2020;
- truly liberalized gas and electricity markets.

Also these decisions will have their impact on emissions of greenhouse gases and will influence the obligations of the Member States in the to be adopted BSA.

[129] Replacing the one in Annex II of Council Decision 2002/358/EC of 25 April 2002 (concerning the approval, on behalf of the European Community, of the Kyoto Protocol to the UNFCCC and the joint fulfillment of commitments thereunder (OJ, L 130, 15 May 2002).

[130] All embedded in the "Energy Action Plan 2007–2009" (basis for the "Energy Policy for Europe"), as adopted by the Council, and based on the "Energy Package" as presented by the Commission on 10 January 2007.

2.3. *The aviation proposal*

On 20 December 2006, the Commission presented a proposal for a Directive aiming at amending Directive 2003/87/EC so as to include aviation activities in the EU Emissions Trading Scheme (EU ETS).[131]

2.3.1. *Context of the proposal*

(a) Grounds for and objectives of the proposal

Air transport has become an integral part of society in the 21st century, enabling both passengers and freight to travel long distances at an unprecedented speed and contributing to European and global integration. Unfortunately, aviation also contributes to climate change. In 2004 greenhouse gas emissions from the Community's share of international aviation increased by a further 7.5% compared with 2003, resulting in a cumulative growth of 87% since 1990. If this continues, there is a risk that growth in the Community's share of international aviation emissions could by 2012 offset more than a quarter of the environmental benefits of the reductions required by the Community's target under the Kyoto Protocol. Since international aviation is not yet covered by the Kyoto Protocol, this growth currently has no legal implications. Nonetheless, this does not lessen its environmental implications which will have to be addressed as part of any effective forward-looking climate policy.

The objective of this proposal is to address the growing climate change impact attributable to aviation by including aviation in the Community emissions trading scheme (the "Community scheme").

(b) General context

On 27 September 2005 the Commission adopted a Communication on Reducing the Climate Change Impact of Aviation. The key conclusion drawn in the Communication was that, in view of the likely

[131] COM(2006)818 final.

future growth in air traffic, further policies and measures are needed to address the climate impact of aviation. Having analysed a number of options, the Commission decided to pursue a new marketbased instrument at Community level in preference to other financial measures such as taxes and charges and considered that *"...the best way forward from an economic and environmental point of view, lies in including the climate impact of the aviation sector in the [Community] scheme"*. On the basis of this conclusion, the Commission announced that it intended to present a legislative proposal to this effect and invited the other Community institutions to consider the policy and design recommendations made in the Communication. The present proposal aims at implementing this key pillar of the strategy without affecting its other means of addressing climate change through a comprehensive approach based on improved technology and utilization of aircraft (including improvements in air traffic management, research, etc).

On 2 December 2005 the Council of Environment Ministers adopted conclusions recognising that the inclusion of the aviation sector in the Community scheme seems to be the best way forward and urged the Commission to bring forward a legislative proposal by the end of 2006. The European Council confirmed the key conclusions reached by the Environment Council. On 21 April 2006 the European Economic and Social Committee adopted an opinion on the Communication expressing its view that inclusion of aviation in the Community scheme could be a very feasible option. On 4 July 2006 the European Parliament adopted a Resolution welcoming the Commission's Communication and recognising that emissions trading has the potential to play a role as part of a comprehensive package of measures to address the climate impact of aviation, provided it is appropriately designed.

The ultimate objective of the United Nations Framework Convention on Climate Change is to stabilise greenhouse gas concentrations in the atmosphere at a level that would prevent dangerous anthropogenic interference with the climate system. The Community has repeatedly stated that the overall global annual mean surface temperature increase should not exceed 2°C above pre-industrial levels.

All sectors of the Community economy should be subject to policies and measures designed to generate the substantial reductions in greenhouse gas emissions compared with 1990 levels which are needed to meet this target.

The limitation and reduction targets adopted under the Kyoto Protocol include emissions from domestic aviation but not emissions from international flights. Instead the Kyoto Protocol places an obligation on the parties to "*pursue limitation or reduction of emissions of greenhouse gases...from aviation...bunker fuels, working through the International Civil Aviation Organization...*". At the sixth meeting of the ICAO Committee on Aviation Environmental Protection in 2004, it was agreed that an aviation-specific emissions trading system based on a new legal instrument under ICAO auspices "*...seemed sufficiently unattractive that it should not be pursued further*". However, Resolution 35-5 of the ICAO Assembly instead endorsed open emissions trading and requested the development of non-binding guidance for use by states, as appropriate, to incorporate emissions from international aviation into their emissions trading schemes. The Commission and Member States are participating in and supporting this work which is scheduled for finalisation by ICAO in 2007. The ICAO Assembly in September 2007 will discuss this issue. In any case, the proposal of directive will not enter into force before that date. The final ICAO guidance will be taken into account, as appropriate, during the co-decision procedure. The objective of this proposal is to provide a model for aviation emissions trading that can be a point of reference in the EU's contacts with key international partners and to promote the development of similar systems worldwide. The Commission also supports the objective of a global agreement aimed at effectively tackling aviation emissions at global level.

2.3.2. *Legal elements of the proposal*

(a) Summary of the proposed action

The proposed Directive would amend Directive 2003/87/EC to include aviation in the Community scheme. As indicated before, the existing Community scheme works by allocating to operators a number of allowances each giving them a right to emit one tonne of carbon dioxide per year. The total number of allowances allocated sets a limit on the overall emissions from participants in the scheme. By 30 April each year operators must surrender allowances to cover their actual emissions. Operators can trade allowances so that emissions reductions can be made where they are most cost-effective.

The requirements to monitor and report emissions will take effect from 2010. From 2011 emissions from the aviation sector will be subject to a cap and aircraft operators will be required to surrender allowances to cover their emissions.

Key aspects are:

- Aircraft operators will be the entities responsible for complying with the obligations imposed by the scheme;
- The scheme will cover all flights arriving at or departing from an airport in the Community as of 1 January 2012. Flights between EU airports will be covered from 1 January 2011;
- Flights by State aircraft, flights under visual flight rules, circular flights, flights for testing navigation equipment or for training purposes, rescue flights and flights by aircraft with a maximum take-off weight of less than 5700 kg will be excluded from the scheme;
- To address other gases, by the end of 2008, the Commission will put forward a proposal to address the nitrogen oxide emissions from aviation after a thorough impact assessment;
- In order to avoid duplication and an excessive administrative burden on aircraft operators, each aircraft operator, including operators from third countries, will be administered by one Member State only;
- In contrast to the existing scheme, the method of allocating allowances will be harmonised across the Community;
- The total number of allowances to be allocated to the aviation sector will be determined at Community level by reference to average emissions from aviation in the years 2004–2006;
- A fixed percentage of the total quantity of allowances will be allocated free of charge on the basis of a benchmark to aircraft

operators which submit an application (the earliest application relating to 2008 data). For the period 2011–2012 this percentage will correspond to the average percentage proposed by the Member States including auctioning in their national allocation plans. Thereafter this will be reviewed in the light of the results of the general review of the emissions trading scheme;

- The details of how auctioning will work such as appropriate design and timing will be set out in a Commission Regulation. Auctioning proceeds should be used to mitigate and adapt to the impacts of climate change and to cover administrative costs;
- Like other participants in the Community scheme, aircraft operators will have to monitor their emissions of carbon dioxide and report them to the competent authority of its administering Member State by 31 March each year. The reports must be verified to make sure that they are accurate. The basic principles for monitoring, reporting and verifying emissions set out in the proposal will be elaborated by guidelines;
- Aircraft operators will be able to buy allowances from other sectors in the Community scheme for use to cover their emissions;
- Aircraft operators will also be able to use ERUs and CERs — from the Joint Implementation or Clean Development Mechanisms (JI/CDM) up to a harmonised limit equivalent to the average of the limits prescribed by Member States in their national allocation plans for other sectors in the Community scheme;
- Domestic aviation will be included in the scheme and treated in the same way as international aviation;
- Special consideration to the treatment of air services to remote or isolated regions, which are particularly dependent on air transport services, can best be given within the framework of existing measures such as public service obligations and aid having a social character under Article 87(2) of the Treaty.

(b) Legal basis

The legal basis for the proposal is Article 175 of the Treaty (which was also the legal basis for Directive 2003/87/EC).

(c) Subsidiarity principle

The subsidiarity principle applies insofar as the proposal does not fall under the exclusive competence of the Community. The objectives of the proposal cannot be sufficiently achieved by the Member States for the following reasons:

- The economic benefits from emissions trading will only be obtained if allowances can be traded and accepted across the Community. To achieve this, the establishment of a common framework is necessary.
- Given the high level of integration in the Community air transport market, distortions of competition could be created if different policies were applied in Member States.

Community action will better achieve the objectives of the proposal for the following reasons:

- Harmonized action to reduce the climate change impact of aviation can best be achieved by adopting legislation at Community level. As all airlines operating from Community airports would be treated in the same way, distortions of competition would be minimised.
- The Council and the European Parliament have called for the Commission to propose Community action and recognised the potential of emissions trading. Directive 2003/87/EC needs to be amended to include aviation in the Community scheme.
- The Community is a major player in global aviation, accounting for about half of the carbon dioxide emissions from international aviation reported by developed countries to the UNFCCC.
- The proposal amends the Community scheme to establish a common framework but will leave implementation and enforcement tasks to Member States which are better placed to perform them.

The proposal therefore complies with the subsidiarity principle.

(d) Proportionality principle

The proposal complies with the proportionality principle for the following reasons:

- Only the elements that are necessary to ensure that the scheme functions properly and to prevent distortions of competition which could result from differences in treatment of aircraft operators between Member States are regulated by the proposal. To ensure consistency and reduce implementation costs, the proposal makes use of the existing architecture of the Community scheme as much as possible. Where the procedures applied in the existing scheme are not appropriate for the aviation sector, special provision has been made whilst preserving the simplicity and environmental integrity of the scheme.
- Emissions trading can be used to achieve an environmental benefit at a lower cost than other means and is therefore the most cost-effective way of addressing greenhouse gas emissions from aviation.

2.3.3. *Evaluation*

The final outcome (i.e. approval or not) of this proposal is still uncertain, due to the heavy protests against it, and this as well inside as outside the Community. One can wonder if this modification of the EU ETS will not needlessly complicate matters. Already now one the biggest complaints about the EU ETS is its complexity and its lack of transparency. It does not look as if this is going to improve with the inclusion of the aviation sector.

Would a tax measure not be a much more sensible measure to tackle aviation emissions? Aviation can already be tackled, be it modestly, through Directive 2003/96/EC of 27 October 2003 restructuring the Community framework for the taxation of energy products and electricity.[132] Although according to this directive, in

[132] OJ, L 283, 31 October 2003.

principle, energy products supplied for use as fuel for the purpose of air navigation other than in private pleasure-flying, are exempt from taxation, Member States may limit the scope of this exemption to international and intra-Community transport. Thus for all air transport within a Member State or between two Member States that have signed a bilateral agreement to this effect, the Member States may apply a level of taxation, be it lower than the minimum that is set by the Directive. In the upcoming revision of this Energy Taxation Directive, a broadening of the scope of this regime can and should be considered.

Also, in the recent Green Paper on market-based instruments for environment and related policy purposes,[133] a plea is made for a more considered and targeted use of tax instruments to tackle greenhouse gas emissions.

2.4. *The maritime sector under EU ETS...*

Same remarks can be made for the maritime sector as for the aviation sector: the maritime sector is highly polluting and not yet covered by EU ETS.

At the moment, air pollutant emissions from ships are covered by Annex VI of the Marine Pollution Convention, MARPOL 73/78, of the International Maritime Organization (IMO). This contains provisions on Sulphur Oxide Emission Control Areas (Baltic Sea, North Sea & English Channel) and nitrogen oxide emissions standards for ships' engines. The EU strategy seeks to implement the SO_x Emission Control Areas set out in Annex VI, and to press for tighter NO_x standards.

In November 2002, the European Commission adopted a European Union strategy to reduce atmospheric emissions from seagoing ships. The strategy reports on the magnitude and impact of ship emissions in the EU and sets out a number of actions to reduce the contribution of shipping to acidification, ground-level ozone, eutrophication, health, climate change and ozone depletion.

---

[133] COM(2007)140final.

In January 2007 a Dutch consultant finalized a report on greenhouse gas for shipping and implementation guidance for the marine fuel sulphur Directive and proposed to integrate the maritime sector into EU ETS. Consequently, in May 2007, the European Commission launched a study, prepared by IIASA with regard to the cost effectiveness of measures with different ambition levels to reduce air emissions from maritime transport. The report and its appendices demonstrate that the reduction of these emissions can be substantially lower than the costs of additional controls for land-based sources with the same effect on health and environment. Additionally the European Commission launched in October 2007 a tender regarding the framework contract for services related to the quality of fuels sold and used in the EU and beyond. The framework contract aims to provide services in respect to the further development of EU policy on the quality of fuels used in the EU and governed by Community legislation. In particular, the framework contract will be used to inform a review of Directive 1999/32/EC on the sulphur content of certain liquid fuels. It seems that the European Commission is intensively investigation to bring the maritime sector also under the scope of EU ETS.

2.5. *The EU ETS global revision...*

2.5.1. *General*

On the basis of article 30, 2 of Directive 2003/87/EC, the Commission[134] had to submit by 30 June 2006 to the European Parliament and the Council, a report on the application of Directive 2003/87/EC, considering the following issues:

> "(a) how and whether Annex I should be amended to include other relevant sectors, inter alia the chemicals, aluminium and transport sectors, activities and emissions of other greenhouse

[134] "On the basis of experience of the application of this Directive and of progress achieved in the monitoring of emissions of greenhouse gases and in the light of developments in the international context...".

gases listed in Annex II, with a view to further improving the economic efficiency of the scheme;

(b) the relationship of Community emission allowance trading with the international emissions trading that will start in 2008;

(c) further harmonisation of the method of allocation (including auctioning for the time after 2012) and of the criteria for national allocation plans referred to in Annex III;

(d) the use of credits from project mechanisms;

(e) the relationship of emissions trading with other policies and measures implemented at Member State and Community level, including taxation, that pursue the same objectives;

(f) whether it is appropriate for there to be a single Community registry;

(g) the level of excess emissions penalties, taking into account, inter alia, inflation;

(h) the functioning of the allowance market, covering in particular any possible market disturbances;

(i) how to adapt the Community scheme to an enlarged European Union;

(j) pooling;

(k) the practicality of developing Community-wide benchmarks as a basis for allocation, taking into account the best available techniques and cost-benefit analysis."

Obviously, the Commission's report could be accompanied by proposals for amending the scheme, but the resulting changes would most likely not take effect until the third trading period beginning in 2013. There are three main reasons for this. First, any amendments have to be adopted by the Parliament and Council in co-decision, a procedure which can take two to three years. Second, operators need regulatory stability, which means they should be given time to adapt before the amendments enter into force. Third, any amendments cannot be reflected in the NAPs for the second trading period since these had to be submitted by 30 June 2006 as well.

The Commission presented its report on 13 November 2006 through Communication 2006/676 — "Building a global carbon market — Report pursuant to article 30 of Directive 2003/87/EC".

Taking into account the aforementioned timing and in order to ensure broader involvement of stakeholders with a high quality input into the review process, the Commission decided to consult further, by means of a separate Working Group on the Review of the EU ETS within the framework of the European Climate Change Programme (ECCP), which contributed to the development of the EU ETS and other climate measures.

The Working Group met in Brussels on four occasions, each time tackling a number of specific topics:

- 8/9 March 2007: scope of the EU ETS Directive,
- 26/27 April 2007: robust compliance and enforcement,
- 21/22 May 2007: further harmonization and increased predictability,
- 14/15 June 2007: linking with emission trading schemes of third countries.

The topics were selected on the basis of experience gathered, and problems signaled, until now with the EU ETS.

2.5.2. *The scope of the Directive*

The original scope of the Directive was driven by the objective of achieving "critical mass" for the introduction of a trading scheme. The selected activities covered large stationary sources where monitoring of emissions could be done with sufficient accuracy. In the review, the Commission is addressing the scope of the EU ETS.

Firstly, it considers streamlining the application of the current scope, notably as regards combustion installations and the smallest installations. There has been significant debate about the fact that Member States have applied different interpretations of a "combustion installation" as mentioned in Annex I to the Directive. The review explores how to give further clarity on specific types of combustion installation and their coverage under the Directive. The cost-effectiveness of covering small installations in the EU ETS is also considered in the review.[135]

Secondly, the review looks at expanding the EU ETS to other sectors and gases including N_2O from the production of ammonia and CH_4 from coal mines. The Commission will also assess to what extent to recognise carbon dioxide capture and geological storage activities in the EU ETS. Moreover, the extension of the EU ETS to other sectors and gases should be part of a comprehensive and coherent policy mix, and it must be ensured that these alternative policies are effective in contributing towards reductions in greenhouse gas emissions.

2.5.3. *Further harmonisation and increased predictability*[136]

In the original design of the EU ETS, most of the initial tasks related to the allocation of allowances are foreseen to be carried out at national level. The Directive created the National Allocation Plans (NAPs) as the tool by which each Member State proposes how many allowances to allocate for a period (setting the cap) as well as how these allowances should be distributed among existing installations and potential new entrants (allocation). In addition to ensuring application of the EC Treaty rules, the Commission is mandated to ensure that NAPs comply with a number of allocation criteria set out in Annex III to the Directive. These criteria, which have been subject to guidance by the Commission, continue to apply for the periods after 2012 unless amended. The result of this nationally-driven process is that the national allocation plans differ from each other and many stakeholders have raised concerns that such differences are having an impact on the internal market. The Commission takes the view that a number of options exist to harmonise the treatment of new entrants, and that having new entrants buy allowances

[135] Cf Ecofys, Small installations within the EU emission trading scheme, July 2007, 27p.
[136] Cf Ecofys, Harmonisation of allocation methodologies, October 2006, 55p.

in the market or in an auction is in accordance with the principle of equal treatment.[137]

Being in theory able to go back to the drawing board prior to each allocation period means that certainty can only be given for up to five years ahead. This is considered by many as too short to give sufficient predictability for investment decisions in sectors which are capital intensive and result in installations intended to be operated for decades. The Commission shares these views and regards further harmonization of the cap-setting and allocation process, as well as increased predictability, as key strategic issues. The review explores the option of a single EU-wide cap and that of separate national caps after 2012 determined by each Member State, and explores specific issues related to auctioning and benchmarking. Another important task is to find cost-effective solutions for providing information to the market on a more structured and regular basis so as to ensure optimal market transparency.

2.5.4. *Robust compliance and enforcement*

Robust compliance and enforcement procedures are important for the good functioning of the scheme, not least in view of linking with schemes in third countries. While initial experience of the learning-by-doing phase of the scheme with respect to compliance and enforcement has been encouraging, there is nevertheless a need to assess further harmonisation requirements.

As regards monitoring and reporting of emissions, it is considered whether the monitoring and reporting guidelines should be laid down in a Regulation in order to aid harmonised application of the legislation. As regards third party verification of emissions reports produced by companies and the accreditation process for approving organisations to act as verifiers, many stakeholders have expressed a preference or even stressed the need for more elaborated provisions at Community level. The need for additional compliance provisions is also considered.

---

[137] Cf Ecofys, The approach to new entrants and closures in the EU ETS, December 2006, 73p.

2.5.5. *Linking with emissions trading schemes in third countries, and appropriate means to involve developing countries and countries in economic transition*[138]

As regards the relationship of the EU ETS to schemes in third countries, the design of third country emissions trading schemes that are in operation or planned, and the possibility of linking third country schemes with the EU ETS is considered. The review considers extending arrangements for linking emissions trading schemes to other mandatory emission trading schemes in third countries capping absolute emissions at national or regional level.

The Directive provides for the continued recognition of credits from the Kyoto Protocol's project-based mechanisms beyond 2012. Regulatory certainty is important for companies, and the Commission is committed to maintaining recognition of these credits. The involvement of developing countries and countries in economic transition in emissions abatement efforts through respectively the Clean Development Mechanism and Joint Implementation and the extent to which emission reduction projects to date are contributing to countries shifting to more sustainable development paths are therefore assessed.

2.5.6. *Institutional and procedural aspects*

The introduction of the EU ETS and the establishment of an international carbon market bring with it new institutional needs both at Community and Member State level. For instance, some Member States have established a specialised agency dedicated to the implementation of the EU ETS. Institutional arrangements give rise to administrative costs, which in principle should be kept as low as possible.

Some of the options to address the strategic issues identified above would result in a considerable number of additional tasks to be

---

[138] Cf IEA, Linking non-EU domestic emission trading schemes with the EU emissions trading scheme, 2004, 37p.

undertaken at Community level to implement the EU ETS after 2012 (e.g. administration of an EU-wide new entrant reserve, approval process for projects not conducted in accordance with the Linking Directive, administration of EU-wide allowance auctions, development of EU-wide benchmarks).

The Commission therefore reviews also the existing institutional arrangements at both Community and Member State level with a view to adapting them to new needs arising out of the review and identifying areas where administrative costs could be further reduced (e.g. a single EU registry) and also invites stakeholder comments in this regard. Procedurally, the Commission considers the need for improvements regarding the notification and assessment of national allocation plans.

2.5.7. *The relationship between the EU ETS and other market-based regulatory instruments*

The relationship between the EU ETS and other policies and measures that pursue the same objective is most obvious in the case of taxes levied on energy products and electricity, which are harmonised by Council Directive 2003/96/EC (cf the before mentioned "Energy Tax Directive"). Greenhouse gas emissions trading and energy taxation are different economic instruments, operating by different legal means but partially pursuing the same objectives, in particular as concerns targeted CO_2 taxes. The Energy Tax Directive foresees that under certain conditions taxation can be fully or partially replaced, in particular for energy intensive companies, by some other instrument, including tradable permit schemes. This possibility is subject to the applicable state aid provisions. The EU ETS limits emissions of covered installations in the EU collectively, and until 2012 the Directive requires most allowances to be allocated free of charge.

In line with its commitments, the Commission considers further the interplay of the EU ETS with other measures pursuing the same objectives, and in particular with energy taxation. The earliest occasions for this will be the discussions on the review of the Emissions

Trading Directive and the before mentioned Green Paper on the use of indirect taxation and other market based instruments for Community policy purposes.

2.5.8. *Meetings*

During the meetings, which as indicated took place between March and June of 2007, an enormous amount of suggestions, proposals, best practices, etc were formulated by all kinds of stakeholders.[139] The general message could be summarized as follows: more harmonization, centralization, simplicity and transparency is needed in the EU ETS. Unfortunately, the European Commission has stayed at the surface until now and has not really indicated yet which direction it is willing to take in the global review of the EU ETS. Wait and see...

---

[139] See: http://ec.europa.eu/environment/climat/emission/review_en.htm

CHAPTER 9

TRENDS IN CARBON TRADING: PRACTICAL LESSONS

Andrew Beatty and Evan Williams

Environment and Environmental Markets Group
Baker & McKenzie, Sydney

1. Introduction

The global carbon market has grown rapidly in recent years, spurred on by the establishment of the EU ETS and a number of other voluntary reduction schemes in developed nations. The estimated value of carbon credit trades made in 2006 was US$30bn, three times what it was in 2005.[1] In the rush to gain exposure to this new commodity or to secure credits for existing or anticipated compliance regulations, new market entrants must be careful not to overlook prudent risk management practices.

A carbon credit can be defined as a tradeable right to GHG emissions reductions. There are a range of carbon credits currently being traded on the global market, which are defined by the legal framework under

The law is stated to be correct as at 1 July 2007. Key terms are defined in the Glossary at the end of the paper.

[1] K. Capoor and P. Ambrosi, *State and Trends of the Carbon Market 2007* (Washington DC: World Bank, May 2007) at 12. Available at http://www-wds.worldbank.org/external/default/WDSContentServer/WDSP/IB/2007/06/06/000020953_20070606144908/Rendered/PDF/399230Carbon1Trends1200701PUBLIC1.pdf (Last Accessed: 5 July 2007), [hereinafter "*States and Trends of the Carbon Market*, 2007"].

which the credit is created. Major types of carbon credits currently being traded include:

- ERUs and CERs created under the Kyoto Protocol;
- EUAs recognized under the EU ETS;
- NGACs traded under the GGAS system in the Australian State of New South Wales; and
- voluntary carbon credits, a catch-all term which describes any carbon credit that is based entirely in contract and is not defined by the rules of a legislative or treaty-based system.

As more emissions trading systems develop, it is likely that more types of carbon credits will arise. The focus of this paper is on trades involving CERs produced through the Clean Development Mechanism and voluntary carbon or VERs,[2] since these two types of carbon credits dominate the current market in Asia.

Risk management in carbon trading is complicated by the immaturity of the carbon market and the lack of established carbon market infrastructure in many of the jurisdictions in which carbon emission reduction projects are conducted. Immature commodity markets are often characterised by a lack of sophistication in many market players who have yet to develop an understanding of the underlying drivers of supply and demand. In the absence of established market history and methods of market analysis, accurate forecasting of market price is difficult. In such an environment, herd behaviour and the need to "feel out" appropriate price levels as the market matures can result in

[2] A class of voluntary carbon credits, Verified Emissions Reductions (VERs) are carbon credits that have been verified by third parties "in line with developing national or international regulatory standards and regulatory schemes, intended to be 'converted' in the future for use in regulatory regimes": M. Wilder and P. Curnow. "Trading Carbon as a Commodity — Sale and Purchase Agreements for Carbon Credits" (2005) *29th Annual AMPLA Conference* (24–27 August 2005) at 6. [hereinafter "*Trading Coupon as a Commodity*"] This paper refers to VERs when discussing voluntary carbon, since there is increasing pressure for all voluntary credits to be subject to some form of third party verification process and companies which purchase voluntary carbon usually seek credits that may be used for compliance purposes if a regulatory system was imposed on them.

greater price volatility and increase the danger that market prices will overvalue or undervalue the commodity.

A clear demonstration of this was the correction in EUA prices in May 2006. After reaching a peak of around €30 in April 2006, EUAs for the first commitment period of the EU ETS fell to €10 within 5 working days. This correction was due to the leaks of verified emissions data for the installations covered by the scheme, which showed that instead of the market prediction of a 50MtCO_2-e shortage in EUAs, the market was going to be over-supplied. Combined with the inability to bank Phase I EUAs into Phase II, the correction led to the price of Phase I EUAs falling to below €1 in early 2007.[3]

The EU ETS price correction illustrates the fact that demand in a cap-and-trade market is driven by the overall cap that is imposed by the regulator.[4] If the cap does not establish an environment in which allowances are scarce, the market price will be negligible. As a result the carbon market will always be subject to an element of uncertainty related to policy change, as regulators tighten caps in successive periods to ensure that scarcity is maintained (although this uncertainty can be mitigated by regulators mapping out clear emissions cap "gateways"). On top of this there is a greater risk that regulation may be brought in to plug market failures. As global demand for carbon credits is driven primarily by the purchase of credits to meet obligations under existing or anticipated emissions trading schemes, regulatory changes in one jurisdiction can have global implications.

---

[3] K. Capoor and P. Ambrosi, *State and Trends of the Carbon Market 2007* (Washington DC: World Bank, May 2007) at 3.

[4] This is true of most markets created for the purposes of regulating environmental outcomes. As J. Salzman and J. Ruhl explain in their article, "Currencies and the Commodification of Environmental Law" (2000) 53 *Stanford Law Review* 607 at 617:

> "In establishing a market, the government first creates a new form of property — legal entitlements to emit pollutants, catch fish, develop habitat — and then imposes a set of rules governing their exchange...Absent legal restrictions on pollutant emissions, fish landings or wetlands development, and the creation of alienable entitlements to these activities, few if any trades would take place."

Despite the risks, there are benefits in getting involved early in the carbon market. Players that enter a market early on get the benefit of first-mover status, enabling them to establish expertise and market presence which can be hard for later entrants to overcome. An obvious advantage of early entry into the carbon market is the ability to secure cheaper credits than will be available at the time that offsets need to be made. In 2003 the average price paid for the right to emissions reductions that were intended to be registered as CERs was around US$4.88.[5] In 2006 the average price for the rights to CERs purchased from a CDM project developer was between US$8 to US$10.[6] A higher price is paid for CERs that have already been issued or have been guaranteed by a seller. Care must be taken in comparing prices paid for CERs over time, as prices are influenced by a range of factors and vary greatly according to project stage and type. However it is clear that the general trend in project-based CERs has been for price rises over recent years.

Within a specific carbon transaction, parties must consider not only the trends of the overall carbon market, but also the risk profile of the specific carbon credits which they are trading. Because of the nature of carbon credits, this will usually involve a consideration of the underlying emissions reduction project or activity which generates the carbon offsets or provides a surplus of emissions allowances. The legal risks associated with a carbon transaction can be divided between:

- country risk — the legal and practical difficulties unique to the jurisdiction in which the carbon credits are generated or traded;
- contractual risk — the project or transaction specific risks that must be addressed on a project-by-project basis; and
- regulatory risk — whether credits will be recognised or accepted by regulators or the market. This risk applies not just to credits in regulated systems but also to VERs in the form of carbon reduction

[5] F. Lecocq and K. Capoor, *State and Trends of the Carbon Market 2003* (Washington DC: World Bank, December 2003) at 16–17. Available at http://www-wds.worldbank.org/external/default/WDSContentServer/WDSP/IB/2006/04/28/000012009_20060428141515/Rendered/PDF/36010.pdf> (Last Accessed: 4 July 2007).

[6] K. Capoor and P. Ambrosi, *State and Trends of the Carbon Market 2007* (Washington DC: World Bank, May 2007) at 3.

efforts not being accepted by a market as genuine or when attempting to convert VERs to meet compliance targets introduced by a mandatory trading system.

This paper explores examples of country risk and contractual risk and the means by which lenders to projects and acquirers of carbon credits can seek to mitigate those risks. With appropriate contractual protections and by undertaking proper due diligence the risks involved in a carbon credit transaction can be minimized.

Key Issues

- While there is some evidence of commoditization in carbon trading contracts, the nature of emissions reduction projects which generate credits and the differences between carbon credit types mean that contracts usually require customisation.
- The voluntary carbon market is expanding. Buyers of voluntary carbon need to be aware of the different risks that are associated with this largely unregulated form of carbon credit and adjust trading contracts accordingly.
- Regulators, consumers, NGOs and investors are increasingly scrutinising carbon reduction activities (including the purchase of credits).
- The carbon market remains an immature market, with many unsettled areas of law which may become the subject of disputes in the near term: these may help clarify areas of uncertainty in these types of transactions.
- While carbon trading is now a permanent feature of the global economy, uncertainty over the shape that future carbon markets will take means that there remains considerable risk in long-term credit delivery arrangements. In this environment of uncertainty, the use of standardized commodity or derivative trading contracts carries with it a considerable degree of risk.
- Competition for carbon credits has led some buyers to relax standard contractual protections against breach.

2. Current Trends in the Carbon Market

2.1. *State of the Global Carbon Market*

(a) Overall trends in price and volume of carbon credits being sold

With an estimated value of US$25bn the EU ETS is by far the largest carbon market in the world. EUAs in the current commitment period remain around the €0.30 mark, a result of the over-allocation of EU permits for the 2005–2007 commitment period.[7] However as the market matures and with the implementation of the new European cap under the 2008–2012 commitment period, the EU market is expected to stabilize.

CDM and JI projects have experienced greater price stability. In 2006, 450MtCO_2-e were traded on the primary CDM market, for a market value of US$5bn.[8] Prices paid for CERs in 2006 generally range between US$8–14 on the primary market and US$14–20 on the secondary market.[9] Investment in JI projects has begun to accelerate with 16.3 MtCO_2-e of ERUs transacted at an average price of US$8.70.[10] The JI market is likely to continue expanding as we approach the start date for ERUs being able to be issued and used as offsets to meet Kyoto targets.

Competition for CDM projects is strong, as the delays in the ITL (see below) have slowed the flow of projects through the pipeline, and more investors enter the market in an effort to secure cheap, high-quality emissions reductions credits. The scarcity of the projects has driven up prices in the EU ETS for the 2008–2012 commitment period, since fewer CERs are being imported into the system, resulting

---

[7] Point Carbon, *Carbon Market Analyst: Outlook for 2007* (26 February 2007) at 5–7; Anon. "European carbon prices spike to 13-month high," *Environmental Finance* (31 May 2007), available at http://www.environmental-finance.com/onlinews/0531eur.htm (Last Accessed: 4 June 2007).

[8] *States and Trends of the Carbon Market 2007* at 4.

[9] *Ibid.*

[10] *Ibid.*

in increased pressure on supply. The result is that in May 2007, EUAs for the December 2008 delivery period reached price levels not seen since April 2006.[11]

(b) Delay in the International Transaction Log

The International Transaction Log (ITL) is the registry that will track trades in Kyoto rights such as CERs and ERUs. Until it is in place it is not possible to forward these credits into national accounts or to transfer them between the national accounts of different parties. While it was initially hoped that the ITL would be in place by the time the first CERs were issued it has experienced a series of delays and it is now predicted that it will not be in place until late 2007. Further delays are predicted before the ITL is linked to the EU registry for EUAs — the Community Independent Transaction Log (CITL). Until the ITL and CITL are linked, it will not be possible to transfer CERs and ERUs between national accounts in the EU or to convert CERs into EUAs.

In the interim CERs will be delivered into the temporary accounts of buyers. CERs can be held in the temporary account until the ITL comes online, at which time the CERs can then be forwarded to national registry accounts and traded on secondary markets.

(c) Growth of the Voluntary Carbon Market

The voluntary carbon market is distinguished from the compliance market by buyer motivation — buyers in the voluntary market purchase carbon credits for reasons other than complying with a mandatory emissions trading scheme. The voluntary carbon market can involve the purchase of any carbon credit type, including credits from statute or treaty based schemes. However, the majority of credits transacted in the voluntary market are VERs.

---

[11] Anon, "European carbon prices spike to 13-month high," *Environmental Finance* (31 May 2007). Available at http://www.environmental-finance.com/onlinews/0531eur.htm (Last Accessed: 4 June 2007).

Monitoring the size of the voluntary market and the prices paid for VERs is difficult due to the lack of central registries and the proliferation of voluntary offset schemes around the world. The World Bank estimates that the voluntary carbon market expanded to US$100m dollars in 2006.[12] The voluntary market is estimated to grow rapidly in coming years, one estimate suggesting that it could grow to 400 $MtCO_2$-e by 2010, which is roughly equivalent to the size of the CDM today.[13]

Prices offered for VERs vary significantly but our experience at Baker & McKenzie is that in some trades in 2007, VER prices have surpassed prices being offered for CERs and EUAs. This has led several CDM projects to sell their CERs as voluntary carbon to take advantage of the higher prices.

The demand for voluntary carbon is being driven by:

(i) Many companies have entered the voluntary carbon market to get experience in carbon trading and operating in a carbon constrained environment in anticipation of greater regulation in the future.

This activity has been increasing in anticipation of the introduction of mandatory trading schemes in a number of jurisdictions.[14] One example of such a company is American Electric Power which

---

[12] *Supra* note 3.

[13] *States and Trends of the Carbon Market 2007* at 41.

[14] A number of jurisdictions have decided to implement or are considering implementing mandatory emissions trading schemes. In the US, mandatory schemes are being developed through the cooperation of States. The Regional Greenhouse Gas Initiative (RGGI), which will impose caps on electricity generators in member states, is to commence in 2009. California has also committed to developing a state-wide emissions cap and an emission trading system under Assembly Bill 32. Australia seems set have a mandatory scheme following the release of the report of the Prime Minister's Task Group on Emissions Trading: Prime Ministerial Task Group on Emissions Trading, *Report of the Task Group on Emissions Trading* (Commonwealth of Australia, May 2007), available at http://www.pmc.gov.au/publications/emissions/index.cfm (Last Accessed: 5 July 2007). The states in Australia have previously committed to implementing a national emissions trading scheme in the absence of action by the federal government: Council for Australian Federation, *Communiqué 9 February 2007* (2007) at 5, available at http://www.emissionstrading.nsw.gov.au/key_documents (Last Accessed: 5 July 2007).

entered an agreement in June this year to purchase 4.6 million credits via the Chicago Climate Exchange for the period from 2010 to 2017, in anticipation of some form of carbon constraining mandate being placed on the company by 2010.[15] Japanese companies who wish to gain exposure to the carbon market may agree to voluntary emissions reduction commitments under the Keidanren Voluntary Action Plan.[16]

(ii) Litigation risk of inaction

A number of high-profile cases have been launched against companies and government agencies for their contribution to climate change or failure to take action to prevent climate change.[17]

[15] Point Carbon, "AEP buys 4.6 million offset carbon credits" (14 June 2007).

[16] The Keidanren Voluntary Action Plan only allows for Kyoto credits, and not VERs to be used as offsets.

[17] The most famous of these cases to date is probably *Massachusetts et al. v Environmental Protection Agency et al.* 549 U.S. (2007) in which the US Supreme Court settled a dispute over whether the US EPA has the authority to regulate greenhouse gases under the Clean Air Act, finding that: (1) greenhouse gases are "air pollutants" under the Act (at 26–30); and (2) the EPA shall regulate greenhouse gases for new automobiles unless "it determines that greenhouse gases do not contribute to climate change or if it provides some reasonable explanation as to why it cannot or will not exercise its discretion to determine whether they do" (at 30). Another case against government agencies is *Friends of the Earth et al. v Mosbacher* in which Friends of the Earth, Greenpeace, and four US cities have alleged that the US Export-Import Bank (Ex-Im) Bank and the Overseas Private Investment Corporation (OPIC) have granted loans to oil fields, pipelines and coal-fired power plants without satisfying a legal obligation to assess the impacts of these projects on global warming or on the US environment as required by the National Environmental Policy Act (NEPA). California has lodged a public nuisance claim against six car-makers (Chrysler, General Motors and Ford and the US subsidiaries of Toyota, Honda and Nissan) for their contribution to global warming: D. Glaister. "California sues car firms for global warming," *The Guardian* (21 September 2006). Another nuisance claim brought by eight US States and the City of New York against the power utility American Electric Power was dismissed in a decision of the US District Court for the Southern District of New York in September 2005: *Connecticut et al. v American Electric Power Company* 04 Civ 5669 (LAP) (SDNY July 21 2004); *Open Space Institute et al. v American Electric Power Company* 04 Civ 5670 (LAP) (SDNY July 21, 2004).

By getting involved in carbon reduction efforts and acquiring voluntary carbon offsets a company can reduce the chance that it will become engaged in these costly cases.

(iii) Consumer based voluntary offset schemes

A number of schemes have arisen under which companies offer to secure carbon offsets for consumers who wish to reduce the impact of their activities on climate change. These are being offered by both specialized carbon offset companies and by existing companies (e.g. airlines who provide an add-on service for consumers who want to offset the impacts of air travel[18]).

(iv) Pledges by major corporations to reduce carbon footprint or become "carbon neutral"

This has led to a significant demand for carbon credits to offset emissions which cannot be avoided through efficiency programs. Companies that have made carbon neutral pledges include HSBC, Swiss Re, IAG and recently NewsCorp.[19] Often such companies choose to meet these targets by purchasing voluntary carbon since it generally offers a more flexible and less expensive source of offsets.

Voluntary carbon has the potential to be more flexible since it is outside the regulatory processes required under Kyoto and the EU ETS. Some forms of offsets are currently incapable of being recognised under regulated schemes, and thus in order to produce a valuable commodity, companies producing those offsets can only do so by creating VERs.

---

[18] A number of airlines including British Airways, Delta and Qantas have announced or introduced schemes whereby customers can pay an additional fee to offset the GHG emissions emitted by their flight. D. Cameron. "Industry to redress imbalance," *Financial Times* (18 June 2007); S. Rochfort. "Qantas offers passengers chance to be greener," *Sydney Morning Herald* (27 June 2007).

[19] NewsCorp, *News Corporation: Global Energy Initiative* (2007). Available at http://www.newscorp.com/energy/carbon_report.pdf (Last Accessed: 5 July 2007).

A major concern over the voluntary market is the "quality" of VERs.[20] Doubts have been raised as to whether these credits represent genuine emissions reductions. One key problem is that there are no national or global registries for VERs. This gives rise to the potential for double counting or sellers offering the same credits for sale multiple times. For companies purchasing VERs it is important that they exercise extreme care in their due diligence process for assessing sellers. The credibility of the VERs is essential, particularly since companies that purchase voluntary carbon are often motivated (at least in part) by pressure from investor groups (e.g. the Carbon Disclosure Project) or consumers.[21] The reputational advantage gained in taking action to reduce the carbon intensity of a business depends on the credibility of the action. Revelations that the offsets being purchased are not backed up by real-life emissions reductions could result in serious damage to the company's reputation. Furthermore, if a company hopes to be able to convert VERs into credits recognisable under future mandatory trading schemes, it is essential that they are based on sound methodology and have requisite safe-guards.

As a result, VER contracts often adopt or mirror the rigorous methodologies and mechanisms of the CDM and issues described in this paper as being related to the CDM rules — such as additionality — are increasingly relevant to VERs.

An alternative means of establishing credibility of voluntary offsets is through NGO developed accreditation programs such as the Voluntary Carbon Standard[22] and the Gold Standard.[23]

---

[20] F. Harvey. "Beware the carbon offsetting cowboys," *Financial Times*, (25 April 2007); F. Harvey and S. Fidler. "Industry caught in 'carbon credit' smokescreen," *Financial Times* (25 April 2007).

[21] J. Lash and F. Wellington, "Competitive Advantage on a Warming Planet," *Harvard Business Review* (March 2007) 94, at 96.

[22] Developed by IETA, the World Economic Forum and the Climate Group. See http://www.v-c-s.org/About_VCS.html (Last Accessed: 29 June 2007).

[23] Developed by the World Wildlife Fund, SouthSouthNorth and Helio International. See http://www.cdmgoldstandard.org/index.php (Last Accessed: 29 June 2007).

2.2. *State of the carbon market in Asia*

(a) Asia hosts a significant proportion of carbon emission reduction projects

Over the last few years there has been a gradual increase in the number of transactions involving the buying and selling of carbon credits and the financing of carbon emission reduction projects throughout Asia. Asia is now one of the main sources of CDM projects. As at the end of June 2007, the registered CDM projects hosted by China and India were expected to produce 58% of expected average annual CERs.[24]

(b) Carbon credit exchanges

In addition to the bilateral deals that have proliferated under the CDM and voluntary carbon projects, the Asia Carbon Exchange (ACX) has opened up the potential for CER holders to access a broad range of buyers in both the primary and secondary markets for emissions reductions. The establishment of such exchanges allows for buyers and sellers to ascertain the prices being paid in recent transactions, and thus ensure that they are themselves getting a deal that is reflective of market rates. Based in Singapore, ACX aims to attract owners and developers of CDM project owners in Asia to use the exchange as the mechanism for locating buyers for their CERs. In doing so, it is argued that sellers will be able to achieve better prices for the credits than if they negotiated with one major buyer. Bids for CERs can be made for a whole project or for an individual vintage (i.e. the CERs produced by a project in a specific verification period). From its launch in mid-2005 to March 2007, ACX was reported to have seen about 2.4 million CERs traded,[25] a fairly small

---

[24] Figure taken from UNFCCC "CDM Statistics" website http://cdm.unfccc.int/Statistics/Registration/AmountOfReductRegisteredProjPieChart.html (Last modified: 29 June 2007; Last Accessed: 2 July 2007).

[25] Yaw Yan Chong, "Asia carbon exchange sees growth after slow start," *Reuters* (23 March 2007).

proportion of the total CERs bought from Asian projects in that time.

(c) Speculation over a regional cap-and-trade system

There has been growing speculation over the development of a regional cap-and-trade system in Asia, fuelled by the policies being suggested by the members of the Asia-Pacific Partnership on Clean Development and Climate (APP) — US, Japan, China, India, Republic of Korea and Australia.[26]

2.3. *Players in the carbon market*

(a) Buyers and sellers

Suppliers are parties that own or develop the emissions reduction projects or activities that generate the carbon credits. These include owners of CDM or VER-generating projects and companies which reduce their emissions below mandated targets enabling them to sell excess allowances. With the commencement of Emissions Trading under Article 17 of the Kyoto Protocol, Annex I countries that have met their targets under the Kyoto Protocol will also be able to sell their excess AAUs.

The end-buyers of carbon credits can be separated into compliance buyers and voluntary buyers. Compliance buyers purchase credits in order to meet a legally defined target under a mandatory emissions trading scheme.[27] They include companies regulated under the EU ETS and Annex I countries that purchase credits to meet their targets under the Kyoto Protocol. We would also include in this category carbon funds that have been established to distribute carbon credits

---

[26] Anon., "Australia plans carbon scheme counting US, China," *Reuters News* (19 May 2007); S. Marris and D. Shanahan. "Key role in global forum on climate," *The Australian* (1 June 2007); K. Green. "Canada joins anti-Kyoto bloc," *National Post* (19 April 2007).

[27] "*Trading Carbon as a Commodity*" at 3–4.

to investors for the purposes of compliance (e.g. the European Carbon Fund). Voluntary buyers are those that purchase credits for reasons other than meeting a target under a mandatory scheme. They may do this in anticipation of regulatory barriers being imposed in the future, as part of a program for achieving corporate responsibility, or in response to market and consumer pressure.

The other major buyers and sellers in the market are intermediaries and speculators. These are parties that purchase credits in the hope of selling them to another buyer for a profit. Within this category we include for-profit carbon funds and investment banks such as Climate Change Capital, which have been active in the carbon market.

Suppliers are generally able to sell to buyers around the world although there are some important segmentations in the market. For example while compliance buyers in the EU ETS can purchase CERs and use them as EUAs, they cannot apply VERs for the same purpose.

(b) Lenders

CDM and JI projects often require large investments to get up and running. The income generated by the sale of carbon credits will often be a small proportion of the total cost of the project. Usually the remaining funds are obtained through conventional project finance. The project finance arrangements for CDM projects may impose additional requirements on the project. Lenders will want to impose conditions and warranties similar to any such transaction to minimise their risk exposure.

In addition lenders may be under a requirement to consider the Equator Principles. By May 2007, 51 institutions had adopted the Equator Principles.[28] In 2005, it was estimated that the 31 institutions which had then signed the Equator Principles represented over 80% of the global project finance market in developing countries.[29]

---

[28] See www.equator-principles.com/index.shtml (Last Accessed: 5 July 2007).

[29] "The Equator banks account for around 85% of the market for project finance in developing countries, which on average is about US$ 90 billion per year during the period 1998–2002": B. Scholtens. "Finance as a driver of corporate responsibility" (2006) 68(1) *Journal of Business Ethics* 19, at 27; see also P. Watchman, "Beyond the

The increasing coverage of the Principles and the practice of syndication means that the vast majority of project finance arrangements now involve a signatory of the Equator Principles. The Principles[30] require any project with a capital cost of more than US$10 million which is categorized as having a certain level of environmental or social risk to undertake an environmental and social impact assessment. Projects may also need to comply with an ongoing Environmental Management Program.

(c) Regulators

The regulation of the carbon market is made up of formal and informal regulators. Formal regulators include:

- scheme regulators that are responsible for developing and administering the rules of emissions reduction and trading schemes. For CERs the most important scheme regulators are the COP/MOP and the CDM Executive Board;
- governments which play a role in establishing national and regional carbon trading schemes. National governments also play a regulatory role in the CDM through the assessment and approval processes associated with the host country Letter of Approval, a step required in the registration of a CDM Project; and
- administrators of carbon market exchanges such as the ECX and ACX, who may impose trading rules for transactions which occur on their exchanges.

Carbon market regulators have a significant impact on the carbon market. As can be seen from the EU ETS, policy decisions such as setting

---

Equator," *Time Magazine* (30 May 2005), 16–17. The ten original Equator Principles banks together arranged loans representing 31.4% of the total project finance market in 2002: B. C. Esty, C. Knoop and A. Sesia. "The Equator Principles: An Industry Approach to Managing Environmental and Social Risks" (2005) (Revised Jan 30, 2007) *Harvard Business School Case Study 9-205-114* at 16.

[30] Available at http://www.equator-principles.com/principles.shtml (Last Accessed: 3 July 2007).

targets for mandatory emissions trading schemes can shift regional and global prices in carbon. The CDM Executive Board and COP/MOP play a very significant role in developing the rules and granting approval for CDM projects. The importance of this work is even greater due to the practice of adopting the work done by the CDM into other carbon credit projects and trading schemes.

Three important actors in informal regulation are:

- Investor Coalitions — investor coalitions that demand information on the greenhouse emissions of a company are generally identified as one of the major drivers behind companies undertaking voluntary carbon reduction initiatives.[31] A striking example of investor pressure on companies to take action on climate change is the Carbon Disclosure Project (CDP). The CDP is a program where institutional investors collectively sign a single global request for disclosure of information on greenhouse gas emissions and climate change risk by the largest companies in the world. The request for information to be compiled into the CDP5 report was sent to companies in February 2007, and was signed by 284 institutional investors, representing US$41 trillion under management.[32] In 2006, 72%, or 360 of the FT500 companies, and 58% of all US companies which were sent the request for information responded.
- Consumers — consumers participate in the carbon market by using services that provide offsets for personal emissions and by exercising consumer choice to favour companies that they perceive to be taking action on climate change.
- NGO Observers — NGOs play an important role in scrutinising the carbon market. In practice NGO observers act as one of the

[31] J. Lash and F. Wellington, "Competitive Advantage on a Warming Planet," *Harvard Business Review* (March 2007) 94, at 96.

[32] Carbon Disclosure Project, "Press Release: World's Largest Investor Coalition representing $41 trillion Seeks Further Disclosure on Climate Change and Shareholder Value From World's Largest Corporations" (1 February 2007). Available at http://www.cdproject.net/pressreleases.asp (Last Accessed: 29 June 2007).

major checks against low-quality voluntary carbon, through their administration of accreditation standards and their mobilisation of publicity and consumer power.

(d) Carbon market services

Specialized services have developed to assist participants in the carbon market, including:

- Carbon credit intermediary services established by organizations which specialize in locating and sometimes aggregating emission reduction projects or carbon credits for companies that are looking to purchase offsets.
- Designated Operational Entities (DOEs) are the experts accredited for assessing emissions reduction projects to be registered as CDM projects. They are obviously a pivotal part of the CDM system. However, they have also played a significant role outside the CDM process including the assessment of voluntary carbon offset schemes. It is important to note that when working on non-CDM projects, there is no requirement for these organizations to meet the CDM methodologies or rules.

There are currently two DOEs based in Asia — Korea Energy Management Corporation (KEMCO) and Korean Foundation for Quality (KFQ). The CDM Executive Board has expressed a desire to encourage more DOEs to set up in non-Annex I countries, in line with the aim of the CDM to involve as many potential host countries as possible and not restrict itself to those non-Annex I countries that have already established a presence in the CDM market.[33]

[33] Statements to this effect can be found at para 14 of Decision 17/CP.7, para 1(g)–(h) of Decision 18/CP.9 and para 4(b)–(c) of the Annex to Decision 3/CMP.1. Additionally, the "Nairobi Framework" announced by the UN Secretary-General at COP/MOP 2 on 15 November 2006 has the "specific target of helping developing countries, especially those in sub-Sahara Africa, to improve their level of participation in the CDM" see http://cdm.unfccc.int/Nairobi_Framework/index.html (Last Accessed: 5 July 2007).

3. Types of Carbon Transactions

There are three main forms of carbon credit transactions — forward sales from emissions reduction projects, carbon portfolio acquisitions, and spot trades.

3.1. *Forward sales from emissions reduction projects*

Most emissions reduction projects enter into contracts to sell the credits they expect to generate — i.e. the credits being traded have not actually come into being yet. Forward sales of CERs and VERs give the project additional future income on which they can rely on when planning and developing the project and to access funding early in the project. Assessing the contractual and country risk issues that face these transactions requires an understanding of the Project Cycle. An outline of the CDM Project Cycle is included in Schedule 1.

It is possible for parties who want to acquire carbon credits to enter a simple off-take arrangement. The parties sign an Emissions Reduction Purchase Agreement (ERPA) under which they contract to buy the carbon credits which will be generated by a project in a future period. The purchaser takes no direct role in the project itself.

Straight purchase agreements were preferred by carbon credit buyers in the early CDM market. However, as the market has matured, many buyers have increased their direct involvement in the underlying project, usually by providing additional services to the project, providing finance, or taking a direct equity stake in the project.[34]

While a number of standard form credit trading contracts have been developed,[35] ERPAs that involve the purchase of credits

[34] M. Wilder and M. Willis. "Carbon Contracts, Structuring Transactions: Practical Experiences" (2007) 2 *Environmental Liability* 101 at 104–106, [hereinafter "*Carbon Contracts, Structuring Transactions*"].

[35] For example IETA has published a standard ERPA and Code of CDM Terms. These can be accessed at http://ieta.org/ieta/www/pages/index.php?IdSitePage=990 (Last Accessed: 5 July 2007).

directly from the owner or developer of the project that generates the credits:

> *are almost invariably individually negotiated for each project, taking into account the specific circumstances of the individual project, the domestic rules of the governing law for the contract and the legal and business constraints binding each of the parties. Many sophisticated buyers tend to prefer to use their own unique contract as a 'calling card' rather than rely on generic templates, and individual sellers often wish to negotiate clauses due to their own internal requirements.*[36]

The standard form contracts do, however, provide a useful indicator of the market norms that have begun to develop in ERPAs.

When drafting ERPAs for primary sales it is important to consider the overall contractual environment for the project. Accompanying contracts involved in project-related carbon credit purchases may include:

- loan agreements and other financing contracts;
- a project development agreement;
- a CDM services agreement; and
- Engineering Procurement and Construction (EPC) contracts.

3.2. *Acquisitions of carbon portfolios*

Another major way in which carbon credits (or the rights to carbon credits) change hands, is through the sale of carbon credit portfolios. In this situation, Company A enters into forward sales agreements with a number of projects, often managed through a special purpose fund. Company B wants greater exposure to the carbon market and purchases the portfolio from Company A. Generally legal title to the

---

[36] "*Carbon Contracts, Structuring Transactions*" at 103.

carbon credits is transferred from Company A to Company B through contract novation.

3.3. *Spot trades*

Spot trades are transactions that involve existing carbon credits. When dealing with well-defined credits under an international, regional or national emissions trading scheme, spot trades are similar to any commodity or derivative trade. As such participants are likely to use standard form agreements. EUAs are often traded using the standard agreements developed by ISDA, EFET or IETA.[37]

Spot trades are not yet possible for CERs or ERUs because of the delays in the ITL. It is possible for a party that has had CERs issued into their temporary account to engage in trading by entering forward sales contracts under which they agree to deliver the CERs to another party once the ITL is established. However until the ITL is in place, these contractual obligations can never be fulfilled, as it is not possible to transfer the ownership of a CER without the ITL. It is also possible for a party that has purchased the right to receive CERs into their temporary account on issuance (which is what a buyer acquires from a seller in the case of forward sales contracts for CERs generated by a CDM project) to transfer that right to another party that has a temporary account. For these transactions, standard form agreements are rarely used because of the strict penalties they impose for failure to deliver.[38]

4. Country Risk

The laws, policies and customs in host countries can have a considerable impact on the way in which carbon credit producing projects are undertaken. For host country jurisdictions that lack mature administrative

---

[37] *Ibid.*

[38] *Ibid.* at 103–104.

mechanisms to enable a ready trade in carbon emissions reductions, lenders to projects and buyers of CERs must factor into their loan agreements or ERPAs a host of contingencies.

4.1. *Legal and policy treatment of carbon credits*

An important consideration in dealing in different countries is whether and how domestic law recognises carbon credits. Carbon credits are a novel property concept in many jurisdictions and as such may require specific legislation in order to recognise the ability to trade them under domestic law. Furthermore some jurisdictions may recognise one type of carbon credit but not another.

CERs Under the Law of Indonesia

The legal nature and status of CERs has not as yet been tested by Indonesian courts nor by other courts or tribunals applying Indonesian law. Indonesian law does not recognise the ability to assign future property. In view of the fact that CERs are a form of intangible right which only come into existence if and when a CDM project passes certain Indonesian and international regulatory hurdles, any agreement governed by Indonesian law which sought to transfer CERs to a person prior to such CERs coming into existence under the UNFCCC regime would likely not be an effective transfer of such CERs.

If domestic law does not have a means of recognising carbon credits, it is advisable to choose as the governing law of the carbon trading contract a system of law which does. It is also important to ensure that the lack of recognition under domestic law does not create any obstacles to the transaction even though it is conducted under a different system of law. Problems may arise if a step in a transaction is governed by a subsidiary contract which is conducted under local law.

Other domestic law and policy factors that may need to be considered include:

- how ownership and possession of the carbon credits are recognized and to whom this is given;
- how carbon credits are regulated under local securities or sale of goods legislation;
- whether there are any restrictions on the sale or export of emissions reductions or the value that can be taken from them;
- whether the carbon credits are subject to sales taxes or duties, capital gains or income tax, and whether there are any concessions or exemptions.

4.2. *Foreign investment rules*

Often carbon credit transactions involve significant foreign investment in countries hosting a project. This is particularly true where buyers enter into forward sales agreements. The CDM is specifically designed as a mechanism of channelling foreign investment into non-Annex I countries. As such it is important to consider any rules which restrict or regulate foreign investment or the repatriation of profits. Some countries may have investment incentives and tax concessions which are designed to promote emissions reduction projects. It is important to engage with the government department responsible for foreign investment in the host country.

4.3. *CDM infrastructure*

The first regulatory hurdle in establishing a CDM project is to obtain a Letter of Approval from the host country. Certain institutional precursors must be in place in a country before it can host

CDM projects and issue Letters of Approval. The country must have[39]:

(a) ratified the Kyoto Protocol; and
(b) established a Designated National Authority (DNA). The DNA is the body responsible for approving CDM projects. Generally it is the DNA which will issue the Letter of Approval, as well as co-ordinating the assessment process.

Although not required, countries may also establish sustainable development criteria for CDM projects. The Letter of Approval is required to state that the project will assist the host country to achieve sustainable development. What the criteria are, how they are interpreted and how they are applied differs from country to country.

The effectiveness of the DNA and its approval procedures is an important consideration when entering contracts to buy or sell CERs from a project which has not yet been approved by the host country. One way of assessing this is by looking at the host country's track record — in particular how many of the projects which it approved have subsequently been registered by the Executive Board and produced CERs.

4.4. *Domestic regulations specific to carbon credit projects*

Local legislation and policy may impose special rules on the manner in which carbon credit producing projects must be developed (e.g. China's new Renewable Energy Law, which provides for preferential feed-in tariffs for renewable energy generation). China has a number

[39] M. Wilder, M. Willis and J. Carmody, *Legal Issues Guidebook to the Clean Development Mechanism* prepared by Baker & McKenzie for the United Nations Environment Programme CD4CDM project (Roskilde, Denmark: UNEP, 2004) at 31–34 and 49. Available at http://cd4cdm.org/publications.htm (Last Accessed: 28 June 2007).

of regulations specific to CDM projects, the most significant of which are outlined in Schedule 2. These rules are in addition to the standard domestic laws governing project development such as environmental and planning legislation which we will consider later under the Environmental Risk section.

4.5. *The impact of domestic regulations on the assessment of additionality*

A CDM project can only be approved if it is shown to achieve emissions reductions which are additional to those that would have occurred under the Baseline Scenario (i.e. the future projection of emissions had the project not been approved under the CDM).[40] How additionality should be tested is the matter of considerable debate.[41] Paragraph 45(e) of the CDM Modalities and Procedures states that in calculating the Baseline Scenario relevant national and/or sectoral policies and circumstances are to be taken into account.[42] However, the CDM Executive Board has since clarified this position.[43] Now:

- policies which give a comparative advantage to more emissions intensive technologies or fuels over less emissions intensive

---

[40] Paragraph 43 of the CDM Modalities and Procedures reads, "A CDM project activity is additional if anthropogenic emissions of greenhouse gases by sources are reduced below those that would have occurred in the absence of the registered CDM project activity." The CDM Modalities and Procedures were adopted by the 1st meeting of the COP/MOP in decision 3/CMP.1 and are contained in the Annex to that decision. The decisions of the COP and COP/MOP which are related to the CDM can be accessed at http://cdm.unfccc.int/Reference/COPMOP/index.html (Last Accessed: 28 June 2006).

[41] M. Wilder, M. Willis and J. Carmody, *Legal Issues Guidebook to the Clean Development Mechanism* prepared by Baker & McKenzie for the United Nations Environment Programme CD4CDM project (Roskilde, Denmark: UNEP, 2004) at 53–55. Available at http://cd4cdm.org/publications.htm (Last Accessed: 28 June 2007).

[42] Paragraph 45(e) of the CDM Modalities and Procedures, see note 41.

[43] The most recent statement from the Executive Board on this issue is contained in Annex 3 to the 22nd Executive Board Meeting. Decisions of the CDM Executive Board can be accessed at http://cdm.unfccc.int/EB/index.html (Last Accessed: 28 June 2007).

technologies or fuels can only be taken into account in the Baseline Scenario if they were in place before 11 December 1997 (i.e. when the COP adopted the Kyoto Protocol).

- policies which give a comparative advantage to less emissions intensive technologies or fuels over more emissions intensive technologies or fuels can only be taken into account in the Baseline Scenario if they were in place before 11 November 2001 (i.e. when the COP adopted the CDM Modalities and Procedures).[44]

5. Contractual Risk

5.1. *Defining the assets and rights being traded*

Properly defining the assets and rights associated with carbon emissions reductions which are capable of being created and traded is an essential element in any carbon trading contract.

(a) Defining different types of carbon credits in contract

Kyoto credits can be defined in contract by reference to the relevant international rules. In the case of VERs, the credits are a construct of the contract and have no legal status or recognition beyond the contract. As such it is essential that the contract provides sufficient definitions, requirements and protections to establish a tradeable contractual right.[45]

Wilder and Curnow outline six critical components of any contractual definition of a carbon credit[46]:

- a reference to the physical reduction in levels of greenhouse gas emissions;

[44] This clarification means countries are not discouraged from implementing legislation to encourage the uptake of cleaner fuels and technologies.

[45] "*Carbon Contracts, Structuring Transactions*" at 101–102. "*Trading Carbon as a Commodity*" at 7.

[46] "*Trading Carbon as a Commodity*", *Ibid.*

- a clear definition of the greenhouse gases to be included (generally being the basket of six gases under the Kyoto Protocol[47]);
- a clear unit of measurement usually expressed in tonnes of CO_2-e;
- a clear definition of the project creating the emissions reductions;
- independent verification and assessment methodology; and
- the baseline scenario against which the reductions are being measured.

(b) Legal title in carbon credits

The seller must have legal title to the carbon credits they are offering. The rules under the Kyoto Protocol make no reference to legal title, but assume that the project participants responsible for a project will be entitled to legal title to the CERs. Difficulties may arise in countries that have not implemented specific legislation dealing with the ownership and possession of carbon credits. The government may claim a right to emissions reductions similar to the claims they make to other natural resources such as air and water.[48] Alternatively disputes could arise between different entities involved in the project such as the landowner, lessee, project developer, contractor, project manager, project owner and financiers.[49]

To ensure that full legal title passes to the buyer, the contract must clearly outline what steps must be taken to "deliver" the carbon credits and transfer all associated rights and interests. In the case of credits governed by a formal scheme, these steps will usually be outlined in the scheme rules and generally involve lodging the transaction with a

[47] CO_2 — Carbon dioxide, CH_4 — Methane, N_2O — Nitrous oxide, PFCs — Perfluorocarbons, HFCs — Hydrofluorocarbons and SF_6 — Sulphur hexafluoride.
[48] M. Wilder, M. Willis and J. Carmody, *Legal Issues Guidebook to the Clean Development Mechanism* prepared by Baker & McKenzie for the United Nations Environment Programme CD4CDM project (Roskilde, Denmark: UNEP, 2004) at 64–66. Available at http://cd4cdm.org/publications.htm (Last Accessed: 28 June 2007).
[49] "*Trading Carbon as a Commodity*" at 13.

central registry.[50] For VERs, the delivery steps required will vary depending on the terms of the original contract. Generally delivery of these contractually based credits is linked either to the receipt of a verification report that details the number of credits created in a specified period or a system of transfer forms which mimics the registry structures of formal schemes.[51]

(c) Quantity of credits

It is important that a contract clearly specifies the quantity of credits being traded. The problem of specifying quantity is of particular concern in contracts dealing with forward sales of credits generated by a project, since the number of credits which will actually be produced in a verification period cannot be known for certain until the end of that period.

This problem can be addressed in the contract by[52]:

- Specifying that the buyer will purchase a percentage of the total credits produced e.g. "90% of all CERs produced" or "all CERs produced by a project" over a certain time period; or
- By using the form, "the first 1 million CERs."

An intermediary party that buys credits and then enters into a secondary ERPA to on-sell the rights to those credits to a third party will generally use two main forms to specify the quantity of credits they are contracting to sell:

- An obligation to deliver a percentage of the CERs which are delivered to the intermediary under the primary ERPA.

---

[50] "*Trading Carbon as a Commodity*" at 3, 14.

[51] "*Trading Carbon as a Commodity*"at 14.

[52] "*Carbon Contracts, Structuring Transactions*" at 102.

- An obligation to deliver a minimum fixed volume, which usually reflects the minimum guaranteed under the primary ERPA.

Many CDM projects have been noted to under-perform against the original estimates of the emissions reductions they will generate.[53] For sellers who "guarantee" the delivery of a minimum number of credits, it is advisable not to specify absolute volumes of credits but rather percentages of total credits, thus minimizing their risk in the event that the project does not produce the number of credits originally expected. In either case sellers should give themselves a buffer by guaranteeing a number of CERs below that which they can reasonably expect a project to generate.[54]

(D) Post-2012

There remains considerable uncertainty as to what will happen to the carbon market upon the expiry of the Kyoto Protocol obligation period on 31 December 2012. It is not yet clear what targets will be imposed (if any) or whether the Kyoto Protocol will be replaced by another framework for emissions trading. Consequently, it is not yet clear whether there will be a formal or informal market for CERs and other Kyoto rights beyond 31 December 2012.

While increasing interest in establishing mandatory carbon trading schemes is being shown by the traditional opponents to the Kyoto Protocol — the United States and Australia — their visions of global or regional approaches to carbon reductions still stand in opposition to Kyoto. The APP countries appear to support regional or global trading schemes which impose voluntary emissions reductions targets on countries rather than the mandatory targets imposed under Kyoto. If a Kyoto style regime continues to exist but no

[53] *Ibid.*

[54] *Ibid.* "*Trading Carbon as a Commodity*" at 13–14.

further emissions reductions targets are imposed on countries then the demand for Kyoto credits is likely to fall. No such targets have yet been agreed and there is currently considerable opposition from the United States and Australia to any further targets in the absence of binding targets on all major emitters and in particular India and China.

Regardless of what the regional or global emissions reduction commitments system will look like, it seems likely that many of the methodologies and processes developed through the Kyoto mechanisms would be retained. However buyers must when entering ERPAs be aware of the risk of being left with an obligation to purchase carbon credits that will not be recognised in the carbon market of the future.

The post-2012 issue can be dealt with in a number of ways:

- A purchaser can elect to purchase credits only for the period up to 2012;
- A purchaser can include an option to purchase credits produced by the project after 2012 in the ERPA or right of first refusal;
- the buyer can agree to purchase post-2012 credits, usually for a reduced price; or
- the ERPA can include an outright trade of all right and interest in the emissions reductions of the project now and in the future.

If the purchaser decides to try and acquire the potential credits generated by the project after 2012, then it is advisable to include wording that is designed to capture any future forms of carbon credit which may exist at that time and thus maximise the opportunity for conversion to future rights.[55] There are obvious difficulties in attempting to create a contractual clause that attempts to cover future and largely unforeseeable contingencies. Generally contracts adopt language that attempts to assign all potential rights or values that

[55] *Ibid.* "*Trading Carbon as a Commodity*" at 7.

could arise from the reductions in GHG emissions. Such broad drafting may have the potential to capture other future rights created in relation to GHGs that are not related to addressing climate change (as we understand the notion today).

5.2. *Delivery risk*

A key question, particularly in the case of forward sales, is who will bear the risk of delivery not occurring. Delivery risk can be conceived by looking at the mechanism by which the contract responds to a delivery shortfall. Common mechanisms for dealing with delivery failure that have been adopted include[56]:

- a right for the buyer to terminate the contract immediately or after a specified notice period;
- a "grace period" during which the seller can make up the shortfall e.g. delivering additional credits in the next delivery period or purchasing replacement credits on the market and delivering them to the buyer. An obligation to use best endeavours to make good the shortfall in later delivery periods may also be imposed;
- an obligation to procure replacement credits;
- an obligation to pay the buyer the difference between the contract price and the prevailing market price;
- payment of the buyer's costs of finding and buying replacement credits. Usually the amount payable will be determined by reference to market price and may include interest on money the buyer paid upfront for the contract credits.

These delivery default mechanisms must be read in conjunction with the definition of a delivery default. Early forward sales contracts

[56] "*Trading Carbon as a Commodity*" at 14; "*Carbon Contracts, Structuring Transactions*" at 102–103.

were often based on the World Bank model ERPA for the Prototype Carbon Fund. This agreement emphasised the importance of building capacity and had weak enforcement provisions against the sellers of the project — most of the breach provisions were restricted to events of "gross negligence", "fraud", or "wilful misconduct". By the time commercial players and governments looking to secure carbon credits for compliance reasons or seeking financial returns began to enter the market in force, sellers relied on the terms of the World Bank contract as being industry standard.

Commercially minded purchasers who wish to have greater certainty of delivery of emissions reductions often request warranties from the seller which guarantee delivery. On the other hand, the intense competition amongst buyers for high quality carbon credit projects has meant buyers have continued to agree to relatively weak enforcement provisions when compared to normal commercial agreements. The majority of ERPAs continue to restrict the more onerous breach provisions to instances of "intentional breach", "wilful default" or "gross negligence". Some purchasers even advertise their lack of enforcement as a selling point. For example the website of the European Carbon Fund (ECF) notes as one of the advantages of entering into an agreement with ECF as that there is no penalty in the case of "project technical failure to deliver CERs/ERUs"[57] or "unintentional delivery default".[58]

The result is that the more stringent mechanisms for dealing with delivery default, such as the buyer paying the seller's replacement costs, often do not come into play unless the seller has deliberately breached the contract, such as by selling the credits to another buyer.

Defaulting sellers may be made to bear the liability for penalties imposed on a purchaser for exceeding their emissions cap under their governing emissions trading scheme. Such liability would be

---

[57] See http://www.europeancarbonfund.com/transactions.php (Last Accessed: 4 June 2007).

[58] See European Carbon Fund Brochure (April 2007) Available at http://www.europeancarbonfund.com/documents/ecf_brochure_0704.pdf (Last Accessed: 4 June 2007).

significant. The current excess emissions penalty in the EU ETS is €40/tCO_2-e, and will rise in the 2008–2012 commitment period to €100/tCO_2-e.[59]

(a) Risk of delay in registration of a CDM project

A key consideration with CDM projects is that emissions reductions cannot be converted into CERs until the project has been registered. Few projects are permitted to generate CERs from emissions reductions achieved before registration (known as retroactive credits).[60] If a contract involves the purchase of CERs produced by a project from 2008 to 2012 and the project is only registered in 2009, then the emissions reductions produced in 2008 will not be convertible to CERs. It is possible to use these emissions reductions as VERs, an option which may be satisfactory for the seller when the VER prices that can be obtained are relatively high. However it is in the interests of a buyer to be able to use them as either CERs or VERs, so that they can maximise the use of the credits for compliance or on-selling purposes.

(b) Dealing with ITL delays in secondary trades

Although the delays in the establishment of the ITL restricts the transfer of CERs between parties, it is still possible to enter into

---

[59] Article 16, Directive 2003/87/EC. In addition to having to pay the penalty, an installation which does not have sufficient EUAs to cover its emissions in one year must also surrender an amount of EUAs equivalent to its excess emissions in the following year.

[60] Paragraph 4 of Decision 7/CMP.1. The Executive Board at its 28th Meeting decided (at para 78) for a project to produce retroactive credits it must have:

(a) commenced between 1 January 2000 and 18 November 2004;
(b) either:
  (i) requested validation by a DOE by 13 December 2005; or
  (ii) submitted a new methodology by 11 January 2006;
(c) submitted the request for registration by 31 March 2007.

forward sales agreements under which a party that has purchased a CER agrees to deliver the credit to another party, either:

- at a set time in the future. In such a case the seller assumes the risk of further delays in the implementation of the ITL. E.g. if they agree to transfer the CERs in July 2008, but the ITL is delayed until December 2008, then the seller will be in default since it would not be possible to transfer title to the CERs in July 2008; or
- at a time when the ITL will be established. Contracts can provide for the CER to be delivered on a certain date or "five days after the ITL is implemented, whichever is the earlier."

5.3. *Dispute resolution*

Given the relative novelty of carbon trading there have been few (publicly recorded) disputes between buyers and sellers of carbon: it is only a matter of time, however, before disputes become public and are the subject of either arbitral awards or judicial determination in the Courts of host countries or, more likely, in those jurisdictions traditionally selected as venues for commercial arbitration (e.g. Singapore, Hong Kong, New York and London).

(a) Common dispute resolution mechanisms in ERPAs

Methods of dispute resolution that can be selected in ERPAs include mediation or conciliation, litigation in the national courts of one of the parties or a neutral country, and expert determination. However, arbitration is by far the most common method of dispute resolution nominated in ERPAs, for the same reasons it is nominated in most international commercial contracts — speed, flexibility, finality and confidentiality. Arbitral awards are also enforceable under the New York Convention 1958[61] — where the country of a party has ratified the

---

[61] *UN Convention on the Recognition and Enforcement of Foreign Arbitral Awards 1958.*

Convention, its courts are obliged to enforce the arbitral awards made in relation to the contract. The Convention specifies some limited exceptions to enforceability. The main exceptions are in Article V of the New York Convention, which outlines the limited mechanisms by which a court may refuse to recognize a foreign award:

(i) where a party was under an incapacity or the agreement was invalid; or
(ii) where a party was not given proper notice of the appointment of an arbitrator or the arbitral proceedings, or a party was not given a proper opportunity to be heard; or
(iii) where the award deals with a dispute that is not within the ambit of the agreed referral to arbitration; or
(iv) where the arbitrator was not appointed in accordance with the agreement, or, failing such agreement, was not appointed in accordance with the law of the country where the arbitration occurred; or
(v) where circumstances exist such that the award has not yet become binding or has been set aside or suspended.

A court may also refuse to recognise the award if the subject matter involved in the dispute is not capable of settlement under the laws of the country or where recognition or enforcement of the particular award would be contrary to public policy.

There is always a risk that a domestic court will consider it has jurisdiction to hear a matter for one reason or another, notwithstanding domestic or international instruments to the contrary. Domestic statute law can also impact on the enforcement of awards. For example, China and Indonesia limit the operation of the Convention to commercial disputes.

Depending on the country, arbitration awards (or the judgments of foreign courts) might also be enforced according to specific foreign judgment enforcement law or because the parties have agreed to the dispute mechanism in the contract and thus are under a contractual obligation to adhere to the award.

The choice of law specified for arbitration, and the choice of arbitral rules, may be those of a jurisdiction other than that specified as the location of the hearing for the arbitration. Complex rules have developed under international law for the purpose of establishing how laws should be applied out of the jurisdiction, so for contracts that might be the subject of arbitration the most important consideration is to ensure that the commercial intention of the parties is clear throughout the document.

(b) Choice of rules for arbitration

The rules chosen to govern arbitration of issues under a carbon trading contract are generally the UNCITRAL Rules, PCA Rules or ICC Rules.

(c) Choice of law in carbon trading contracts

Carbon trading contracts normally include an explicit choice of law provision. The most common selection is English law, reflecting the preference for English or US law in international transactions (particularly in commodity trading) and the involvement of the UK in both the Kyoto Protocol and EU ETS. Because of the prevalence of English law in the carbon trading market, selecting English law is preferable when the intention is to on-sell carbon credits since it reduces the potential for problems to arise from differences in the drafting or enforcement of the contracts governing the primary and secondary trades.

(d) Types of disputes that can arise from carbon trading transactions

In 2001, Kalas and Herwig outlined types of disputes that may arise in relation to a CDM transaction, including[62]:

(i) Disputes arising out of the financial failure of a project. Similar considerations apply here as for any investment;

---

[62] P. R. Kalas and A. Herwig. "Dispute Resolution Under the Kyoto Protocol" (2000) 27 *Ecology Law Quarterly* 53 at 111.

(ii) Disputes over registration of the projects and certification and issuance of CERs;
(iii) Disputes over the retrospective invalidation of CERs;
(iv) Disputes over changes in the rules of a trading or compliance system;
(v) Disputes brought by third parties (NGOs and other Annex I countries) who challenge the validity of a project or transaction.

Although the strict requirements for registration and issuance are likely to minimise the number of CER revocations that will occur, if the CERs issued for a project are not backed up by genuine emission reductions, then the CERs may in some circumstances be revoked. This will give rise to a range of disputes over the liabilities arising for parties who have relied on the CERs to meet emissions caps.

(e) Interpretation of breach provisions

The majority of ERPAs continue to restrict penalties to events of "intentional breach", "wilful default" or instances of "gross negligence". It is unclear how such clauses will be interpreted when they begin to be tested in dispute resolution. Some of the phrases used are without established legal meanings under certain systems of law. For example, the phrase "gross negligence" while having a meaning in some jurisdictions, appears not to be a term of art which has a distinctly different meaning to "negligence" under English commercial law.[63] The English courts have been prepared to attach a separate meaning to "gross negligence" when the phrase is used in contracts on the basis that in choosing to use the phrase the parties have clearly intended to refer to something other than just mere negligence — they do not intend to refer to all conduct which would satisfy the elements of the tort of negligence. To give effect to the intention of the parties,

[63] *Armitage v Nurse and Others* [1997] 2 All ER 705 per Millet LJ; *Tradigrain SA and Others v Intertek Testing Services (ITS) Canada Ltd and Another* [2006] All ER (D) 127 (Apr) Langley J at [97].

the courts will, on the merits of each case, attempt to distinguish between "mere" negligence and "gross" negligence.[64]

Under the *contra proferentem* rule an exclusion clause is construed strictly against the party at whose instigation it was included in the contract and who seeks to rely on it.[65] Ambiguous words in exemption clauses are construed in the way least favourable to the party relying on them.[66] This risk may be mitigated by providing a clear definition within the contract of the terms used in an exclusion clause.

(f) Misleading representations

Parties that purchase carbon credits for the purpose of providing carbon-offsetting services must be careful in the way that they describe the products they offer, as inaccurate descriptions may expose them to liability for misleading representations in jurisdictions where such an offence exists. Before making statements that may have a technical meaning (e.g. describing a product or a business as a whole as "carbon neutral") companies should seek advice. Those using voluntary carbon to support carbon initiatives are at the greatest danger, again reinforcing the need for robust contractual frameworks for VERs.

(g) Potential for State Immunity to bar action

Disputes arising out of carbon trading transactions are likely to involve States as parties, either as a buyer or seller of the carbon

---

[64] See *Red Sea Tankers Ltd v Papachristidis (The Hellespont Ardent)* [1997] 2 Lloyd's Rep 547. Further support for this approach can be found in *Fellus v National Westminster Bank* (The Times, June 28, 1983 cited in Walton, Cooper and Wood (eds), *Charlesworth & Percy on Negligence,* 11th Ed (London: Sweet & Maxwell, 2006) at 1–15, note 37) in which it was held that where the phrase "undue negligence" appeared in a contract for the purchase of traveller's cheques, it meant "excessive negligence" and was a question of fact and degree to be decided in each case.

[65] G. Treitel, *The Law of Contract,* 11th Edition (London: Sweet & Maxwell, 2003) at 221.

[66] *Ibid.* citing *Morris v C W Morris & Sons* [1966] 1 QB 716.

trades, an investor in a project or in their regulatory capacity. This raises the question of whether such a party can rely on principles of sovereign immunity to avoid obligations being enforced against them.

5.4. *Environmental risk*

Lenders to projects and buyers of CERs need to be satisfied that the projects generating carbon credits not only have all necessary environmental approvals to carry on business but that they do not, in their operation, breach any domestic environmental or safety laws thereby potentially exposing third parties (such as those lenders or those buyers) to liability particularly where "step in" rights are exercised or where buyers take equity positions in a project. The high level of involvement some credit buyers have through the provision of services to the project producing the carbon credits can potentially expose them to claims that they are exercising control over the site, and are therefore directly liable for environmental liabilities.

However, this will depend not only on the domestic laws of each individual jurisdiction and their capacity to enforce those laws, but also the extent of the lender's or buyer's involvement in the project. For example, some jurisdictions make project managers, occupiers and others liable for breaches of the environmental law in addition to the direct polluter or owner of the site.

The extent to which lenders and purchasers can mitigate liability will depend on a range of factors including:

(a) the availability of insurance coverage, either inside or outside of the host country, for those risks;
(b) the contractual arrangements in place between the parties involved in the project; and
(c) indemnification arrangements in place between the parties involved in the project.

Domestic law will also specify what environmental impact assessment processes are required for the project. In some jurisdictions certain types of CDM project may be given more favourable treatment whether by way of lower taxes or charges or a reduced level of rigour in the process that must be undertaken. In other jurisdictions, domestic law or CDM project approval policy may identify certain project types as always acceptable or never acceptable in terms of environmental assessment.

5.5. *Market price risk*

Buyers and sellers must be wary of the potential for the price of carbon credits to fluctuate over the period of delivery. Such movements can significantly impact on the value of the benefits obtained by each party under the contract. The price volatility in the carbon market and the long-term nature of many carbon credit contracts means that market price risk is an essential consideration in any carbon credit transaction.

(a) Price flexibility mechanisms

The response to price volatility taken in many ERPAs has been to introduce price adjustment mechanisms, whereby the agreed sales price is varied to reflect drastic changes in the market price between the time of contracting and the time of transfer. The price can either be pegged to the prevailing market price or a discount of market price, or, alternatively or in combination, a floor or ceiling price can be imposed on a market-indexed price.

(b) Advance payments under ERPA

Generally payment is on delivery of the credits. However, in an effort to secure supplies of carbon credits in the increasingly competitive CDM market, buyers are agreeing to pay in advance for a proportion of the credits that are anticipated to be delivered under the contract. This obviously offers advantages to the seller. The buyer may wish to include a clause by which the advance payment is reclaimed if the delivery fails to recover it.

(c) Hedging risk through carbon derivatives markets

Participants in the carbon market often diversify their risk exposure through a portfolio of carbon credit sources or by utilizing financial instruments such as options, swaps and futures that have been developed for the carbon market to hedge the risk parties have from their primary exposure to the carbon market.

6. Future Developments

A number of law reform measures which can be taken now may assist in supporting the carbon trading market. Implementing legislation that gives legal recognition to carbon credits and establishes the rules for determining ownership and possession of the credits will encourage investors and project developers to establish carbon credit projects in a jurisdiction. In doing so, it is important to try and achieve consistency with other jurisdictions, since credit purchasers will generally be operating in another jurisdiction. Removing barriers to foreign investment which discourage carbon credit generating projects, will also make developing a carbon credit project in a country more favourable.

Establishing the legal framework and institutions to support the generation of CERs is a necessary step for countries that wish to access the CDM market. This infrastructure required for the CDM process is likely to become a feature of future legislative or treaty-based trading systems and may become part of the regulatory measures for ensuring the integrity of the voluntary carbon market. Establishing national or regional voluntary credit registers to ensure integrity of VERs generated within the area, will be an advantage in a market where VER buyers are being subjected to increased scrutiny over their emissions reduction activities. The integrity of the voluntary market would also be strengthened by establishing global assessment standards for voluntary carbon, whether by government intervention or the continuing cooperation of NGOs and industry.

Glossary

AAU	Assigned Amount Unit. A unit issued pursuant to the relevant provisions on registries in the Marrakesh Accords and is equal to one metric ton of CO_2-e.
ACX	Asia Carbon Exchange or ACX-Change
Additionality	The reduction in Greenhouse Gas emissions by sources or removals by sinks that is additional to any that would occur in absence of the CDM Project activity. The Marrakesh Accords state that a project activity is additional if anthropogenic emissions of Greenhouse Gases are reduced below those that would have occurred in the absence of the CDM project.
Annex B Country	One of the countries listed in Annex B to the Kyoto Protocol, being a list of Annex I Countries that have committed to a quantitative emission reduction target under Article 3.1 of the Kyoto Protocol.
Annex I Country	Countries that have committed to emission restraints under Article 4.2(a) and (b) of the UNFCCC as listed in Annex I of the UNFCCC (generally developed countries and countries undergoing the process of transition to a market economy).
APP	Asia-Pacific Partnership on Clean Development and Climate. Group of 6 countries (US, Japan, China, India, Republic of Korea and Australia) that have formed a partnership for developing responses to climate change.
APEC	Asia-Pacific Economic Co-operation
Assigned Amount	Amount of Greenhouse Gases emissions that an Annex B Party can emit during the Commitment Period taking into account the quantified emission limitations reduction commitments of Annex B of the Kyoto Protocol.
Baseline Scenario	The scenario that reasonably represents the anthropogenic emissions by sources of Greenhouse Gases that would occur in the absence of the proposed project activity.
Carbon Neutral	Phrase used to describe an activity that produces no net Greenhouse Gas emissions. Prominent companies including IAG, HSBC and BSkyB have adopted the

(Continued)

	goal of attaining carbon neutral status. A company can achieve carbon neutral status by reducing their overall emissions and purchasing carbon offsets. (e.g. carbon credits) or investing directly in activities that reduce the amount of CO_2-e in the atmosphere.
CDM	Clean Development Mechanism. Mechanism under Article 12 of the Kyoto Protocol designed to facilitate investment by Annex I countries in emissions reduction projects in non-Annex I countries.
CDM Executive Board	The formal governance body established under Article 12 to oversee the implementation and administration of the CDM, under the authority and guidance of the COP/MOP.
CDM Methodology	Methodology that has been approved by the CDM Executive Board, which can be used to establish a baseline and prove additionality for a CDM Project. A large number of Methodologies have been approved for different types of emissions reduction projects.
CDM Modalities and Procedures	The basic set of rules that govern the CDM. The Modalities and Procedures were adopted by COP7 as part of the Marrakesh Accords (Decision 17/CP.7) and were subsequently adopted by the COP/MOP at its first meeting (Decision 4/CMP.1). The rules established by the Modalities and Procedures have been expanded and clarified by subsequent decisions of the COP and COP/MOP and by the decisions and practice of the CDM Executive Board.
CDM Project	An emission reduction project which is intended to be registered with the CDM Executive Board and ultimately realise the delivery of CERs.
CDP	The Carbon Disclosure Project is a program where institutional investors collectively sign a single global request for disclosure of information on greenhouse gas emissions and climate change risk by the largest companies in the world. The first request was sent in 2002, resulting in the publication of the CDP1 report in 2003. The CDP4 report was released 2006. The request for information to be compiled into CDP5 was sent to companies in February 2007.

(*Continued*)

CERs	Certified Emissions Reductions. Carbon credits issued in relation to emissions reductions achieved by CDM Projects. It is equal to one metric ton of CO_2-e.
CITL	Community Independent Transaction Log. An add-on to the ITL which is checks and registers Kyoto credits that are introduced into the EU ETS.
CCX	Chicago Climate Exchange. A voluntary trading community for GHGs established in 2003, the CCX now has over 300 members.
CO_2-e	Carbon dioxide equivalent. There are a range of Greenhouse Gases that contribute to the greenhouse effect and climate change. However, each contributes to different degrees (i.e. they have a different Global Warming Potential). For example one molecule of methane has the same effect as 21 molecules of carbon dioxide. For ease of calculation emissions of GHGs other than CO_2 are described in terms of the equivalent number of molecules of CO_2 that they represent. Global Warming Potentials for the purpose of the UNFCCC and Kyoto Protocol are defined in decision 2/CP.3 of the Marrakesh Accords or as subsequently revised in accordance with Article 5.
COP	Conference of Parties to the UNFCCC, held on a regular basis to establish the rules to implement the UNFCCC.
COP/MOP	Conference of the Parties serving as the meeting to the Parties to the Kyoto Protocol, being the Kyoto Protocol's supreme body. The sessions of the COP and COP/MOP will be held during the same period.
DNA	Designated National Authority. The authority for CDM designated by a Party to the Protocol. DNAs are generally responsible for the assessment and approval of CDM Projects in Host Countries.
DOE	Designated Operational Entities. An independent legal entity accredited by CDM Executive Board that can validate proposed CDM Projects and verify and certify Greenhouse Gas emission reductions.
ECX	European Carbon Exchange. The ECX is the platform on which the majority of EUA trades take place for the EU ETS.

(*Continued*)

EFET	European Federation of Energy Traders.
Equator Principles	A framework for banks to manage the environmental and social issues of projects for which they are giving project financing. Signatories to the Principles agree not to loan money to a project with a total capital cost exceeding US$10 million unless the sponsors of the project comply with requirements to undertake environmental and social assessments and, in some cases, to develop management plans for the environmental and social impacts of the project.
ERPA	Emissions Reduction Purchase Agreement.
ERUs	Emissions Reduction Units. A unit issued under the JI mechanism pursuant to Article 6 and all other relevant Kyoto Protocol requirements and which is equal to one metric ton of CO_2-e.
EUAs	EU Allowance. Permits which give right to holders to emit pollutants covered by the EU ETS.
EU ETS	European Union Emissions Trading Scheme.
First Commitment Period of the Kyoto Protocol	The period between 2008–2012 during which Annex B countries are required to reduce their emissions of Greenhouse Gases to the levels established in the Kyoto Protocol.
GGAS	NSW Greenhouse Gas Reduction Scheme. Commencing on 1 January 2003 GGAS is a mandatory greenhouse gas emissions trading scheme administered by the State of NSW in Australia in which energy generators and large energy consumers must participate.
GHG	Greenhouse Gas. Gas that contributes to the Greenhouse Effect by preventing the escape of radiant heat from the Earth's atmosphere. Six Greenhouse Gases include carbon dioxide (CO_2), methane (CH_4), nitrous oxide (N_2O), ozone, water vapour and chlorofluorocarbons (CFCs). The targets established under the Kyoto Protocol cover emissions of six GHGs, which are listed in Annex A to the Protocol — carbon dioxide, methane, nitrous oxide (N_2O),

(*Continued*)

	hydrofluorocarbons (HFCs), perfluorocarbons (PFCs) and sulphur hexafluoride (SF_6). CFCs are addressed under the Montreal Protocol.
Host Country	Non-Annex I country in which a CDM project is located.
ICC	International Chamber of Commerce.
IETA	International Emissions Trading Association.
ISDA	International Swaps and Derivatives Association.
ITL	International Transaction Log. Once established the ITL will verify the validity of all transactions involving Kyoto Protocol rights including CERs within or between registries (including between a national registry and the CDM registry). The log will cover the issuance, transfer, acquisition, cancellation, retirement or carryover into the next commitment period of any Kyoto Protocol rights.
JI	Joint Implementation Mechanism. Flexible mechanism under Article 6 of the Kyoto Protocol with the purposes (1) to assist Annex I Parties in achieving sustainable development and (2) to contribute to the ultimate objective to the UNFCCC and (3) to assist Annex I Parties to achieving compliance with their quantified emission limitation and reduction commitments.
Kyoto Protocol	The Protocol to the UNFCCC signed at the third COP meeting that requires developed countries to reduce their GHG emissions by an average of 5.2% below 1990 levels by 2008–2012. It was adopted by all Parties to the UNFCCC in Kyoto, Japan, in December 1997. To enter into force the Protocol had to be ratified by 55 parties representing 55% of industrial nations' Greenhouse Gas emissions in 1990. Despite the US and Australia not yet ratifying the treaty, it entered into force in February 2005.
Letter of Approval	A letter issued by the Designated National Authority of the host country to a CDM Project

(*Continued*)

	confirming that the project, as proposed, will assist the host country to achieve its goals of sustainable development. This written approval is a necessary step in the process of getting a CDM Project registered.
Marrakesh Accords	Decisions 2/CP.7 through to Decision 24/CP.7 (inclusive) of the seventh session of the COP/MOP.
Methodologies Panel	Body responsible for assessing proposed new CDM Methodologies. On the basis of this assessment the Panel makes a recommendation to the Executive Board as to whether the Methodology should be approved or not.
National Board for the Examination of CDM Projects	Body established by Article 13 of the "Measures for Operation and Management of Clean Development Mechanism Projects in China" (2005), which is responsible for reviewing CDM projects in China including the terms of the agreement for the sale of CERs produced by a project.
NCCCC	The PRC's National Coordination Committee on Climate Change.
NDRC	The PRC's National Development and Reform Commission. The NDRC is the PRC's Designated National Authority.
NGAC	Greenhouse Abatement Credits created under the GGAS scheme in NSW, Australia.
NGO	Non-Governmental Organization.
Non-Annex I Country	A country not listed in Annex I of the UNFCCC.
PCA	Permanent Court of Arbitration.
Programmatic CDM	CDM Project where emissions reductions are achieved not by one single investment but rather by multiple actions executed over time as a result of a government measure or private sector initiative.
PDD	Project Design Document. The document to be prepared and submitted by CDM Project Participants to an accredited DOE for validation of a proposed project activity.
Registration	The formal acceptance by the CDM Executive Board of a validated project as a CDM Project.

(*Continued*)

	Registration is the prerequisite for verification, certification and issuance of CERs related to that project.
RGGI	Regional Greenhouse Gas Initiative. Cooperative effort by North-eastern and Mid-Atlantic states in the US to reduce carbon dioxide emissions by implementing a cap-and-trade emissions trading system for electricity generators. Member states include Connecticut, Delaware, Maine, New Hampshire, New Jersey, New York, and Vermont. Legislation was signed in April, 2006, that requires Maryland to become a full participant in the process by June 30, 2007. In addition, the District of Columbia, Massachusetts, Pennsylvania, Rhode Island, the Eastern Canadian Provinces, and New Brunswick are observers in the process.
UNCITRAL	United Nations Commission on International Trade Law.
UNFCCC	United Nations Framework Convention on Climate Change, signed at the 'Earth Summit' in Rio de Janeiro in May 1992.
Validation	The process of independent evaluation of a project activity by a designated DOE against the requirements of the CDM as set out in the Marrakesh Accords on Article 12 and on the basis of the Project Design Document.
Verification	The periodic independent review and ex post determination by the designated DOE of the monitored reductions in anthropogenic emissions by sources of Greenhouse Gases that have occurred as a result of a registered CDM Project activity during the verification period.
VERs	Verified Emissions Reductions. A carbon credit generated outside of a formal scheme but which has been verified by a third party, usually in line with the rules of a pending scheme.

Schedule 1: CDM Project Cycle[67]

PROJECT PARTICIPANTS COMPLETE A PROJECT DESIGN DOCUMENT (PDD)

The PDD must be in accordance with the requirements of the CDM Executive Board. It functions to describe the project activity, the proposed baseline methodology and the project participants. It must also explain how the Additionality requirements will be met and how the emission reductions will be monitored. The PDD must also establish a "crediting period", for the project, which (for standard projects other than sinks projects) can be either seven years (with the possibility of 2 renewals) or ten years with no renewal option. Note that the crediting period for sinks projects is different from other CDM Projects.

Issues to be considered:

- The project participants must prove that the CDM Project will reduce or sequester Greenhouse Gases in accordance with the Additionality requirements of the Marrakesh Accords
- Baseline and monitoring methodologies must be approved by the CDM Executive Board.
- What is the "starting date" of the project? (i.e. for what period can CERs be created?)
- Who will be the project participants and who is entitled to the CERs from the project?

↓

DESIGNATED NATIONAL AUTHORITY ISSUES LETTER OF APPROVAL

The Host Country must confirm (generally through its DNA) that it approves a CDM Project for the purposes of the Kyoto Protocol and that the project assists it to achieve its sustainable development goals.

↓

DESIGNATED OPERATIONAL ENTITY VALIDATES PROJECT ACTIVITY

DOE reviews the PDD to confirm that the project activity meets the following requirements:

- participation is voluntary and Parties have established a DNA for the CDM;
- any non-Annex I countries participating in the project are parties to the Kyoto Protocol;

---

[67] This diagram is taken from M. Wilder, M. Willis and J. Carmody, *Legal Issues Guidebook to the Clean Development Mechanism* prepared by Baker & McKenzie for the United Nations Environment Programme CD4CDM project (Roskilde, Denmark: UNEP, 2004) at 32–34.

- stakeholder views have been considered;
- environmental impacts have been analyzed;
- baseline methodology and monitoring plan are appropriate and comply with those approved by the CDM Executive Board;
- PDD contains provisions for monitoring, verification and reporting in accordance with Article 12; and
- project activity conforms to all other requirements for CDM Projects.

After inviting and considering stakeholder comments on the PDD, the DOE determines whether to validate the project.

Issues to be Considered:

- Choosing a DOE — Who should bear the costs of Validation?
- To what extent will stakeholder comments be taken into account?

↓

REGISTRATION BY THE EXECUTIVE BOARD

Registration will automatically occur within 8 weeks after the DOE has submitted the request to the CDM Executive Board, unless a Party involved in the project or three members of the CDM Executive Board request a review in relation to the Validation requirements.

↓

MONITORING BY PROJECT PARTICIPANTS

Project participants are required to implement the monitoring plan in the PDD. This will require:

- collection and archiving of all data needed for the measurement and estimation of Greenhouse Gas emissions or sequestration;
- identification and attribution of the emissions to the project activity;
- assessment of environmental impacts; and
- quality assurance and control procedures for monitoring and calculating Greenhouse Gas emission reductions (simplified monitoring procedures can be used for small scale CDM projects).

Issues to be Considered:

- Project participants must comply with monitoring plan to provide DOE with sufficient information to undertake verification (although DOE can also seek its own information).

↓

VERIFICATION AND CERTIFICATION BY DESIGNATED OPERATIONAL ENTITY

Periodic independent review and written assurances by a DOE of the monitored enhanced reductions in emissions during the verification period. This requires the DOE to:

- make the monitoring reports publicly available;
- calculate the enhanced Greenhouse Gas reductions;
- ensure the project complies with the PDD and CDM rules; and
- complete a Verification Report to confirm the amount of Greenhouse Gas emission reductions, and a Certification Report that certifies to the CDM Executive Board that the project activity actually achieved those reductions.

Issues to be Considered:

- Which DOE should undertake verification and certification? Who will bear the costs?
- Will the DOE indemnify the project participants from any liability owing to inaccurate verification or certification?

↓

ISSUANCE OF CERs BY THE CDM EXECUTIVE BOARD

The CDM Executive Board will automatically issue CERs within 15 days after it receives the Certification Report (unless there is a request for review). CERs will be issued by electronic registry created and maintained by the CDM Executive Board.

The CDM Executive Board registry administrator will:

- issue the CERs and distribute them to the relevant national registries and/or accounts of the project participants as requested in the signed statement contained in the PDD; and
- deduct a share of proceeds to cover administration costs and meet the costs of adaptation.

Issues to be Considered:

- How will CERs be delivered?
- Do the Host Country participants have access to a registry account?

Schedule 2: Case Study — Domestic Regulation of CDM Projects in the PRC

From 12 October 2005, CDM projects in the PRC have been governed by the "Measures for Operation and Management of Clean Development Mechanism Projects in China" (the Measures). In addition to the Measures the PRC regulators have adopted a number of policy positions and practices which impose restrictions on how a CDM project can be carried out.

Project Owners Must be PRC Controlled

Only enterprises in the PRC which are wholly owned by PRC persons or entities or those in which the PRC party or parties holds a controlling interest (i.e. 51%) may undertake CDM projects with foreign parties.[68] This rule means that initially all CERs produced by CDM projects in the PRC are owned by a PRC-controlled company, which may then be under a contractual obligation to sell the CERs to a foreign entity. This places some restrictions on the types of joint venture arrangements that a foreign company might wish to enter into with PRC project owners and increases the importance of having robust contractual arrangements for the sale of the CERs to the foreign buyer.

Foreign Buyer Must Agree to Purchase All of the CERs

Article 15-2 provides that if a foreign buyer for CERs has not been found by the time a project is submitted for approval, the CERs must be transferred into the PRC's national account in the CDM registry and can only be transferred out with the authorisation of the PRC's DNA. In practice the NDRC will only be willing to approve a project when the foreign buyer agrees to purchase all of the CERs produced by the project up until 2012. Otherwise the CERs will be issued to

[68] Article 11 of the "Measures for Operation and Management of Clean Development Mechanism Projects in China." Recently, the NDRC has agreed in principle that a "PRC party" can include entities established in the Special Administrative Region of Hong Kong, however, it is still unclear under what circumstances this will be allowed.

the Chinese government until a buyer can be found after registration. This provision ensures that a project will only be approved once it is confirmed that a price above the floor price has been secured.

The Imposition of a Floor Price for CERs

Article 13 of the Measures establishes a National Board for the Examination of CDM Projects,[69] which is responsible for reviewing CDM projects including the terms of the agreement for the sale of CERs produced by the project, including the price.[70] In practice the NCCCC, together with the Board, determines a floor price for CERs through these reviews. The NDRC, in its capacity as DNA, will only issue a Letter of Approval to a project if it feels the price stipulated in the ERPA is sufficient. The price it will accept appears to vary with the current international market price, the number of CERs produced by a project and across project types. The current floor price which is being enforced by the PRC is €8 or US$10.[71]

In negotiating a contract to purchase CERs in the PRC it is advisable to seek advice on the prices agreed under projects which have been recently approved by the NDRC. A foreign purchaser should also be aware of the impact that currency movements may have on the floor price.

There is anecdotal evidence that the NDRC is unwilling to approve projects that have a price flexibility mechanism which can bring the price below the floor price. On top of this, the desire to ensure that CERs generated within the PRC are getting a fair price may lead to the NDRC looking more favourably on projects which have an ERPA that provides for price increases when the market price for CERs at the time of delivery and payment exceeds the contract price.

---

[69] Article 15 specifies that the Board is led by the NDRC and Ministry of Science and Technology. Other members of the Board include the Ministry of Foreign Affairs, the State Environment Protection Administration, the China Meteorological Administration, the Ministry of Finance and the Ministry of Agriculture.

[70] Article 15–1 of the "Measures for Operation and Management of Clean Development Mechanism Projects in China."

[71] Point Carbon, "China approves 72 projects in latest update" (13 June 2007).

The PRC Government Takes a Share of the CER Revenues

The share of the revenues taken by the Government varies according to the type of project[72]:

- 65% of the CER revenues from HFC and PFC destruction projects;
- 30% of the CER revenues from nitrous oxide (N_2O) destruction projects, and
- 2% of the CER revenues from carbon sequestration projects and projects in the priority areas of energy efficiency improvement, development and utilisation of new and renewable energy sources and methane recovery and utilization (which includes landfill and waste mine gas methane).

NDRC Ruling on Consultancy Fees

This ruling prohibits contracts under which a consulting or intermediary company performs services in return for a percentage of CERs or benefits derived from a transfer of CERs. However, our understanding, based on previous discussions with the NDRC, is that it is possible for the CDM project developer to own the CDM rights to a project without owning or operating the underlying project — this is actually quite common in projects such as coal mine methane or landfill gas projects — and as such the ruling would only apply in instances where a consultant was advising and assisting the PRC company in return for a percentage share of the CERs. On this understanding it seems possible to avoid the consultancy fee ruling by restructuring any CER sharing arrangement into a CER sales arrangement.

72 Article 24 of the "Measures for Operation and Management of Clean Development Mechanism Projects in China".

THEME IV

EFFECTIVE IMPLEMENTATION OF THE KYOTO PROTOCOL IN ASIA

CHAPTER 10

SINGAPORE'S NATIONAL CLIMATE CHANGE STRATEGY

K. Suresh

Climate Change Unit
(Resource Conservation Department)
National Environment Agency
Singapore

Singapore is a city-state with limited natural resources and is totally dependent on imported fossil fuels to meet our energy needs. As an export-oriented economy, much of the energy used by industry is used to make products for export.

The government is committed to addressing climate change in an environmentally sustainable manner that is compatible with economic growth and has developed a National Climate Change Strategy (NCCS). Singapore acceded to the Kyoto Protocol in 2006.

The key carbon mitigation strategies for Singapore are promoting energy efficiency and promoting the use of clean, less carbon intensive energy.

Singapore will continue to seek sustainable energy solutions to reduce the environmental impact of energy production and use. The government is committed to involving and working with both the community at large as well as businesses to achieve a less carbon intensive economy.

1. Introduction

Singapore is a city-state with limited natural resources. It is totally dependent on imported fossil fuels to meet our energy needs. As an export-oriented economy, much of the energy used by our industry is used to make products for export.

Singapore acceded to the Kyoto Protocol in 2006. The government is committed to addressing climate change in an environmentally sustainable manner that is compatible with economic growth and has developed a National Climate Change Strategy (NCCS).

The NCCS comprises the following 4 key areas:

(a) Mitigation

It's main mitigation measure is to improve energy efficiency and use less-carbon intensive fuels.

(b) Vulnerability and Adaptation

As a small island state, it is vulnerable to the impact of climate change. Early this year, it embarked on a 2-year comprehensive study of the impacts of climate change on Singapore to study the most appropriate adaptation measures to take.

(c) Competency Building

Singapore is set to promote climate-related R&D such as low-carbon technologies, facilitate transfer of technologies, develop local expertise in energy management to improve energy efficiency and promote Clean Development Mechanism (CDM) projects.

(d) Public Awareness

It has to raise public awareness on the threat of climate change and educate the public on specific measures that they can take to help reduce greenhouse gas emission.

2. Greenhouse Gas Emissions

Currently, Singapore's annual GHG emissions, mainly carbon dioxide (CO_2) from the combustion of fossil fuels, are about 40 000 kilotons[1]; which is about 0.14% of global annual GHG emissions based on latest available data published by the International Energy Agency (IEA). The table below shows the carbon dioxide contributions of the various sectors. On the supply side, the power generation sector accounts for about 48% of total emissions. On the demand side, industry, transport, buildings and households sectors account for about 54%, 20%, 15% and 9% of total emissions respectively after combining fossil fuel combustion and secondary electricity consumption.

	Power Generation (%)	Industry (%)	Transport (%)	Buildings (%)	Households (%)	Others (%)
Fuel Consumption	48	33	17	1	1	—
Electricity Consumption	—	21	3	14	8	2
Overall		54	20	15	9	2

In the Singapore Green Plan 2012,[2] Singapore has set a target to reduce carbon intensity (GHG emissions per dollar of GDP) to 25% below 1990 levels by 2012. To date, Singapore's carbon intensity is about 26% below 1990 levels and, based on latest available data

[1] 1000 metric tons.

[2] http://www.mewr.gov.sg/sgp2012/index.html

published by the International Energy Agency (IEA), compares favourably against developed nations[3]:

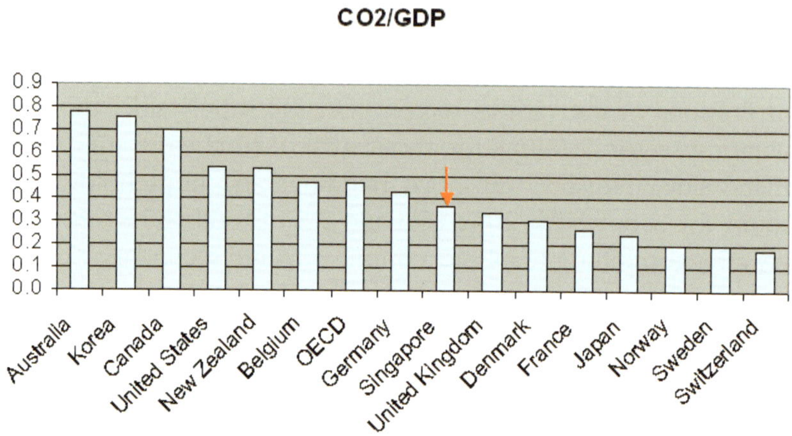

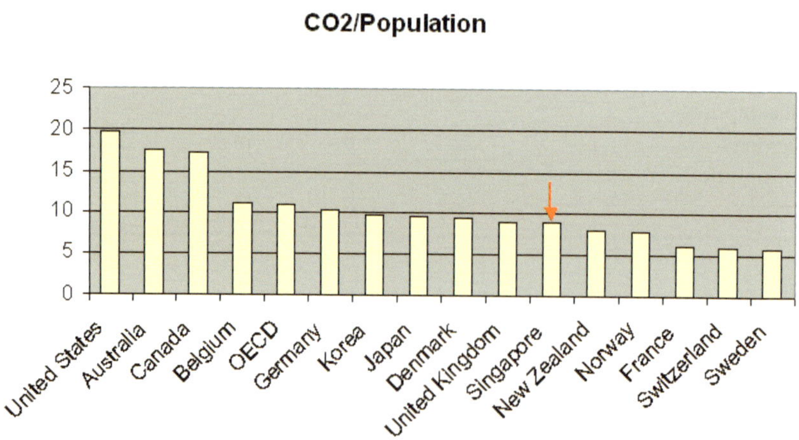

Source: IEA, 2004 provided in para 5.

[3] Singapore's economic and demographic structure is more comparable with those of other developed nations. Nations such as the United Kingdom, Denmark, France, Japan, Norway, Sweden and Switzerland have a lower carbon intensity than Singapore as they have recourse to nuclear and renewable energy.

3. Mitigation, Public Awareness and Competency Building Strategies and Programmes

Singapore's strategy to minimise greenhouse gas emission from each of the energy-consuming sectors is to promote:

(a) energy efficiency throughout the energy cycle and in all sectors of energy use without affecting the quality of life; and
(b) the use of clean or less carbon-intensive energy sources, e.g. natural gas. With natural gas, for each unit of energy produced, less CO_2 is released into the atmosphere.

Some of the key programmes are as follows:

(a) Power Generation Sector

- Since Jan 2003, the electricity market was liberalized and power generation companies (Gencos) are required to sell electricity through a competitive bidding process. The Gencos have over the past several years sought greater generation efficiencies to maintain their commercial viability. Most have opted to switch to natural gas and combined cycle generation technologies to enhance their operational efficiency and environmental performance. The proportion of electricity generated by natural gas has increased from 19% in 2000 to 80% in 2006.
- A cogeneration facility built on Jurong Island in September 2001 generates electricity and recovers waste heat to generate steam. Co-generation is able to achieve generation efficiency of up to 75%. Promoting cogeneration will help to reduce emission of greenhouse gas emission.
- The National Environment Agency encourages private and public sector partners to test-bed and demonstrate innovative clean energy technologies such as solar photovoltaics and hydrogen fuel cells through the Innovation for Environmental Sustainability (IES) Fund. These demonstration projects will help promote early adoption of new and innovative technologies.

For example, in 2006, two trigeneration projects implemented by pharmaceutical companies were supported under the IES Fund. These projects will help reduce carbon emissions by 23 200 tonnes annually and result in annual savings of over SGD $900 000 for the companies.

(b) Industry Sector

- A S$10 million Energy Efficiency Improvement Assistance Scheme (EASe) was introduced in April 2005. The Scheme funds up to 50% of the cost of engaging accredited energy service companies (ESCOs) to conduct energy audits. To-date, manufacturing companies and building owners/operators participating in EASE (provided in first line of this para) have projected annual energy savings of $13 million and reductions of 112 kilo-tonnes of CO_2 emissions for Singapore.
- A one-year accelerated depreciation allowance for energy efficient equipment and technology[4] was introduced in January 1996. Companies that replace old, energy-consuming equipment with more energy efficient ones or invest in energy-saving equipment could depreciate the qualifying capital equipment in one year instead of three.

(c) Transportation Sector

- A key strategy for this sector is to promote the use of public transport. In October 2006, the Ministry of Transport announced plans to increase the modal share of public transport during morning peak-hours from 63% to at least 70% over the next 10–15 years.[5]
- A voluntary Fuel Economy Labelling Scheme was launched in June 2003.[6] The main objective is to raise consumer awareness of the fuel economy in cars. Information provides on the fuel economy labels would help motorists make better-informed

---

[4] http://www.nccc.gov.sg/incentive/home.shtm
[5] http://app.mot.gov.sg/data/s_08_01_18.htm
[6] http://www.nccc.gov.sg/transport/transport.shtm

decisions on fuel efficiency when they are considering the purchase of a new car.

- The Government announced the enhanced green vehicle tax rebates in December 2005. The aim of the rebates is to promote the use of green vehicles such as petrol-electric hybrid cars and vehicles that use cleaner fuels such as compressed natural gas (CNG) and hydrogen. Since the enhancement of the rebate, the number of green vehicles has gone up substantially from about 200 to 700. The rebate has been extended to December 09.

(d) Building Sector

- The Government has established a Building Control Regulation[7] for air-conditioned buildings to stipulate minimum energy efficiency standards. Requirements on the maximum permissible Envelope Thermal Transfer Value (ETTV) and Roof TTV (RTTV) levels, minimum efficiency requirements for air-conditioning systems exceeding 30 kW and maximum lighting power budget have been set.
- The Singapore Standard (SS530) on Energy Efficiency for Building Services and Equipment[8] was implemented from 1 July 2007. SS530 provides the minimum energy-efficiency standard of building equipment such as air-conditioning equipment, heat rejection equipment, water heaters, electric motors and high efficiency lightings. Requirements for air-conditioners are aligned to international standards and the latest technologies. For example, air-conditioning chillers that comply with the new SGD $530 will use up to 30% less energy, compared to the performance requirement in the previous standard (CP 24).
- An Energy Smart Building Labelling Scheme[9] was launched in December 2005 to accord recognition for buildings with good

[7] http://www.bca.gov.sg/PerformanceBased/others/ETTV.pdf

[8] http://www.spring.gov.sg/Content/ModulePage.aspx?group=nw&id=ecbdfd81-0776-4313-8667-70da22f03c0c#SS530

[9] http://www.nccc.gov.sg/energysmart/office.shtm

energy performance, while maintaining a healthy and productive indoor environment. A similar scheme will be introduced for hotels in 2007. The National Environment Agency (NEA) will consider extending the Scheme to other types of buildings e.g. shopping centres and hospitals.

- The public sector is taking the lead on energy efficiency. All large air-conditioned Government office buildings, as well as tertiary institutes, will conduct energy audits to improve their energy efficiency. In addition, all new large government buildings and new schools will meet the Singapore Green Mark certification standard. Some agencies, including the Ministry of the Environment and Water Resources, Ministry of Finance and Ministry of Manpower, have already completed their energy audits, and are projected to reap combined savings of $240 000 annually.

(e) Household Sector

- A voluntary Energy Labelling Scheme[10] was launched in April 2002 for refrigerators and air-conditioners. The aim is to guide consumer-purchasing behaviour towards sustainability in energy use. Mandatory labelling will be introduced for refrigerators and air-conditioners in Jan 2008.
- A Climate Change Awareness Programme[11] (CCAP) was launched in April 2006 to encourage households and motorists to make simple lifestyle changes/choices that would save them money, reduce energy wastage and reduce greenhouse gas emissions. The CCAP is led by the Singapore Environment Council, an NGO, with support from the NEA.

4. Clean Development Mechanism

Under the Kyoto Protocol, there are 2 categories of Parties — Annex 1 and non-Annex 1. Annex 1 Parties are those countries that have

---

[10] http://www.nccc.gov.sg/Households/Households.shtm
[11] http://everydaysuperhero.sg/

greenhouse gas emission reduction targets to meet between 2008–2012. The non-Annex 1 countries do not have such targets. Singapore is a non-Annex 1 country.

The Clean Development Mechanism or CDM is a mechanism under the Kyoto Protocol that allows greenhouse gas emissions reductions from projects implemented in non-Annex I countries to be used to offset emissions of Annex I countries. Such projects, if implemented in Singapore, would earn emission reduction credits known as Certified Emissions Reductions or CERs, which are tradable. Hence, any company in Singapore that is able to reduce its greenhouse gas emissions may be able to benefit from the CDM.

CDM is not only a means to generate cash flow, but also a means for companies to demonstrate their commitment towards taking action against global warming by implementing projects that result in measurable and long-term reductions in greenhouse gas emissions.

The National Environment Agency (NEA) has been appointed Singapore's Designated National Authority (DNA). NEA's role is to ensure that CDM projects to be implemented locally meet national sustainable development criteria. NEA will support sustainable CDM projects to the fullest possible extent and provide clear guidelines to ensure an efficient evaluation and approval process.

5. Vulnerability and Adaptation

Like other nations, Singapore will not be spared from the effects and impacts of climate change. We have put in place various measures to adapt to the impacts of climate change.

Notwithstanding this, Singapore needs to ensure that we address all the key risks. To do that, the NEA has commissioned a 2-year study to get a better understanding of the likely long-term impacts of climate change on Singapore arising from the projected effects of climate change such as changes in rainfall patterns, sea levels and extreme weather conditions. This study, which is conducted by a team of local and foreign experts, will incorporate the latest scientific research on climate change such as the latest report from the Intergovernmental Panel on Climate Change (IPCC). A number of the foreign experts who have been actively involved in the drafting of the IPCC reports have also been co-opted in the study.

6. Conclusion

Singapore will continue to seek sustainable energy solutions to reduce the environmental impact of energy production and use. The Government is committed to involving and working with both the community at large as well as businesses to achieve a less carbon intensive economy.

Singapore's key carbon mitigation strategies are promoting energy efficiency and promoting the use of clean, less carbon intensive energy.

CHAPTER 11

JAPAN: ACHIEVING ITS KYOTO TARGET

Hiroji Isozaki
Meiji Gakuin University, Japan

The first commitment period of the Kyoto Protocol will begin in 2008. However, the United States, the world's largest emitter of GHGs, has not ratified the Kyoto Protocol, and developing nations such as China and India are not required to reduce emissions. The Kyoto Protocol also lacks a mechanism for promoting the development of the innovative technology required to fundamentally resolve the global warming problem. Further, major emitters such as the United States and Australia are not expected to ratify the Kyoto Protocol, and developing countries are not expected to make substantial efforts to reduce GHGs. Taken together, all this means that the Kyoto Protocol is not an effective framework for addressing global warming. As the host country of the Conference of Parties in Kyoto ("COP3"), Japan is still working to develop the best possible proposal for combating global warming.

Current Status of the Debate on a Post-Kyoto Framework — Overview of the Main Actors' Positions on a Post-Kyoto Framework

The EU wishes to maintain the Kyoto Protocol framework (setting binding, country-specific, quantitative targets) and to make EU ETS a global standard.

The US calls for promoting a strong and transparent system, distinct from the Kyoto Protocol, under which each country sets different targets, and focuses on technological solutions.

The major developing countries continue to insist that responsibility lies with the developed nations, but there are some signs of a more proactive approach.

Former Prime Minister Shinzo Abe of Japan has proposed a "flexible and diverse framework" for the post-Kyoto framework. The outcome of the G8 summit in Heiligendamm represents the current middle ground and a starting point for negotiations.

At the subsequent APEC Economic Leaders' Meeting, common quantitative targets for energy efficiency were agreed upon in the form of a non-binding declaration. This declaration is significant in that it demonstrates that the concept of energy efficiency can be given a central role in measures against global warming and that major emitters, excluding the EU, can join in an agreement between developed and developing nations when based on the concept of energy efficiency.

Proposal for a Post-Kyoto Framework — Commitment and Action Plan

(1) Necessary elements for a post-Kyoto framework

- (I) Is premised on a long-term perspective and sustainability, in the sense that it ensures the continued efforts of both governments and the domestic entities responsible for emissions
- (II) Identifies the potential reductions in the various countries (particularly major emitters of GHGs) through scientific methods and an objective process, and shares information on these potential reductions among countries
- (III) Includes internationally legally-binding commitments made to policies and actions that governments can definitely implement

(2) Six principles of the new framework

- (I) Environmental effectiveness — Ensure that GHGs are actually reduced

(II) Science-based analysis — Data on potential for reductions; costs should be based on scientific analysis
(III) Equity — In accordance with "common but differentiated responsibilities and respective capabilities of countries and their social and economic conditions" (UNFCCC)
(IV) Inclusiveness — Broaden participation to include non-governmental entities as actors in the new agreement
(V) Political feasibility — Required to enable all countries to participate in the framework
(VI) Sustainable, long-terns perspective — A reasonable amount of lead time should be provided for innovative technology development and diffusion

A new protocol should be agreed on in discussions based on Article 4, paragraph 2 (d) of the UNFCCC. It should be examined based on Article 9 of the Kyoto Protocol, and should replace the Kyoto Protocol after the first commitment period expires at the end of 2012 (or when a COP decision is made to shift to a new protocol). In 2013, the new protocol, comprised of the following three categories as major elements in its structure, should go into effect based on Article 17 of the UNFCCC.

The commitment period should last 50 years beginning in 2013, and every five years negotiations should be held on revising Category I commitments and updating Categories II and III based on the most recent scientific, technical, economic and social information.

Category I: Shared Commitments to Binding Actions by Major Emitters' Governments

(a) Internationally legally-binding provisions are stipulated for major emitters including developing countries (the contents are determined by negotiating a series of policy templates through the "request and offer" negotiation process in order to build an agreed-upon policy matrix)
(b) Provisions to ensure compliance and deal with countries in violation

Category II: Individual Commitments to Non-binding Actions by All Governments

(a) Individual governments make political commitments to policies and measures (the new protocol stipulates the kind of items that should be included in the commitments)
(b) Measures to ensure implementation

Category III: Participatory Commitments to Individual Actions by Private Sector Entities

(a) Commitments to actions to combat global warming by private sector entities that are in accord with the object of the new protocol
(b) Procedures to register, validate and verify those actions and their achievements

(3) International cooperation for innovative technology development

An international cooperation program should be established and an appropriate system developed for handling intellectual property rights and assigning public-private roles. This program should distinguish between short-term practical research on energy conservation and the development and use of new energy and long-term basic research that will lead to innovative technology development.

(4) Financial issues in developing countries' measures against global warming (mitigation and adaptation)

Under the new framework, a system could be designed to encourage independent efforts by the major emitters with steadily developing economies. Specifically, conditions on use of the original CDM should be imposed on major emitters among the developing countries in the aforementioned Category I, by requiring that those countries commit to internationally legally-binding measures to reduce GHGs and actually implement them.

With this as a prerequisite, developed countries should preserve financial mechanisms used for CDM, and the governments of developed countries should commit the funds used to purchase emissions credits generated by CDM to Category 1 policies through negotiations. In this case, one option would be to require that developed countries purchase enough emission credits to compensate for their shortfall in meeting the Kyoto Protocol targets.

Structural Problems of the Kyoto Protocol

(1) The greatest failure of the Kyoto Protocol is its lack of adequate consideration for technology.
(2) A regulatory approach based on setting quantitative targets, as required by the Kyoto Protocol, makes negotiations a diplomatic game rather than a science-based dialogue. Further, developing countries can evade requirements to reduce GHGs in perpetuity as long as the Kyoto Protocol remains in force.
(3) There is such a wide range of GHG-emitting entities that the quantitative targets that the Kyoto Protocol imposes will be very difficult to achieve without regulating a broad range of national economic activity and daily activity.
(4) The Kyoto Protocol places obligations on developed countries but does not require developing countries to reduce emissions, leaving no guarantee of its effectiveness
(5) The Kyoto Protocol sets penalties for non-compliance with quantitative targets committed to by the developed countries that have ratified the Kyoto Protocol, but does not provide any type of disincentive for developed countries that do not ratify the Kyoto Protocol.

1. Reduction Targets for Japan under the Kyoto Protocol

In Japan, the average temperature has risen by approximately 1 degree Celsius during the 20th Century.[1] Climate change will

---

[1] *Japan's Fourth National Communication Under the United Nations Framework Convention on Climate Change*, at p. 3. (http://www.env.go.jp/en/earth/cc/4th_framework.html)

have a significant impact on ecosystems, agriculture, social infrastructure, and human health, possibly leading to drastic lifestyle changes. Under the Kyoto Protocol, Japan has to reduce its greenhouse gas (GHG) emissions by 6% compared to the base year level of 1990.

Japan produces approximately 5% of the total world's carbon dioxide (CO_2) emissions, which makes it the 4th largest emitter in the world after the USA (about 22%), China (about 18%) and Russia (about 6%) in 2004.[2] Developed nations produce larger CO_2 emissions per capita than developing nations.

2. GHG emissions from Japan

Japan's total GHG emissions in Fiscal Year (FY) 2003 was 1339 million tons.*[3] This is an increase of 8.3% compared to emissions in the base year (1990) (1255 million tons*) under the Kyoto Protocol.[4] For emissions of HFCs, PFCs and SF6, the base year adopted for the purposes of the Kyoto Protocol is FY 1995.

There was no significant decrease in Japan's GHG emissions levels in FY2004, in which the country emitted 1355 million tones*, which is 8% higher than the total emissions in the 1990 base year.[5] A sector-by-sector analysis of the country's GHG emissions profile shows that, using the 1990 base year, the industrial sector's emissions had decreased by 3.4%, emissions of the transport sector had increased by 20.3%, emissions of the commercial and remaining sectors had increased by 37.9%, and those of the residential sector had increased by 31.5%.[6]

---

[2] Ministry of the Environment, *White Paper on the Environment and the Sound Material-Cycle Society 2007*, p. 111.

[3] Hereinafter, figures marked with * represent data for CO_2 equivalents.

[4] At p. 41, *op. cit.* note 1.

[5] Ministry of the Environment, *Quality of the Environment in Japan 2006* (White Paper), p. 26.

[6] At p. 26, *Ibid.*

3. Measures Decided by the Japanese Government to be Taken for the Kyoto Protocol

Following the adoption of the Kyoto Protocol, the Japanese government has promulgated a number of measures to address global warming. These measures include:

(a) the Law concerning the Promotion of Measures to Cope with Global Warming (1998, amended 2002).
(b) the Action Program to Arrest Global Warming (1990).
(c) the Basic Policy on Measures to Tackle Global Warming (1999).
(d) the Outline for Promotion of Efforts to Prevent Global Warming (1998, 2002).
(e) the Kyoto Protocol Target Achievement Plan (2005).
(f) the Law on Energy Conservation (1979, amended 1998, 2002).
(g) the Law on Promotion of Use of New Sources of Energy (1997).
(h) the Law concerning the Recovery and Destruction of Fluorocarbons (2001).

In March 2002, the Outline for Promotion of Efforts to Prevent Global Warming was revised following promulgation of the Marrakesh Accords in November 2001 (2002 Revised Outline). In May 2002, the Law concerning the Promotion of the Measures to Cope with Global Warming was also revised.

The 2002 Revised Outline sets out Japan's strategy which is essentially a gradual and incremental one that regularly evaluates the progress of mitigation measures and policies, and that implements any additional requisite measures and policies. Specifically, the period between 2002 and the end of the first commitment period under the Kyoto Protocol (2008–2012) was divided into three phases (2002–2004, 2005–2007 and 2008–2012). Reviews of the existing measures and policies were taken in 2004 and 2007. The 2002 Revised Outline was further revised after the 2004 review process.

4. The Kyoto Protocol Target Achievement Plan, 2005

On April 28, 2005, in response to the further revision of the Outline in 2004, Japan's Cabinet Meeting approved a Kyoto Protocol Target Achievement Plan.[7] This plan contains various policy actions, such as introducing low-emission vehicles, energy-saving equipment, and solar power generation systems. Carbon dioxide arises from every aspect of human activity. Therefore, achieving a 6% GHG reduction commitment and implementing energy-saving measures will require a collective effort by all levels of society, from the national and local governments, to business owners and individual citizens.

The main details of the Kyoto Protocol Target Achievement Plan are as follows[8]:

(a) *Direction of Global Warming Countermeasures*

* Steadily achieve the 6% GHG reduction commitment under the Kyoto Protocol.
* Further pursue long-term and continuous reduction of GHG emissions on a global scale.

(b) *Basic Philosophy*

* Compatibility between the environment and the economy: Addressing global warming may also create opportunities for sustainable development. In this regard, Japan will strive for technological innovation and seek to develop technologies for the low-carbon economy of the future. The research and development, and production of such technologies also present opportunities for the development of Japan's high-technology sector.
* Promotion of Technological Innovation: The Government of Japan will accelerate technological innovation in areas

[7] http://www.env.go.jp/en/press/2005/0427b.html
[8] http://www.env.go.jp/en/press/2005/0427b-01.pdf

such as energy conservation, utilization of unused energy, etc. The dissemination of such technologies abroad will foster Japan's endeavour to be a global leader of environmental technology.

* Promotion of Participation and Collaboration of all Stakeholders (National Campaign and Sharing of Information): The Japanese government will promote the active participation of all stakeholders including central and local governments, corporations and citizens, in GHG mitigation measures and policies, and strengthen multi-stakeholder collaboration by actively providing and sharing information on the efficacy of the mitigation measures as well as additional actions that may be required if existing measures prove inadequate.
* Utilization of Diverse Policy Instruments: The Japanese government will effectively employ diverse policy instruments, such as voluntary methods, restrictive methods, economic methods, and informational methods in order to harness the strengths of various policy tools to address climate change.
* Strengthening the Evaluation and Review Process: In 2007, the Japanese government undertook a comprehensive evaluation of the progress of existing GHG mitigation measures and policies. The intention was to implement additional or alternative measures the following year if the review process indicated the necessity of such actions. An annual review process would be undertaken to ensure that the policies are working as intended.
* Ensuring International Collaboration: The Japanese government reiterated its commitment to active participation in the international climate change negotiations so as to create a global rule-based architecture that will facilitate the involvement of all countries, including the US and developing countries, in the global effort to address climate change. Japan will also take a leading role in the world's

efforts to combat global warming through international cooperation.

5. Additional Measures Taken

While the measures outlined above represent a commendable effort to implement a GHG mitigation strategy, total GHG emissions in FY 2006 increased by 7.8% from the 1990 base levels.[9] As a result, the total amount of GHG emissions that Japan will have to reduce in order to meet its Kyoto Protocol obligations represent a 13.8% decrease from its 1990 base level. This represents a significant challenge that led to the following measures being taken.

By partial amendment to the Law concerning the Promotion of the Measures to Cope with Global Warming that went into effect on April 1, 2006, major emitters are now required to calculate their GHG emission volumes and report them to the government. The government will compile the data submitted and make this information available to the public.

All types of GHGs are subject to this requirement. Businesses that are already required to report energy usage under the Law concerning the Rational Use of Energy (primarily those in the energy and transportation sectors) are now also required to report their CO_2 emissions derived from energy consumption.

For other types of GHGs, businesses with more than 20 full-time employees are required to report the aggregate amounts of their emissions by type of GHG at each business site where emissions exceed 3000 tons.

These measures are aimed at encouraging businesses to tackle the issue more concretely by first requiring them to accurately determine their emission levels. Further, by making the emissions profile of such companies available to the public, consumers and the general public will be better able to put pressure on such companies to improve their environmental performance.

[9] Ministry of the Environment, *White Paper on the Environment and the Sound Material-Cycle Society 2007*, p. 111.

6. Measures under the Kyoto Mechanisms

(a) *Emissions trading*

The Voluntary Emissions Trading Scheme started in FY 2005,[10] and the Ministry of the Environment has selected 34 companies and corporate groups, including INAX Corp., Nippon Electric Glass Co., Mitsubishi Gas Chemical Co. and Yamazaki Baking Co., to participate in the scheme.

Under the scheme, the Ministry subsidizes the costs of installing CO_2 emissions reduction equipment by businesses that are actively attempting to reduce their GHG emissions. In exchange for the subsidy, the participants are required to commit to a certain reduction in their CO_2 emissions. The scheme also allows them to trade CO_2 emission quotas to meet their reduction targets.

The participants in this scheme have committed to a total reduction of 276380 tons for FY2006, or 21% of their average annual CO_2 emissions in the base years, FY 2002 to 2004. The reduction in CO_2 emissions over the officially-recognized service life of the subsidized equipment is estimated to be about 3.7 million tons.[11]

In February 2007, Chuo Mitsui Trust Bank and Mitsubishi UFJ Trust Bank announced that they would accept Emission Trade Rights as Trust Property. The Bank can either take charge of the Right and divide it into small lots (which makes it easier for companies to purchase these rights) or the Bank can keep the money in trust and, under the direction of the entruster, buys and administers the Emission Trade Rights.

(b) *Joint Implementation projects*

On June 1, 2005, the governments of Japan and Romania signed a Statement of Intent on cooperation in the facilitation of Joint Implementation (JI) projects. This is the first JI document that Japan

---

[10] http://www.ets-japan.jp/dms/dms_3_2.html (in Japanese).

[11] http://www.env.go.jp/press/press.php?serial=5979 (in Japanese).

has signed.[12] On December 20, 2005, Japan and Bulgaria signed a Memorandum on cooperation to facilitate the implementation of JI projects.[13]

(c) *Clean Development Mechanism (CDM)*

In the Kyoto Protocol Target Achievement Plan, the Japanese government declared its intention to promote the utilization of the CDM to meet its Kyoto Protocol target. Major Japanese power companies have been entering into agreements for the sale and purchase of emission credits generated by CDM projects in China.[14] These transactions include:

Tokyo Electric Power Co. and Xinjiang Uighur Autonomous Region
650 000 t, 7 years, wind power project 30 000 kW
Tokyo Electric Power Co. and Guizhou Province
840 000 t, 6 years, hydropower project, 80 000 kW
Chubu Electric Power Co. and Zhejiang Province
2 million t, 6 years, chlorofluorocarbon collection project
Kansai Electric Power Co. and Gansu Province
560 000 t, 6 years, hydroelectricity power plants

In addition, the CDM Executive Board recently approved a CDM project proposed by Ricoh Company. In cooperation with Conservation International (CI), this is a reforestation project in Ecuador for an abandoned pasture area of about 5 million square meters.[15] This activity would also help to restore the ecosystem by assisting local people in collecting, nursing, and providing and managing the seeds of original wild species.

---

[12] http://www.env.go.jp/en/press/2005/0601a.html
[13] http://www.env.go.jp/en/press/2005/1220a.html
[14] Tokyo Electric Power Co. (http://www.tepco.co.jp/cc/press/06062901-j.html) (in Japanese); Chubu Electric Power Co. (http://www.chuden.co.jp/corpo/publicity/press2006/0619_2.html) (in Japanese); Kansai Electric Power Co. (http://www.kepco.co.jp/pressre/2006/0711-2j.html) (in Japanese).
[15] http://www.ricoh.co.jp/ecology/topic/02/index.html#ecuador (in Japanese).

7. Local or Private Initiatives

Local authorities and private entities are active in combating global warming, especially as the need for stronger action was made evident by the rising trends in CO_2 emissions in densely populated Tokyo. One measure that the Tokyo metropolitan government has implemented is a program asking companies to submit GHG reduction plans. These companies are those that consume the equivalent of at least 1500 kiloliters of petroleum a year. So far, about 1000 companies have submitted emission reduction plans.

The Tokyo metropolitan government further plans to require large companies and factories to meet GHG reduction goals under a new set of standards from FY 2008. Companies or factories that fail to meet their emission reduction targets would be required to buy carbon emissions credits.[16] The metropolitan government is also considering financial support measures that would allow small businesses to acquire funds more easily for measures to fight global warming.

On September 22, 2006, the Federation of Electric Power Companies of Japan (FEPC) announced the results of a follow-up review on the electric power industry's Environmental Action Plan. It found that CO_2 emissions in FY 2005 were 375 million tons, up 3% from the previous year. The CO_2 emission intensity (CO_2 emissions per kilowatt hour of end use) was 0.425 kg, up from the previous year's figure of 0.421 kg.[17]

There are two reasons for this increase in CO_2 emission intensity. First, thermal generation had to be increased to make up for a decline in hydro generation due to water shortages. Secondly, despite improvement in the capacity utilization rates of nuclear power plants (which resulted in an increase of power generation), record cold winter temperatures caused a spike in electricity demand.

To reduce CO_2 emission intensity to 0.34 kg by 2010 (about 20% down from 1990 levels), the FEPC intends to increase nuclear power

16 "Tokyo to set CO_2 emission targets on major companies" The Asahi Shimbun, June 1, 2007 (in Japanese).

17 Press Conference, the Federation of Electric Power Companies of Japan, September 22, 2006 (in Japanese).

capacity utilization rates, to introduce Liquid Natural Gas (LNG) combined cycle thermal power plants, and to build highly efficient coal-fired generating facilities. It is predicted that, in FY 2010, CO_2 emission intensity will be about 0.36 kg (a little higher than the reduction target) since the number of new nuclear power plants expected to be built has decreased.[18] The FEPC is also promoting biomass generating and methane recovery operations.

8. Further Measures for Kyoto Target Achievement Plan

At the Global Warming Prevention Headquarters meeting on July 7, 2006, the Japanese government evaluated the progress of the Kyoto Protocol Target Achievement Plan. It was found that the measures taken so far have led to concrete progress but further measures would have to be taken in order for Japan to meet its Kyoto Protocol target. The total emissions of GHGs in Japan in FY 2004 increased by 8% from 1990 base levels and have also grown by 1.8% compared with FY 2003. Accordingly, between FY 2004 and FY 2010, it is necessary to achieve reduction targets of 10.4% from FY 2004 levels in the energy conversion sector, 6.6% in the industrial sector, 4.4% in the transportation sector, 27.2% in other sectors, including offices and other business facilities, and 18.3% in the household sector. In the case of forest carbon sinks, if forest management remains at current levels, the amount of the reduction is estimated to fall far below the target of 3.9%.[19]

Quantitative evaluation and revision of the plan was made in FY 2007. As that evaluation was considered to directly influence the amount of emission and absorption during the first commitment period beginning in FY 2008, the government needs to strictly assess the progress of measures and policies to ensure that the 6% reduction commitment is achieved. Based on such evaluation, the government

[18] *Ibid.*

[19] The Report on the Progress of the Kyoto Protocol Target Achievement Plan, Ministry of the Environment, 7 July 2006 (in Japanese). http://www.env.go.jp/press/file_view.php?serial=8254&hou_id=7303 (in Japanese).

is in the process of amending the Law concerning the Promotion of Measures to Cope with Global Warming and the Law on Energy Conservation.[20]

Wide ranges of drastic initiatives are insisted by environmental NGOs in order to tackle the global warming agenda. These initiatives include strengthening policies in all areas of our socio-economic system, re-structuring the various policies, and re-engineering our current socio-economic system of mass production, mass consumption and mass disposal, in the future.

In particular, the following measures are required:

(a) Fuel conversion from coal to natural gas

Some measures, such as restriction of further construction of coal-fired power generation and a heavier taxation on coal, may be required to promote the fuel conversion.

(b) System to promote industry's efforts

Introduction of a carbon tax, a domestic trade emission system and a commitment for reduction targets will be useful in order to raise awareness on global warming among industries and suppliers.

(c) Strengthen policies and measures on housings and buildings

Stronger measures for energy-efficiency standards to be applied for new housing and buildings, as well as subsidies or supporting systems, will be necessary.

(d) Raising the target of renewables/Introduction of feed-in-tariff system

An accelerated spread of renewable energy supported by further amendment of related laws and regulations will be useful. In particular, the feed-in-tariff system, which fixes the price of renewables on electrical power, may lead to widespread use of renewable energy.

(e) Introduction of carbon-tax as a cross-cutting measure

A carbon tax may have an impact on the prevention of climate change.

---

[20] The Amendment Law (Law No. 67) was adopted on 13 June, 2008.

(f) Region-Led Transportation Demand Management (TDM)
A Region-Led TDM based on the framework of local transportation needs will improve the efficiency of transportation and reduce CO_2 emissions.

9. Beyond the First Commitment Period of the Kyoto Protocol

Moving away from the focus on domestic measures taken by Japan to address climate change, this section discusses the post-2012 climate change regime based on a recent movement in Japan.[21] After all, in order to stabilize the atmospheric GHG concentrations, it is necessary for countries to curb their emissions on a long-term, continuous basis, after the first commitment period of the Kyoto Protocol is over. To that end, measures should be taken not only from the perspective of meeting short-term goals, such as the numerical targets stipulated by the Kyoto Protocol, but also from a mid-to-long term perspective. States began discussing the post-2012 options at the 11th Conference of the Parties (COP11) to the United Nations Framework Convention on Climate Change (UNFCCC), and the first Conference of the Parties/Meeting of the Parties (COP/MOP1) to the Kyoto Protocol (held in Montreal in November–December 2005).

Although a variety of policy actions have been proposed for the Post-Kyoto Framework, few of them reflect the basic principles of UNFCCC and those of general environmental law. Those principles and criteria that are commonly contained in international treaties on nature conservation and sustainable use of living resources may be categorized into a few groups such as those principles that set out basic rules that regulate the use of living resources, the basic procedures that must be followed in the application of the basic principles, principles governing the proper management of natural resources and, finally, principles governing the creation and

[21] This section was originally prepared for the panel discussion of the last session of the APCEL Conference, "Crucial Issues in Climate Change and the Kyoto Protocol: Asia and the World".

improvement of legal and institutional frameworks governing environmental protection.

These fundamental principles include equity, ecosystem approach, preventive measures, precautionary approach, multi-disciplinary approach, public awareness and international cooperation. The basic procedures include access to information, prior assessment and monitoring. Resource management measures include appropriate spatial and time frames, decentralization, adaptive management and participatory management. Institutional Improvement include the improvement of laws and regulations, measures that promote public participation, respect for customary rights and traditional knowledge, compliance and dispute avoidance, access to judicial settlement, internalization of environmental costs and responsible commercial practices.

General Principles of International Law, especially the principle of sovereign equality (Article 2(1) of the UN Charter) are closely related to those basic principles set out in Article 3 of UNFCCC. First, there is the debate about the procedural aspect of the principle of sovereign equality, particularly the concepts of absolute equality, relative equality, proportional equality and functional equality. These concepts have been insisted upon by international organizations, especially international financial organizations. Second, the substantial aspect of equality has been basically insisted upon by the concept of absolute equality; but from the 1960's, the concept of substantial equality has been widely supported by developing countries.

The principle of common but differentiated responsibility (CBDR) is set out as a basic principle under the UNFCCC and has been interpreted in two ways: a difference in the obligations undertaken by each country; or a difference in the obligations of developed countries and those of developing countries and the inclusion of financial assistance from the former group to the latter. While these composite elements of the CBDR have received heightened attention in the context of global warming, they are not unique to it and have actually been incorporated in many other international legal regimes governing the international environment. Therefore, in applying the CBDR, the concepts of substantial equality and functional equality may well help to guide and reinforce the development of the post-Kyoto framework.

Finally, the Polluter Pays Principle (PPP) is not explicitly stated in the UNFCCC and its Kyoto Protocol. However, for the purposes of internalizing an environmental externality and in light of justice and fairness concerns, the PPP is expected to play a role in providing a basis for a global warming burden-sharing agreement, especially in the context of financing climate change adaptation.

10. Post-2012 Proposals

Japanese organizations have submitted several proposals for the post-Kyoto framework at various UNFCCC/Kyoto Protocol meetings. In 2007, a legal group and the economic sector submitted two new proposals. The first proposal was developed by leading Japanese environmental law professors, headed by Professor Otsuka at Waseda University. The second was developed by the 21st century Public Policy Institute, Keidanren, headed by Professor Sawa at the University of Tokyo.

Post-2012 Proposals Based on Various Principles[22]

(1) Mitigation — 3 system proposals

Proposal 1: Based on the polluter pays principle (PPP), the ability to pay principle and the principle of CBDR (Fig. 1)

(i) First, we should list those countries with a certain percentage (e.g. 70%) above the average in worldwide cumulative GHG emissions (mainly carbon dioxide) in the past (after 1990) shall be determined (EU is considered as an entity). This section applies the PPP.

(ii) Among these countries, those with a certain level of national income and other indices (purchasing power parity is a possible criterion) should reduce their emissions by a uniform rate of X% from 1990 levels (or from the year in which the above indices reached certain levels for developing countries) ("X" here is the target value that policymakers should aim for in

[22] By a group of environmental law professors.

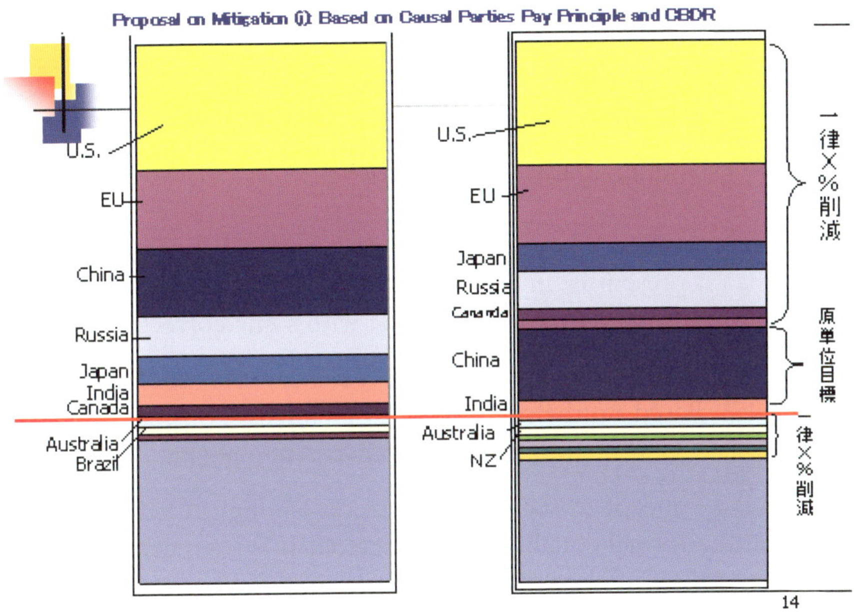

Fig. 1.

order to halve GHG emissions from 1990 levels for 70% of the world while considering iii)). It is based on the CBDR principle (which is a combination of the polluter pays principle (as discussed above in (i)) and the ability to pay principle).

(iii) Countries that are not included in (ii) shall have CO_2 emission intensity targets imposed on them. These targets can be either legally binding or voluntary (combined with the option of emissions trading). This concept is premised on the ability to pay principle.

(iv) Amongst the countries comprising the remaining 30%, countries with national income and other indices at certain levels should be obliged to reduce emissions from the viewpoint of the ability to pay principle. Countries that accept obligations with the hope of participating in emission trading shall also be acclaimed. These countries should be treated in the same manner as those addressed under (ii).

(v) The countries that remain should be given financial assistance by advanced countries, a revised CDM, etc. (based on the CBDR principlecausal party-pays-principle (= *i*) + ability to pay principle]). It is also possible to provide support for developing countries in (iii) above.

These 2 tracks are based on the concept that countries with high emissions (at present or in the past) should reduce more as indicated by the above causal party-pays-principle with some corrections by the application of the CBDR principle.

Proposal 2: Based on the principle of equitable responsibility of each Individual — Contraction & Convergence proposal (Fig. 2)

This is a concept to set up the emission amount target per capita based on the formal equality of each person. It is a concept that pursues one version of the equity principle. A problem of this design is

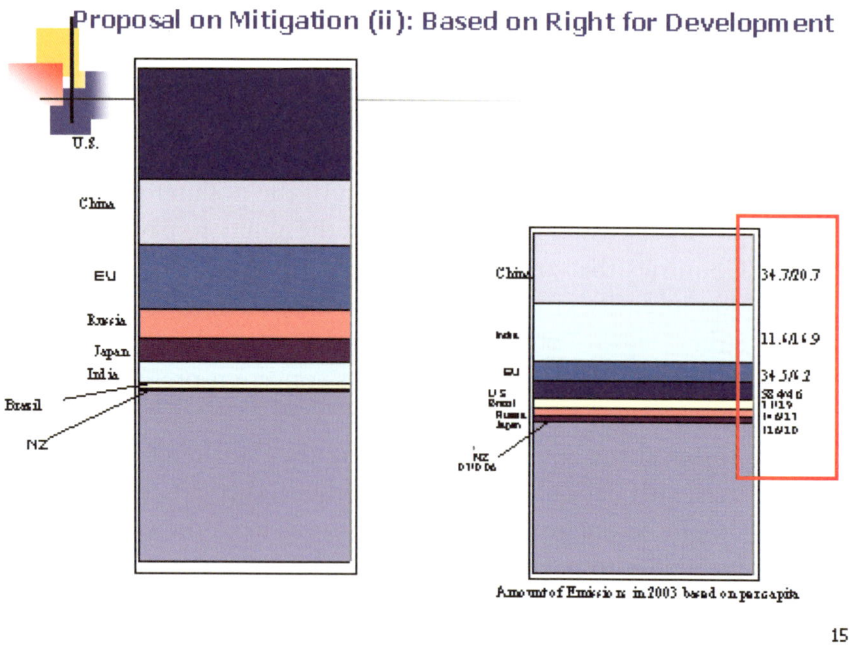

Fig. 2.

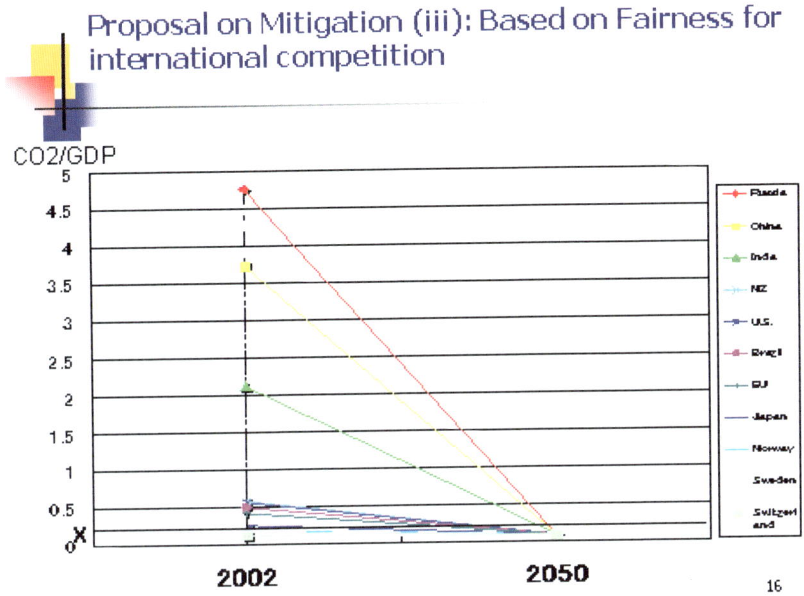

Fig. 3.

that it makes light of the fact that population increase is itself an environmental load and a cause of higher GHG emissions.

Proposal 3: Based on the principle of fairness for international competition (Fig. 3)

This method, based on the concept of fairness for international competition, seeks to establish the intensity target per GDP. This aims to equalize the GHG emission amount per certain volume of economic activity. This, however, cannot be measured with accuracy.

How we should think of the fact that the reduction of total emissions cannot be achieved when GDP increases is also problematic. It is possible to address this problem by setting up a intensity target such that the total level of emissions stays within a certain range by taking production growth into account in such a model. The concept of fairness for international competition should be applied not only to advanced countries but also to developing countries where trade and investment levels are high. In this case, all of those countries with high levels of trade and investment should attain the same target.

For developing countries among the countries with major emission levels, it may also be possible to allow certain transitional periods or to allocate a little larger emission amount. This will allow the CBDR principle to be taken into account while retaining the principle of fairness for international competition as the basis.

This proposal is considered to have the following advantages. First, it may secure the agreement of certain countries that are willing to accept the obligations set out above. Secondly, this proposal avoids the GHG emissions leakage issue. Thirdly, the presence of a concrete reduction target provides certainty and an incentive for action. Fourthly, it promotes energy efficiency and reduced consumption.

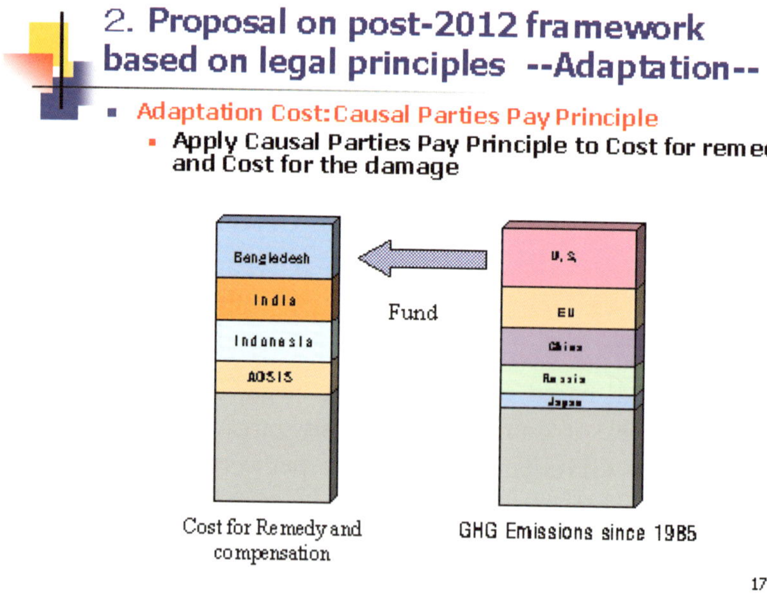

Fig. 4. Proposal for a Post-Kyoto framework

By the 21st Century Public Policy Institute[23]

[23] http://www.21ppi.org/english/pdf/071112.pdf.

(2) Proposals on adaptation (Fig. 4, Appendix 1)

It is arguable that the polluter pays principle provides the strongest justification for a fair burden-sharing agreement when it comes to climate change adaptation. However, the major difficulty with applying the polluter pays principle is the difficulty of proving the causation between the cumulative GHG emissions and the consequent damage. Though there are those who argue that it is possible to demonstrate such causation and therefore allocate liability amongst countries deemed responsible, only a few international law scholars support this. It is therefore difficult to apply the polluter pays principle to the case of adaptation.

CHAPTER 12

COMPLIANCE UNDER THE KYOTO PROTOCOL AND ITS IMPLICATIONS FOR THE ASIAN REGION

Maria Socorro Manguiat[1]

Programme Officer, Legal Affairs Programme (Compliance) Climate Change Secretariat

The strong compliance regime of the Kyoto Protocol (KP), characterized by rigorous reporting and strict eligibility rules for participation in its flexibility mechanisms, constitutes a central element for the Protocol's environmental integrity. The Kyoto Protocol's Compliance Committee and its facilitative and enforcement branches work to facilitate, promote and enforce compliance with the commitments under the Protocol. The facilitative branch aims at providing advice and facilitation of financial and technical assistance and at promoting compliance and providing for early warning of potential non-compliance. The enforcement branch (EB), on the other hand, focuses on *non-compliance* by Parties included in Annex I (KP) in relation to their obligations under the Protocol.

The work undertaken by the two branches of the Compliance Committee is inextricably linked to the reporting and review process, as consideration of any question of implementation comes at the end of the reporting, review and compliance chain. The reporting, review and compliance chain seeks to ensure that the building blocks of Protocol implementation, namely, the assigned amount, the

[1] The views expressed in this paper are personal to the author and do no represent the view of the United Nations.

national system for the estimation of anthropogenic emissions by sources and removals by sinks of greenhouse gases, and the national registry responsible for ensuring the accurate accounting of the issuance, holding, transfer, acquisition, cancellation and retirement of Kyoto units are in place. The decisions of the EB will affect the eligibility of Annex B Parties (KP) to use the flexibility mechanisms, as the eligibility criteria are linked to these building blocks and seek to ensure that any Party that wishes to use the mechanisms is in a position to accurately account for its emissions and its Kyoto units.

Facilitation is not meant to draw from resources available to developing country Parties. There are those who hold the view that it is possible to bring a question of implementation relating to a non-Annex I Party before the facilitative branch, notwithstanding some practical difficulties in doing so. Actions beyond the first commitment period may entail a system of verification that would require reporting and review and logically, a compliance mechanism. At the very least, the facilitative function will continue to be relevant, especially if less developed Parties begin to take on commitments or undertake voluntary actions.

1. Introduction

> "*[E]nvironmental integrity is to be achieved through sound modalities, rules and guidelines for the mechanisms, sound and strong principles governing land use, land-use change and forestry activities, and a* ***strong compliance regime***" (5th preambular paragraph, decision 2/CMP.1, principles, nature and scope of the mechanisms pursuant to Articles 6, 12 and 17 of the Kyoto Protocol; emphasis supplied).

The Conference of the Parties serving as the meeting of the Parties to the Kyoto Protocol (CMP) has recognized the centrality of a strong compliance regime in maintaining the environmental integrity of the Kyoto Protocol by the adoption of decision 2/CMP.1. Interestingly, it is the decision on the principles, nature and scope of the mechanisms that points to the importance of the compliance regime.

One legal scholar explains how a strong compliance regime can contribute to achieving environmental integrity:

> "*...to ensure the integrity, and thus the effective operation of emissions trading mechanisms, the Kyoto Protocol needed a compliance regime that could backstop their abuse. In short, the Kyoto Protocol's emphasis on rigorous reporting, its creation of eligibility requirements for participation in Kyoto mechanisms, and the [non-compliance procedure's] inclusion of consequences such as suspension of eligibility or deduction of excess emissions are all shaped at least in part by the Protocol's reliance on the trading mechanism.*"[2]

Under Article 18 of the Kyoto Protocol, the CMP was required to approve appropriate and effective procedures and mechanisms to determine and address cases of non-compliance with the provisions of the Protocol. In response to this mandate, the CMP adopted decision 27/CMP.1, to which was annexed the Procedures and mechanisms relating to compliance under the Kyoto Protocol (hereinafter, the "Procedures and mechanisms").[3] The objective of the Procedures and

[2] Jutta Brunnée, Enforcement Mechanisms in International Law and International Environmental Law in Ulrich Beyerlin, Peter-Tobias Stoll and Rüdiger Wolfrum, "Ensuring Compliance with Multilateral Environmental Agreements: A Dialogue between Practitioners and Academia" (Martinus Nijhoff Publishers, Leiden, 2006), p. 22.

[3] The second sentence of Article 18 requires any procedures and mechanisms "entailing binding consequences" to be adopted by means of an amendment to the Kyoto Protocol. At the first session of the CMP in Montreal, Canada, the CMP was asked to agree on the legal form of the procedures and mechanisms relating to compliance. In decision 27/CMP.1, the CMP decided to approve and adopt the procedures and mechanisms relating to compliance under the Kyoto Protocol as annexed to the decision, as well as to commence consideration of the issue of an amendment to the Kyoto Protocol in respect of procedures and mechanisms relating to compliance in terms of Article 18, with a view to making a decision by CMP 3, which will take place in Bali. The decision to adopt the procedures and mechanisms relating to compliance was "without prejudice to the outcome" of the process of consideration of the issue of an amendment to the Kyoto Protocol in respect of procedures and mechanisms relating to compliance. While the item entitled "Amendment of the Kyoto Protocol

mechanisms is "to facilitate, promote and enforce compliance with the commitments under the Kyoto Protocol" (Sec. I, Procedures and mechanisms).

This paper seeks to provide an overview of the compliance mechanism of the Kyoto Protocol, to demonstrate how it can contribute to the environmental integrity of the Kyoto Protocol, and to reflect on its implications for the Asian region.

As this paper will demonstrate, while the compliance mechanism is intimately linked to the effective operations of the mechanisms, its value goes well beyond the mechanisms and into the very core of the systems required for the credible implementation of the provisions of the Kyoto Protocol. The paper begins with an overview of the institutional and procedural aspects of the Kyoto Protocol compliance mechanism. It then demonstrates the link between the reporting, review and compliance systems, and discusses the subject of compliance and the mechanisms. The final section deals with the implications of the compliance mechanism on the Asian region both for the first commitment period and beyond.

---

in respect of procedures and mechanisms relating to compliance" has been in the agenda of the Subsidiary Body for Implementation since the 24th session of the Subsidiary Body for Implementation (May 2006), no substantive discussions have been held relating to this issue. Prof. Massimiliano Montini argues that the competence of the CMP is confined to "the approval of appropriate 'procedures and mechanisms' in order to promote compliance with the Kyoto Protocol". After analyzing the consequences listed under the annex to decision 27/CMP.1, Prof. Montini maintains that there are at least two types of consequences that the enforcement branch can impose that entail binding consequences. He goes on to observe that in addition to not having commenced any amendment procedure relating to compliance, "the States' practice subsequent to the approval of the compliance regime by the COP/MOP1 seems to point in the opposite direction and contradicts the text of the Kyoto Protocol itself." The fact that procedures and mechanisms relating to compliance entailing what some legal scholars consider to be binding consequences has been adopted through a decision of the CMP rather than through an amendment of the Kyoto Protocol is referred to as the "amendment dilemma". Massimiliano Montini, The Compliance Regime of the Kyoto Protocol in The Kyoto Protocol and Beyond (Legal and Policy Challenges of Climate Change). W. Th. Duoma, L. Massai and M. Montini (eds.), 2007.

2. The Kyoto Protocol Compliance Mechanism

2.1. *The Compliance Committee*

The Compliance Committee of the Kyoto Protocol is composed of 20 members elected by the CMP, 10 of who are elected to serve in the facilitative branch and the other 10 who are elected to serve in the enforcement branch.[4] Each member has an alternate who is entitled to participate in the proceedings of the Committee but may only cast a vote if he or she is serving as a member.[5] While nominated by groups defined under the Procedures and mechanisms,[6] members and their alternates serve in their individual capacities,[7] and are bound by rules relating to confidentiality and conflict of interest.[8]

The Committee functions through a bureau, a plenary and its two branches, the facilitative and enforcement branches. The bureau is composed of the chair and vice-chair of both branches.[9] The main responsibilities of the bureau are to allocate questions of implementation to the appropriate branch[10] and to provide guidance to the secretariat in drafting the provisional agenda for each meeting of the plenary.[11] The plenary takes care of the general business of the Committee, including reporting

---

[4] Section II, paragraph 3, Procedures and mechanisms.

[5] Rule 3, Rules of procedure of the Compliance Committee of the Kyoto Protocol, annex to decision 4/CMP.2 (hereinafter, "Rules of procedure").

[6] Both the facilitative and enforcement branches are composed of one member from each of the five regional groups of the United Nations (Western Europe and Others, Africa, Asia, Latin America and the Caribbean, and Eastern Europe) one member from the small island developing States, two members from Parties included in Annex I of the United Nations Framework Convention on Climate Change and two members from Parties not included in Annex I. *See* Section IV, paragraph 1 and Section V, paragraph 1, Procedures and mechanisms. Alternates are also nominated in the same manner.

[7] Section II, paragraph 6, Procedures and mechanisms.

[8] *See* rule 5, Rules of procedure.

[9] Section II, paragraph 4, Procedures and mechanisms.

[10] Section VIII, paragraph 1, Procedures and mechanisms.

[11] Rule 7, paragraph 7, Rules of procedure.

to the CMP on the activities of the Committee and approving rules of procedure that are forwarded to the CMP for its adoption.[12] The plenary may also develop further working arrangements to complement and give effect to the Committee's rules of procedure.[13] The facilitative and enforcement branches consider questions of implementation of the type described below.

2.2. *The facilitative branch*

The overall mandate of the facilitative branch is to provide advice and facilitation to Parties in implementing the Protocol, and for promoting compliance by Parties with their commitments under the Protocol.[14] It is responsible for addressing questions of implementation relating to:

(1) Response measures taken by Annex I Parties[15] aimed at mitigating climate change in a way that minimizes their adverse impacts on developing countries[16]; and

---

[12] *See* Section III, paragraph 2, Procedures and mechanisms.

[13] Paragraph 11, Annual report of the Compliance Committee to the Conference of the Parties serving as the meeting of the Parties to the Kyoto Protocol, FCCC/KP/CMP/2006/6 (hereinafter, "First annual report"). At its fourth meeting, the plenary of the Compliance Committee agreed on working arrangements to enable those who would like to observe the meetings of the Committee to register or follow a meeting on the Internet. *See* paragraphs 15 to 17, Annual report of the Compliance Committee to the Conference of the Parties serving as the meeting of the Parties to the Kyoto Protocol, FCCC/KP/CMP/2007/6 (hereinafter, "Second annual report").

[14] Section IV, paragraph 4, Procedures and mechanisms.

[15] Parties included in Annex I or "Annex I Parties" are defined as Parties included in Annex I to the United Nations Framework Convention on Climate Change, as may be amended, or a Party which has made a notification under Article 4, paragraph 2(g), of the Convention, which refers to Parties not listed in Annex I to the Convention that have taken on commitments that would ordinarily pertain only to Annex I Parties. *See* Article 1, paragraph 7, Kyoto Protocol.

[16] Section IV, paragraph 5(a), Procedures and mechanisms in relation to paragraph 32, annex to decision 15/CMP.1, Guidelines for the preparation of the information required under Article 7 of the Kyoto Protocol.

(2) The use by Annex I Parties of the mechanisms as "supplemental" to domestic action.[17]

With the aim of promoting compliance and providing for early warning of potential non-compliance, the facilitative branch is also responsible for providing advice and facilitation for:

(1) Compliance with quantified emissions limitation and reduction targets under Article 3, paragraph 1 of the Kyoto Protocol (hereinafter, "Article 3.1 target" or "Article 3.1 commitment") prior to and during the commitment period[18];
(2) Methodological and reporting commitments relating to greenhouse gas inventories prior to the beginning of the first commitment period as set out in Article 5, paragraphs 1 and 2 of the Kyoto Protocol and related decisions[19]; and
(3) Commitments on reporting supplementary information in a Party's annual inventory as indicated in Article 7, paragraph 1 and 4 of the Protocol and related decisions, prior to the first commitment period.[20]

The facilitative branch may apply the following consequences,[21] taking into account the principle of common but differentiated responsibilities and respective capabilities:

(1) Provision of advice and facilitation of assistance to individual Parties regarding the implementation of the Protocol;

[17] Section IV, paragraph 5(b), Procedures and mechanisms in relation to decisions 2/CMP.1, Procedures, nature and scope of the mechanisms pursuant to Articles 6, 12 and 17 of the Kyoto Protocol 15/CMP.1 and 22/CMP.1, Guidelines for review under Article 8 of the Kyoto Protocol. See subsequent discussion on the flexibility mechanisms.
[18] Section IV, paragraph 6(a), Procedures and mechanisms.
[19] Section IV, paragraph 6(b), Procedures and mechanisms.
[20] Section IV, paragraph 6(c), Procedures and mechanisms.
[21] The term "penalty" does not appear in the Procedures and mechanisms.

(2) Facilitation of financial and technical assistance to any Party concerned, including technology transfer and capacity building from sources other than those established under the United Nations Framework Convention on Climate Change ("Convention") and the Kyoto Protocol for developing counties;
(3) Facilitation of financial and technical assistance, including technology transfer and capacity building, taking into account Article 4, paragraphs 3, 4 and 5 of the Convention[22]; and
(4) Formulation of recommendations to the Party concerned, taking into account Article 4, paragraph 7, of the Convention.[23]

One example of an area where a Party could seek facilitation is if an adjustment of any key source category (as defined in chapter 7 of the Intergovernmental Panel on Climate Change (IPCC) *Good Practice Guidance and Uncertainty Management in National Greenhouse Gas Inventories*)[24] of the Party concerned that accounted for 2% or more of the Party's aggregate emissions of the gases from the sources listed in Annex A of the Kyoto Protocol was calculated during the inventory review in three subsequent years. In this instance, a Party will be considered to fail to meet the methodological and reporting requirements under Article 7, paragraph 1 of the Kyoto Protocol,

[22] Article 4, paragraph 3 of the Convention speaks of the financial obligation of developed country Parties and other developed Parties included in Annex II to the Convention (Annex II Parties) in relation to developing country Parties. Article 3, paragraph 4 of the Convention requires the same Parties to assist developing country Parties that are particularly vulnerable to the adverse impacts of climate change. Finally, Article 3, paragraph 5 of the Convention lays out obligations of developed country Parties and Annex II Parties relating to technology transfer to enable Parties, particularly developing country Parties, to implement the provisions of the Convention.

[23] Section XIV, Procedures and mechanisms. Article 4, paragraph 7 of the Convention ties the effective implementation by developing country Parties of their commitments under the Convention to the effective implementation by developed country Parties of their commitments related to financial resources and transfer of technology, taking "fully into account that economic and social development and poverty eradication are the first and overriding priorities of the developing country Parties."

[24] Available at http://www.ipcc-nggip.iges.or.jp/public/gp/english/

unless the Party has requested assistance from the facilitative branch of the Compliance Committee in addressing this problem, prior to the beginning of the first commitment period, and assistance is being provided.[25]

The window for facilitation prior to the first commitment period, which will begin on 1 January 2008, is quickly closing. When the Procedures and mechanisms were first adopted by the Conference of the Parties to the Convention at its seventh session in Marrakesh in 2001, Parties could not have anticipated that it would take four more years for the Kyoto Protocol to enter into force. While the facilitative branch has not yet elaborated on how it would apply the consequences listed above, one could speculate that more Parties may have tried to avail of facilitation if the Kyoto Protocol had entered into force much earlier than 16 February 2005.

2.3. *The enforcement branch*

The enforcement branch focuses on *non-compliance* and does not determine compliance. It is responsible for determining whether a Party included in Annex I (Annex I Party) is not in compliance with:

(1) Its Article 3.1 target[26];
(2) The methodological and reporting requirements for greenhouse gas inventories under Article 5, paragraphs 1 and 2, and Article 7, paragraphs 1 and 4, of the Kyoto Protocol and the related decisions[27]; and
(3) The eligibility requirements under joint implementation (Article 6 of the Kyoto Protocol), the clean development mechanism (Article 12 of the Kyoto Protocol) and emissions trading (Article 17 of the Kyoto Protocol).[28]

In case of disagreements between a Party and an expert review team (ERT) under Article 8 of the Kyoto Protocol, the enforcement

---

[25] Paragraph 3(e), decision 15/CMP.1.
[26] Section V, paragraph 4(a), Procedures and mechanisms.
[27] Section VI, paragraph 4(b), Procedures and mechanisms.
[28] Section VI, paragraph 4(c), Procedures and mechanisms.

branch will determine whether to apply adjustments to greenhouse gas inventories or to correct the compilation and accounting database (CAD)[29] for the accounting of assigned amounts (i.e. a Party's allowable level of emissions over the Protocol's first commitment period (2008–2012)).[30] In considering adjustments, the enforcement branch may take into account the principle of achieving conservative estimates, which requires that the calculations of adjustments for a commitment period year should not result in an emission estimate that is lower or a removal estimate that is higher than that originally submitted by the Party, and an adjustment for an estimate of the base year should not result in an emission estimate that is higher or a removal estimate that is lower than the originally submitted estimate.[31]

In the case of the enforcement branch, there is a one-to-one correspondence between the consequences that it can apply and the type of non-compliance:

(1) When the enforcement branch determines that a Party is not in compliance with Article 5, paragraph 1 or paragraph 2, or Article 7, paragraph 1 or 4 of the Kyoto Protocol, it will declare the Party to be in non-compliance and require the Party to submit a plan within three months from the determination of non-compliance.[32] The three-month period for submitting a "compliance action plan" may be extended by the enforcement branch, as it considers appropriate.[33] This appears to be one of the few instances when the enforcement branch can take the cause,

[29] The CAD was established to facilitate the assessment of the compliance of each Annex I Party with its Article 3.1 commitment. See paragraph 50, annex to decision 13/CMP.1, Modalities for the accounting of assigned amounts under Article 7, paragraph 4, of the Kyoto Protocol.

[30] Section VI, paragraph 5, Procedures and mechanisms.

[31] Paragraph 53, annex to decision 20/CMP.1, Good practice guidance and adjustments under Article 5, paragraph 2, of the Kyoto Protocol.

[32] Section XV, paragraph 1, Procedures and mechanisms.

[33] Section XV, paragraph 2, Procedures and mechanisms.

type, degree and frequency of the non-compliance into account in its application of consequences.[34]

(2) Failure to meet one or more of the eligibility requirements under Articles 6, 12 and 17 of the Kyoto Protocol leads to suspension of eligibility of that Party,[35] a power that only the enforcement branch has. Suspension has slightly differing effects depending on the flexibility mechanism involved, as will be discussed in Sec. 4 of this paper. A suspension can be lifted through a reinstatement procedure that will be described below.

(3) When the enforcement branch determines that a Party has exceeded its assigned amount, in addition to declaring the Party to be in non-compliance, the enforcement branch will require the Party to make up the difference between its emissions and its assigned amount during the second commitment period, plus 30%. In addition, it will require the Party to submit a compliance action plan and suspend the eligibility of the Party to make transfers under emissions trading until the Party is reinstated.[36]

The 130% per cent deduction is what some claim to be the "teeth" of the Kyoto Protocol compliance mechanism. The bite of such a consequence depends, however, on at least two important future circumstances:

(a) Commitments beyond 2012 of Annex I Parties with commitments listed in Annex B of the Kyoto Protocol (hereinafter, "Annex B Parties") will have an identical structure to those of the Kyoto Protocol. Otherwise, revised rules for applying the 130% deduction would have to be devised. On the other hand, if subsequent commitments follow the same structure, a Party in non-compliance could, theoretically, keep carrying the deduction over to subsequent compliance periods.[37]

---

[34] Section XV, paragraph 1, Procedures and mechanisms.

[35] Section XV, paragraph 4, Procedures and mechanisms.

[36] Section XVI, paragraph 5, Procedures and mechanisms.

[37] *See* Andries Nentjets and Ger Klaassen, On the quality of compliance mechanisms in the Kyoto Protocol, Energy Policy 32 (2004) 531 at 538 and 543.

(b) The Party declared to be in non-compliance remains a Party to the Kyoto Protocol or is a Party to the successor agreement.

A compliance action plan will include an analysis of the causes of non-compliance of the Party, measures that that Party intends to implement in order to remedy the non-compliance, and a timetable for implementation of such measures.[38] In the case of non-compliance with the requirements of Article 5, paragraph 1 or 2 or Article 7, paragraph 1 or 4, the Party in non-compliance must implement the plan within a time frame not exceeding twelve months.[39] In the case of non-compliance with a Party's Article 3.1 target, the plan can have a maximum time frame of three years or up to the end of the subsequent commitment period, whichever occurs sooner.[40] Progress reports are required to be submitted to the enforcement branch on a regular basis.[41]

The enforcement branch may refer a question of implementation to the facilitative branch for consideration at any time.[42] Such a referral would be made through a decision by the enforcement branch with a statement identifying the question of implementation and the information on which the question is based.[43] One form or assistance that the facilitative branch could give to a Party concerned is in preparing or implementing its compliance action plan.

2.4. *Operations of the Committee*

At least three-fourths of the members of the Committee need to be present for it to adopt a decision.[44] However, there is no quorum requirement for holding a meeting. Thus, at its third meeting, despite the absence of a quorum to adopt a decision, the facilitative branch

[38] Section XV, paragraphs 2 and 6, Procedures and mechanisms.
[39] Section XV, paragraph 2, Procedures and mechanisms.
[40] Section XVI, paragraph 6, Procedures and mechanisms.
[41] Section XV, paragraphs 3 and 7, Procedures and mechanisms.
[42] Section IX, paragraph 12, Procedures and mechanisms.
[43] Rule 23, paragraph 1, Rules of procedure.
[44] Section II, paragraph 8, Procedures and mechanisms.

decided to continue its meeting. It could not, however, elect a vice-chairperson and instead invited a member of the branch to become a friend of the chair.[45]

The Committee seeks to reach agreement on any decision by consensus. As a last resort, a decision may be adopted by a majority of three-fourths of the members present and voting. In addition, decisions adopted by the enforcement branch must also be adopted by a majority of members from Annex I Parties and a majority of members from Parties not included in Annex I to the Convention (hereinafter, "non-Annex I Parties") present and voting.[46] The difficulty of achieving this three-fourths majority is illustrated by the facilitative branch's failure to adopt either a decision to proceed or a decision not to proceed in relation to the submission by South Africa as Chairman of the Group of 77 and China on behalf of the Group of 77 and China entitled "Compliance with Article 3.1 of the Kyoto Protocol."[47]

The Committee can only be seized of a matter through by a report of an ERT indicating a question or questions of implementation,[48] or a submission by a Party with respect to itself or with respect to another Party,[49] which is likely to be primarily based on, but not limited to, an ERT report. All final review reports, whether questions of implementation are indicated by the ERT or not, including status reports on initial check on annual inventories, will be published and

---

[45] *See* the Report on the third meeting of the facilitative branch, CC/FB/3/2006/2, available at http://unfccc.int/kyoto_protocol/compliance/facilitative_branch/items/3786.php.

[46] Section II, paragraph 9, Procedures and mechanisms.

[47] A detailed account of the proceedings of the facilitative branch on this matter is found in Annex IV of the First annual report.

[48] Section VI, paragraph 1, Procedures and mechanisms. Only an unresolved problem pertaining to language of a mandatory nature as found in the annex to decision 22/CMP.1, the guidelines for review under Article 8 of the Kyoto Protocol, that still exists after the Annex I Party has been provided with opportunities to correct the problem during the review, will be listed as a question of implementation. An unresolved problem pertaining to language of a non-mandatory nature will be noted in the final review report, but will not be listed as question of implementation (paragraph 8, annex to decision 22/CMP.1).

[49] *Id.*

forwarded by the secretariat, together with any written comments on the final review report by the Party which is the subject of the report, to the CMP, the Compliance Committee, and the Party in respect of which a question of implementation is raised ("Party concerned"), as applicable.[50]

After receipt of an ERT report indicating a question of implementation or a Party submission, the bureau must allocate, i.e. assign, the question of implementation to the facilitative or enforcement branch within seven days from receipt thereof.[51] Upon receipt by the branch of a question of implementation, it has three[52] (and in the case of expedited procedures described below, two[53]) weeks to complete its preliminary examination. Completion means deciding to either proceed or not to proceed with the question of implementation, after determining whether the question before it:

(1) Is supported by sufficient information;
(2) Is not *de minimis* or ill-founded; and
(3) Is based on the requirements of the Protocol.[54]

These criteria do not apply to questions raised by a Party with respect to itself,[55] which would likely relate to facilitation. A referral of a question of implementation by the enforcement branch to the facilitative branch in the exercise of its powers under section IX, paragraph 12 of the Procedures and mechanisms would likewise not require a preliminary examination.[56]

In the first 15 and so far only questions of implementation considered by the facilitative branch, the branch took the view that its members must vote on two separate decisions, i.e. a decision to proceed *and* a decision not to proceed. Failure of one type of decision to

---

[50] Paragraph 49, annex to decision 22/CMP.1, Section VI, paragraphs 2 and 3, Procedures and mechanisms.

[51] Section VII, paragraph 1, Procedures and mechanisms and rule 19, Rules of procedure.

[52] Section VII, paragraph 3, Procedures and mechanisms.

[53] Section IX, paragraph 1(a), Procedures and mechanisms.

[54] Section VII, paragraph 2, Procedures and mechanisms.

[55] *Id.*

[56] Rules 23, paragraph 2, Rules of procedure.

garner the required number of votes does not lead to the conclusion that the other type of decision is adopted.[57]

While the secretariat is required to make a question of implementation submitted with respect to the Party concerned availability to that party, there is no provision requiring the Party to be advised of the decision of the bureau to allocate the question of implementation. The Party concerned could therefore only learn of the action taken by the bureau on the question of implementation after the branch to which the question of implementation is assigned completes the preliminary examination.[58]

After the preliminary examination, no fixed timelines apply to the proceedings of the facilitative branch, which may engage in a dialogue with the Party concerned, either on its initiation[59] or the at the request of the Party concerned.[60] In contrast, the work of the enforcement branch follows well-defined steps and is governed by strict timelines set out in Secs. 9 and 10 of the Procedures and mechanisms.[61] If the enforcement branch decides to proceed with the consideration of question of implementation, it may decide, at the request of the Party concerned, to hold a hearing,[62] after which it must adopt a preliminary finding that the Party concerned will have an opportunity to comment on[63] and thereafter adopt a final decision.[64]

---

[57] See Annex IV, First annual report.

[58] See section VII, paragraphs 4 to 6, Procedures and mechanisms. On the other hand, there is also no prohibition against providing such information.

[59] Rule 24, paragraph 1, Rules of procedure.

[60] Rule 24, paragraph 2, Rules of procedure.

[61] When the circumstances of an individual case warrant, the enforcement branch may extend any time frames in the Procedures and mechanisms. Section IX, paragraph 11, Procedures and mechanisms. This power does not appear to extend to expedited procedures, which fall under Section X of the Procedures and mechanisms.

[62] Section IX, paragraph 2, Procedures and mechanisms.

[63] Section IX, paragraphs 4 to 8, Procedures and mechanisms.

[64] Section IX, paragraphs 7 and 8, Procedures and mechanisms. These are the steps that apply to ordinary proceedings of the enforcement branch as well as expedited procedures under Section X, paragraph 1 of the Procedures and mechanisms. It remains to be seen whether the enforcement branch will apply the same steps to other types of questions of implementation requiring the use of expedited procedures, e.g. adjustments and corrections.

Application of normal procedures means that the enforcement branch would have about 40 weeks from its receipt of the question of implementation to adopt a final decision. Expedited procedures could entail anywhere between 12 weeks in the case of adjustments and corrections under Sec. 10, paragraph 5 of the Procedures and mechanisms to close to 40 weeks in the case of a request for reinstatement of eligibility under Sec. 10, paragraph 2 of the Procedures and mechanisms. In the case of a suspension of eligibility to participate in the mechanisms under Sec. 15, paragraph 4, of the Procedures and mechanisms, the enforcement branch will decide a request for reinstatement that is submitted to it directly "as soon as possible."

As a general rule, decisions of the facilitative and enforcement branch are final. Copies of these decisions are forwarded to the CMP only for its information. However, a Party concerned may appeal a decision of the enforcement branch relating to is Article 3.1 target, if the Party believes that it has been denied due process.[65] Any appeal must be filed within 45 days after the Party has been informed of the decision of the enforcement branch and will be decided by the CMP at its first session after the lodging of the appeal.[66]

The compliance mechanism is an integral part of the broader Kyoto Protocol system. It is inextricably linked to the reporting and review process, as consideration of any question of implementation by the two branches of the Compliance Committee comes at the end of the reporting, review and compliance chain. At the other end, the decisions of the enforcement branch will affect the eligibility of Annex B Parties to use the flexibility mechanisms.

The next two sections discuss how the compliance mechanism links to two very important components of the Kyoto Protocol system: the reporting and review system and the flexibility mechanisms.

3. Reporting and Review

Unless a Party uses material other than ERT reports to raise a question of implementation, the work of the Compliance Committee will

---

[65] Section XI, paragraph 1, Procedures and mechanisms.

[66] Section XI, paragraph 2, Procedures and mechanisms.

depend to a large degree on reviews conducted by ERTs.[67] ERT reviews are based, in turn, on information that Annex B Parties are required to submit under the Kyoto Protocol.

Under the Kyoto Protocol, ERTs review four basic types of information submitted by Parties:

(1) The initial report required under decision 13/CMP.1, which was due on 1 January 2007[68];
(2) The supplementary information required under Article 7, paragraph 1, of the Kyoto Protocol, together with the greenhouse gas inventory for each year of the commitment period, due on 15 April 2010 and every year thereafter until 2014[69];
(3) the supplementary information required under Article 7, paragraph 2, of the Kyoto Protocol, which was required to be submitted together with the national communication due on 1 January 2006[70]; and

[67] In its second annual report, the Committee highlighted that "the Committee relies on the quality and timeliness of reports of the expert review teams." *See* paragraph 23, Second annual report.

[68] Paragraph 2, decision 13/CMP.1. Alternatively, if one year after the entry into force of the Kyoto Protocol for an Annex B Party were later than this date, the later date would apply. This is the case for Croatia and Monaco, which became Parties to the Kyoto Protocol on 28 August 2007 and 28 May 2006 respectively.

[69] Article 7, paragraph 3, Kyoto Protocol and paragraphs 15 and 51, annex to decision 22/CMP.1. The supplementary information will be included with the inventory submission due under the Convention for the first year of the commitment period after the Kyoto Protocol has entered into force for that Party. The first annual report under the Kyoto Protocol is due on 15 April 2010, which is the year for the submission of annual inventories for the first year of the commitment period. However, a Party may start reporting this information from the year following the submission of its initial report, or in 2007. *See* paragraph 2, decision 15/CMP.2.

[70] Article 7, paragraph 3 of the Kyoto Protocol provides that Annex I Parties must submit the supplementary information required under Article 7, paragraph 2 of the Kyoto Protocol as part of the first national communication due under the Convention after the Protocol has entered into force for that Party and after the adoption of the guidelines for the preparation of the supplementary information required under Article 7, paragraph 1 and 2 of the Kyoto Protocol. Fourth national communications from Annex I Parties were due to be submitted to the secretariat by 1 January 2006 (paragraph 3, decision 4/CP.8).

(4) the true-up period report, due at the expiration of the period for fulfilling the Article 3.1 commitment.[71]

In addition to preparing reports that are forwarded to the Compliance Committee, ERTs can provide technical advice to the Compliance Committee upon request.[72] To date, 16 reports of the centralized in-depth review of the fourth national communication and 18 reports of the review initial reports of Annex I Parties have been prepared by ERTs and forwarded to the Compliance Committee. None of these reports indicate questions of implementation.[73]

ERT review of these reports aims at a thorough, objective and comprehensive technical assessment of all aspects of the implementation of the Kyoto Protocol by Annex I Parties.[74] The building blocks of Kyoto Protocol implementation, and the role of ERTs in assessing these, are explained below.

3.1. *Building blocks of implementation*

3.1.1. *Assigned amount*

Before participating in the flexibility mechanisms, an Annex I Party's assigned amount must be established and recorded in the CAD.[75]

---

[71] Paragraph 3, decision 13/CMP.1, and paragraph 49, annex to decision 13/CMP.1. For purposes of fulfilling its Article 3.1 commitment, an Annex B Party may, until the hundredth day after the date set by the CMP for the completion of the ERT review process under Article 8 of the Kyoto Protocol for the last year of the commitment period, continue to acquire, and other Parties may transfer to such Party, emission reduction units, certified emission reductions, assigned amount units and removal units under Articles 6, 12 and 17 of the Kyoto Protocol, from the preceding commitment period, provided the eligibility of any such Party has not been suspended in accordance with section XV, paragraph 4 of the Procedures and mechanisms (section XIII, Procedures and mechanisms). The date that triggers the 100 days, known as the "true-up period", has not yet been set by the CMP.

[72] Paragraph 5, annex to decision 22/CMP.1.

[73] Copies of these in-depth review reports can be downloaded through this link: http://unfccc.int/documentation/documents/advanced_search/items/3594.php?such=j&symbol="/IDR"#beg

[74] Paragraph 2, annex to decision 22/CMP.1.

[75] Paragraphs 10 and 52, annex to decision 13/CMP.1.

A Party's proposed calculation of its assigned amount, i.e., in tonnes of CO_2 equivalent, is set out in its initial report.[76] An ERT verifies this calculation during the in-country review of the initial report.[77] Any unresolved issues between the Party and the ERT relating to the calculation of a Party's assigned amount is referred to the enforcement branch of the Compliance Committee,[78] which seizes of the matter through the ERT's report. The enforcement branch will have 12 weeks to decide on the question of an adjustment.[79] Once a question of implementation relating to the calculation of a Party's assigned amount is resolved, the assigned amount will be recorded in the CAD.[80] The Party may thereafter issue assigned amount units (AAUs)[81] in its national registry equivalent to its assigned amount.[82] Once fixed, the assigned amount of a Party cannot be changed.[83] Any adjustments applied to the base year inventory estimates of an Annex I Party, which will be reviewed by the ERT at the same time as the initial report, will be used in the calculation of the Party's assigned amount.[84]

3.1.2. *National system*

Under Article 5, paragraph 1, of the Kyoto Protocol, each Annex I Party is required to have a national system for the estimation of

---

[76] Paragraph 7(d), annex to decision 13/CMP.1.

[77] Paragraph 87, annex to decision 22/CMP.1.

[78] Paragraph 10, decision 20/CMP.1.

[79] Paragraph 5, Procedures and mechanisms.

[80] Paragraphs 10 and 52, annex to decision 13/CMP.1.

[81] A Party's AAU is based on the aggregate anthropogenic carbon dioxide equivalent emissions of the greenhouse gases and sources listed in Annex A to the Kyoto Protocol in the base year, multiplied by five, taking into account the rules listed under paragraph 5 of the annex to decision 13/CMP.1. The proper calculation of a Party's AAU assumes that the ERT assigned to review its inventory for the base year has found its inventory to meet the guidelines under Article 5, paragraph 2 and Article 7, paragraph 1 of the Kyoto Protocol.

[82] Paragraph 24, annex to decision 13/CMP.1.

[83] Paragraph 10, annex to decision 20/CMP.1 and paragraph 54, annex to decision 22/CMP.1. Thus, any adjustment to estimates of emissions and removals for purposes of establishing the assigned amount will only be applied during the initial review.

[84] Paragraph 8, decision 20/CMP.1.

anthropogenic emissions by sources and removals by sinks of all greenhouse gases not controlled by the Montreal Protocol in place no later than one year prior to the start of the first commitment period. A national system is defined as including:

> "*all institutional, legal and procedural arrangements made within a Party included in Annex I for estimating anthropogenic emissions by sources and removals by sinks of all greenhouse gases not controlled by the Montreal Protocol, and for reporting and archiving inventory information.*"[85]

A national system is intended to enable Annex I Parties to estimate anthropogenic greenhouse gas emissions by sources and removals by sinks, and to report these, to assist Parties included in Annex I in meeting their commitments under Articles 3 and 7 of the Kyoto Protocol, to facilitate review of information submitted under Article 7, as required by Article 8, and to assist Parties included in Annex I to ensure and improve the quality of their inventories.[86] An ERT reviews each Annex I Party's national system to assess the capacity of the national system and the adequacy of the institutional, legal and procedural arrangements to produce an inventory of anthropogenic emissions by sources and removals by sinks in conformity with Article 5, paragraph 2 of the Kyoto Protocol.[87] The review is also meant to assess the extent to which guidelines for national systems under Article 5, paragraph 1 of the Kyoto Protocol have been adhered to, and to assist Annex I Parties in meeting their commitments under Article 5, paragraph 1.[88] Finally, the review is meant to provide the CMP and the Compliance Committee with reliable information on the national systems.[89]

---

[85] Paragraph 2, annex to decision 19/CMP.1, Guidelines for national systems under Article 5, paragraph 1 of the Kyoto Protocol.
[86] Paragraph 5, annex to decision 19/CMP.1.
[87] Paragraph 96(a), annex to decision 22/CMP.1.
[88] Paragraph 96(b), annex to decision 22/CMP.1.
[89] Paragraph 96(c), annex to decision 22/CMP.1.

A thorough and comprehensive review of the national system is undertaken during the in-country review,[90] while subsequent reviews, which will be done during the review of the national inventory,[91] will focus on any significant changes in national systems.[92]

3.1.3. *National registry*

A national registry is responsible for ensuring the accurate accounting of the issuance, holding, transfer, acquisition, cancellation and retirement of emission reduction units (ERUs), certified emission reductions (CERs), AAUs and removal units (RMUs)[93] and the carry-over of ERUs, CERs and AAUs[94] (collectively known as "Kyoto credits"). ERT review of an Annex I Party's national registry seeks to confirm adherence by the national registry to technical standards for data exchange between registry systems,[95] which is necessary to ensure accurate, transparent and efficient exchange of data between national registries, the clean development mechanism registry and the international transaction log.[96]

A lack of capacity of the national registry to ensure the accurate accounting, issuance, holding, transfer, acquisition and retirement, replacement and carry-over of Kyoto credits could result in discrepancies

---

[90] Paragraph 100, annex to decision 22/CMP.1.

[91] Paragraph 108, annex to decision 22/CMP.1.

[92] Paragraph 101, annex to decision 22/CMP.1.

[93] RMUs refer to the units issued by a Party on the basis of its activities under Article 3, paragraph 3 and the activities elected under Article 3, paragraph 4. *See* paragraphs 3 and 25, annex to decision 13/CMP.1.

[94] Paragraph 17, annex to decision 13/CMP.1.

[95] Paragraphs 110 and 115, annex to decision 22/CMP.1. The technical standards for data exchange are laid out in the data exchange standards (DES), which have been developed by the secretariat in consultation with registry developers. Version 1.1 of the DES, which is currently in use, is available at http://unfccc.int/files/kyoto_mechanisms/registry_systems/application/pdf/des_techspec_v1_1.pdf. *See* Kyoto Protocol Reference Manual on Accounting of Emissions and Assigned Amounts, UNFCCC Secretariat (February 2007), pp. 45, 46 and 48 (hereinafter, KP Reference Manual) and decisions 24/CP.8 and 16/CP.10.

[96] Paragraph 115, annex to decision 22/CMP.1.

in transactions initiated by the Party.[97] Another problem that could arise from defects in the national registry could be the failure to replace temporary or long-term CERs (tCERs or lCERs).[98]

As in the case of national systems, a thorough review is conducted as part of the initial review, and a desk or centralized review of any change of the national registry is conducted in conjunction with the annual review.[99] In addition, another thorough review will be conducted if one is recommended by the ERT.[100]

3.1.4. *Adjustments*

Adjustments are applied for a variety of reasons, including avoiding double counting.[101] Adjustments will be calculated, documented and recommended under the collective responsibility of the ERT reviewing the report.[102] The will apply to the emissions and removals for purposes of establishing the assigned amount, or to individual inventory years, i.e. the base year or the latest year of the commitment period under review.[103]

Adjustments under Article 5, paragraph 2 of the Kyoto Protocol are applied by an ERT when inventory data submitted by an Annex I Party is found to be incomplete and/or prepared in a way that is not consistent with the Revised 1996 IPCC Guidelines for National Greenhouse Gas Inventories as elaborated by the IPCC good practice guidance and any good practice guidance adopted by the CMP.[104] Problems for which adjustments would be appropriate will be identified by an ERT during the individual inventory review. The ERT will officially notify the Annex I Party of the reason why an adjustment is

---

[97] Paragraph 88(j)(v), annex to decision 22/CMP.1.

[98] Paragraph 88(k)(v), annex to decision 22/CMP.1.

[99] Paragraph 111, annex to decision 22/CMP.1 Part of the material to be provided to each ERT is an independent assessment report for each national registry.

[100] Paragraph 112, annex to decision 22/CMP.1.

[101] Paragraph 62, annex to decision 20/CMP.1.

[102] Paragraph 5, annex to decision 20/CMP.1.

[103] Paragraphs 10 and 11, annex to decision 20/CMP.1.

[104] Paragraph 3 decision 20/CMP.1 and paragraph 79, annex to decision 22/CMP.1.

considered necessary and provide advice on how the problem could be corrected.[105] If an Annex I Party does not adequately correct the problem through the provision of a revised estimate within the given time frame, despite being given an opportunity to do so, the ERT will initiate procedures for the calculation of adjustments[106] by officially notifying the Party concerned of the calculated adjustments. The notification will describe the assumptions, data and methodologies used to calculate the adjustment(s), as well as the value of the adjustment(s).[107] The Party concerned will have four weeks to indicate whether it accepts or rejects the adjustment(s), indicating its rationale.[108] Failure to respond within the four weeks would be considered an acceptance of the adjustment, which would then be applied for the purpose of compilation and accounting of emissions inventories and assigned amounts.[109] If the Party disagrees with the proposed adjustments(s), it must send a notification to the ERT, indicating its rationale. The ERT would then send the Party's notification and the recommendation for an adjustment to the Compliance Committee.[110]

The adjustment procedure is intended to result in estimates that are conservative for the Party concerned.[111] Adjustments are intended to correct inventory problems for the purpose of accounting emissions inventories and assigned amounts of the Annex I Parties.[112] The Annex to decision 20/CMP.1 provides technical guidance on methodologies for adjustments to ensure consistency and comparability and the use of similar methods for similar problems as far as possible across all inventories reviewed under Article 8.[113] The aim is for emission estimates for the base year not to be overestimated and

[105] Paragraph 80(a), annex to decision 22/CMP.1.
[106] Paragraphs 68 and 80(b), annex to decision 22/CMP.1.
[107] Paragraph 80(d), annex to decision 22/CMP.1.
[108] Paragraph 76 and 80(3), annex to decision 22/CMP.1.
[109] Paragraph 80(e)(i), annex to decision 22/CMP.1.
[110] Paragraph 80(e)(i), annex to decision 22/CMP.1.
[111] Paragraph 5, decision 20/CMP.1.
[112] Paragraph 6, decision 20/CMP.1.
[113] Paragraph 7, decision 20/CMP.1.

emission estimates for a year of the commitment period not to be underestimated relative to the likely true value of the emissions of the Party concerned.[114]

3.1.5. *True-up period*

On the basis of an Annex B Party's true-up report, a compliance assessment will be conducted by an ERT, which will compare the ERUs, CERs, AAUs and/or RMUs valid for the commitment period in question, retired by the Party with the emissions of that Party during the commitment period with that Party's aggregate anthropogenic carbon dioxide equivalent emissions of gases and sources listed in Annex A.[115] Credits not retired or cancelled during the commitment period in question that are over and above that Party's emissions may, with the exception of RMUs, tCERs and lCERs, be carried over to the subsequent commitment period.[116] The review of the true-up report must be completed within 14 weeks of the due date for the submission of information, including a two-week period within which the Party concerned may comment on the draft ERT report.[117] All such review reports would then be forwarded to the Compliance Committee. In case an ERT raises a question of implementation relating to non-compliance with an Annex B Party's Article 3.1 commitment, the enforcement branch will consider the report and determine whether the Party concerned is not in compliance.

3.2. *Establishment, suspension, and reinstatement of eligibility*

An Annex I Party will be considered to meet the eligibility requirements under Articles 6, 12 and 17 of the Kyoto Protocol after 16 months

---

[114] Paragraph 51, annex to decision 20/CMP.1.

[115] Paragraph 14, annex to decision 13/CMP.1 and paragraph 92, annex to decision 22/CMP.1.

[116] Paragraphs 41 and 45, annex to decision 5/CMP.1, Modalities and procedures for afforestation and reforestation project activities under the clean development mechanism in the first commitment period of the Kyoto Protocol.

[117] Paragraph 94, annex to decision 22/CMP.1.

have elapsed since the submission of its initial report to facilitate the calculation of its assigned amount.[118] After an in-depth review of the initial report of an Annex I Party, which must be completed within 12 months from its initiation,[119] the enforcement branch would consider any questions of implementation relating to eligibility. Should the enforcement branch decide not to proceed with a question of implementation relating to eligibility, the Party's eligibility could be established even earlier than the 16 months. On the other hand, if, before the end of the 16-month period, the enforcement branch determines that the Party does not meet the eligibility requirements, the Party's eligibility will not be established, i.e., information that the Party is eligible cannot be recorded in the CAD. However, even if the enforcement branch is considering a question of implementation relating to eligibility, if it is unable to decide on this question before the lapse of the 16-month period, then the Party's eligibility will be established at the end of the 16 months.[120] Eligibility to transfer and/or acquire ERUs, CERs, AAUs and RMUs and to use CERs to contribute to its compliance under Article 3, paragraph 1 is recorded in the CAD.[121]

The Kyoto Protocol presumes that each Party meets the eligibility criteria for participation in the Kyoto mechanisms.[122] Thus, after eligibility is established, this eligibility is maintained until the enforcement branch suspends the Party's eligibility during the course

---

[118] Paragraph 32(a), annex to decision 3/CMP.1, paragraph 22(a), annex to decision 9/CMP.1 and paragraph 3(a), annex to decision 11/CMP.1, Modalities, rules and guidelines for emissions trading under Article 17 of the Kyoto Protocol. To date, all Annex B Parties to the Kyoto Protocol have submitted their initial reports. Information on when each initial report was submitted, and a copy of the text of these reports are available at Information is available at http://unfccc.int/national_reports/initial_reports_under_the_kyoto_protocol/items/3765.php.

[119] Paragraph 2, decision 22/CMP.1 and paragraph 1, decision 26/CMP.1, Review process during the period 2006–2007 for Parties included in Annex I to the Convention that are also Parties to the Kyoto Protocol.

[120] The enforcement branch could, however, continue with its proceedings and thereafter suspend the eligibility of the Party concerned.

[121] Paragraph 53, annex to decision 13/CMP.1.

[122] KP Reference Manual, p. 33.

of the commitment period. This would happen if the enforcement branch finds, based on an ERT report on its review of the supplementary information required under Article 7, paragraph 1, of the Kyoto Protocol, or a submission from a Party, that a Party does not meet one or more of the eligibility requirements relating to the mechanisms and suspends the Party's eligibility.[123] Expedited procedures will apply where a question of implementation relates to eligibility requirements under Articles 6, 12 and 17 of the Kyoto Protocol.[124]

The expedited procedures of the enforcement branch are complemented and linked to the expedited procedures of the ERT. Expedited procedures are applied when an Annex I Party's eligibility is suspended, and the Party is asking for reinstatement. The review for reinstatement of eligibility to use the mechanisms will be limited to the review of the matter or matters that led to the suspension of the eligibility.[125] The annex to decision 22/CMP.1 outlines the procedure that a Party must observe to work for the reinstatement of its eligibility. During its review, the ERT will assess whether the question or questions of implementation that led to suspension of eligibility have been addressed and resolved.[126] Expedited review by an ERT would take a maximum of 15 weeks from the time that information is received from the Party concerned.[127] At the end of the review, the ERT will produce a final review report on the reinstatement of eligibility, which would indicate whether there is still a question of implementation with respect to the eligibility of the Party concerned to use the mechanisms.[128]

3.3. *ERTs and the facilitative branch*

3.3.1. *Article 3, paragraph 14 of the Kyoto Protocol*

Information on how a Party is striving, under Article 3, paragraph 14, of the Kyoto Protocol, to implement its Article 3.1 commitments in

---

[123] Paragraph 32(b), annex to decision 3/CMP.1, paragraph 22(b), annex to decision 9/CMP.1 and paragraph 3(b), annex to decision 11/CMP.1.

[124] Section X, Procedures and mechanisms.

[125] Paragraph 148, annex to decision 22/CMP.1.

[126] Paragraph 153, annex to decision 22/CMP.1.

[127] Paragraphs 156 and 157, annex to decision 22/CMP.1.

[128] Paragraph 159, annex to decision 22/CMP.1.

such a way as to minimize adverse social, environmental and economic impacts on developing country Parties, particularly those identified in Article 4, paragraphs 8 and 9,[129] of the Convention will be part of the supplementary information in the annual inventory starting in 2010.[130] In addition, Parties included in Annex II of the Convention and other Parties included in Annex I that are in a position to do so are required to incorporate information on how they give priority, in implementing their commitments under Article 3, paragraph 14, of the Kyoto Protocol, to the actions listed in paragraph 24 of the annex to decision 15/CMP.1.

3.3.2. *Advice and facilitation*

In the course of conducting a review, the ERT could also offer advice to Annex I Parties on how to correct problems that they identify, taking into account the national circumstances of the Party.[131] An ERT could, for instance, recommend possible ways for improving the estimation and reporting of inventory information[132] and make recommendations on how deficiencies in the functioning of the national system could be improved.[133] An ERT may also recommend how problems relating to the performance of the functions of the national registry and the adherence to technical standards for data exchange between the registry systems could be addressed.[134] The ERT may also recommend possible ways to improve the reporting of information relating to the minimization of adverse effects under

---

[129] Article 4, paragraph 8 of the Convention lists criteria that would make developing country Parties particularly vulnerable to the adverse effects of climate change and/or the impacts of the implementation of response measures. Countries meeting these criteria are recognized to have specific needs and concerns that Parties are required to give full consideration to. Article 4, paragraph 9 asks Parties to take full account of specific needs and special situations of the least developed countries with regard to funding and transfer of technology.

[130] Paragraph 23, annex to decision 15/CMP.1 and paragraph 3, decision 15/CMP.1.

[131] Paragraph 5, annex to decision 22/CMP.1.

[132] Paragraph 65(h), annex to decision 22/CMP.1.

[133] Paragraph 106, annex to decision 22/CMP.1.

[134] Paragraph 118, annex to decision 22/CMP.1.

Article 3, paragraph 14.[135] In documenting potential problems in its final review report, an ERT is required, among other things, to describe any potential problem in, and factors influencing the fulfilment of, commitments identified during the review, any recommendations provided by the ERT to solve the potential problem, and any assessment of the efforts by the Annex I Party to address potential problems identified by ERTs during the current review or during previous review that have not been corrected.[136] Information on circumstances surrounding the problems identified, as well as the advice given by an ERT may prove useful for the facilitative branch, should Party request facilitation, and could also form an important part of any compliance action plan required from the Party by the enforcement branch.

4. Compliance and the Flexibility Mechanisms

4.1. *Eligibility and the compliance mechanism*

Eligibility to participate in the flexibility mechanisms will depend on compliance with the methodological and reporting requirements under Article 5, paragraphs 1 and 2 and Article 7, paragraphs 1 and 4, of the Kyoto Protocol. Eligibility is determined through the application of six criteria that are derived from these provisions and the relevant decisions of the CMP. The eligibility criteria have been developed "to ensure that a Party is accounting accurately for its emissions and assigned amount, so that the use of the Kyoto mechanisms will not jeopardize the Party's compliance with its Article 3, paragraph 1 commitment."[137] The enforcement branch of the Compliance Committee will provide oversight of compliance with this provision.[138]

[135] Paragraph 123(e), annex to decision 22/CMP.1.
[136] Paragraph 48, annex to decision 22/CMP.1.
[137] KP Manual, p. 29.
[138] Paragraph 5, decision 2/CMP.1 and Section V, paragraph 4, Procedures and mechanisms.

The six basic criteria for eligibility to participate in the flexibility mechanisms are:

(1) The Party is a Party to the Kyoto Protocol;
(2) The Party's initial assigned amount has been recorded in the CAD;
(3) The Party's national system is in compliance with the requirements established under Article 5, paragraph 1 of the Kyoto Protocol;
(4) The Party's national registry is in compliance with the requirements established under Article 7, paragraph 4 of the Kyoto Protocol;
(5) The Party has submitted its inventory for the most recent year, and this inventory meets the requirements under Article 7, paragraph 1 of the Kyoto Protocol; and
(6) The Party has submitted information on its assigned amount under Article 7, paragraph 1 of the Kyoto Protocol.[139]

As demonstrated in the previous section, determination of a Party's eligibility is part of the review of the initial report required under decision 13/CMP.1. This review establishes the eligibility of a Party to participate in the Kyoto mechanisms. Subsequent annual reviews of the inventories of Parties can then result in suspension of a Party's eligibility, which can be reinstated through a process outlined in the Procedures and mechanisms.

Being a Party to the Kyoto Protocol, having one's assigned amount fixed and recorded in the CAD, and having a national registry that complies with the requirements established under Article 5, paragraph 2 of the Kyoto Protocol and Article 7, paragraph 4 of the Kyoto Protocol are the core requirements for eligibility to participate in the Kyoto mechanisms. Failure to meet the eligibility requirements relating to the national system, the annual inventory, and the submission of information on its assigned amount (the "other requirements") has different consequences depending on the mechanism involved.

---

[139] *See* paragraph 31, annex to decision 3/CMP.1, paragraph 21, annex to decision 9/CMP.1, and paragraph 2, annex to decision 11/CMP.1.

In the case of the clean development mechanism (CDM), a finding of non-compliance with *any* of the six requirements listed above prevents an Annex I Party from using CERs generated from CDM project activities to contribute to compliance with part of its Article 3.1 commitment.[140] This means, in practical terms, that those CERs cannot be retired and counted for purposes of demonstrating that Party's compliance with its quantified emissions limitation or reduction commitment,[141] a detail to be picked up during the automated check to be done by the international transaction log for proposed retirement of CERs.[142] This will affect public and/or private entities authorized by such a Party to participate in CDM project activities as they may only transfer and acquire CERs if the authorizing Party is eligible to do so at that time.[143]

If the Party hosting a joint implementation (JI) project does not meet these other requirements, it can only issue and transfer ERUs for JI project activities that have been verified by independent entities accredited by the Joint Implementation Supervisory Committee (JISC), or the so-called "independently-verified, track 2 or CDM like" JI for which it is the host Party.[144] A host Party that is considered to meet all the eligibility requirements may, on the other hand, use the "Party-verified, track 1 or ET like" JI, which entitles the host country Party to verify reductions in anthropogenic emissions by sources or enhancements of anthropogenic removals by sinks from a JI project and thereafter issue the corresponding ERUs, without need to resort to the verification procedure for track 2 JI.[145] Such a Party may, however, opt to use the verification procedure at any time.[146] Legal entities authorized by an Annex I Party to participate in JI project activities under its responsibility would,[147]

---

[140] Paragraph 31, annex to decision 3/CMP.l.
[141] Paragraph 13, annex to decision 13/CMP.1.
[142] Paragraph 42(d), annex to decision 13/CMP.1.
[143] Paragraph 33, annex to decision 3/CMP.1.
[144] Paragraph 24, annex to decision 9/CMP.1.
[145] Paragraph 23, annex to decision 9/CMP.1.
[146] Paragraph 25, annex to decision 9/CMP.1.
[147] Article 6, paragraph 3, Kyoto Protocol.

as a consequence, also be prevented from transferring and/or acquiring such credits.[148]

For emissions trading (ET), a finding by the enforcement branch that a Party does not meet *any* of the six requirements enumerated above would prevent that Party from transferring and/or acquiring ERUs,[149] CERs, AAUs or RMUs.[150] Any legal entity authorized by such Party to transfer and/or acquire credits under ET would also lose such right if the authorizing Party does not meet the eligibility requirements or has been suspended.[151]

Eligibility of the Parties involved in the transaction to participate in the mechanisms is one of the conditions required for the international transaction log to clear transfers between registries.[152]

4.2. *Supplementarity*

Observance of the principles relating to mechanisms is ensured through the compliance mechanism. The facilitative branch is responsible for addressing questions of implementation with respect to the provision of information on the use by an Annex I Party of the mechanisms as supplemental to domestic action.[153] Decision 2/CMP.1 requires the use of mechanisms to be supplemental to domestic action, and further requires domestic action to constitute a significant element of the effort made by each Annex I Party to meet their Article 3.1 targets.[154] Decision 15/CMP.1 requires Annex I Parties to include information in its national communication on how their use of mechanisms is supplemental to domestic action and how its domestic action constitutes a significant element of the effort made to meet

---

[148] Paragraph 29, decision 9/CMP.1.

[149] AAUs and RMUs may also be converted into ERUs. *See* paragraph 29, annex to decision 13/CMP.1.

[150] Paragraph 2, annex to decision 11/CMP.1.

[151] Paragraph 5, annex to decision 11/CMP.1.

[152] Paragraph 42(b), annex to decision 13/CMP.1.

[153] Section IV, paragraph 5(b), Procedures and mechanisms.

[154] Paragraph 1, decision 2/CMP.1, Principles, nature and scope of the mechanisms pursuant to Articles 6, 12 and 17 of the Kyoto Protocol.

its Article 3.1 target.[155] ERTs examine supplementarity relating to mechanisms pursuant to Articles 6, 12 and 17 through a review of the supplementary information incorporated by an Annex I Party into its national communication.[156]

While there are substantive, reporting and review requirements on supplementarity relating to mechanisms, there is no methodology available to define supplementarity in a quantitative way. It therefore remains to be seen how the ERTs will ascertain fulfilment of this requirement. So far the 16 in-depth review reports of fourth national communications prepared by ERTs and forwarded to the Compliance Committee do not shed light on this issue.

4.3. *Institutional collaboration*

The JISC is encouraged by the CMP to collaborate with the Compliance Committee, in particular with regard to the list of Parties that meet the eligibility requirements and those that have been suspended.[157] The details of such collaboration have yet to be worked out.

5. Implications for the Asian Region

5.1. *The present*

The consequences that the facilitative branch apply make it clear that facilitation provided by the facilitative branch is not meant to draw from resources available to developing country Parties. The provision of Sec. XIV paragraph (b), of the Procedures and mechanisms is explicit on this point.

While it is clear that the mandate of the enforcement branch applies only to Annex I Parties with a commitment inscribed in Annex B of the Kyoto Protocol, there is room to argue that facilitation may

[155] Paragraph 33, annex to decision 15/CMP.1.
[156] Paragraphs 132 and 135, annex to decision 22/CMP.1
[157] Paragraph 27, annex to decision 9/CMP.1.

be available to non-Annex I Parties. This interpretation is based on a close reading of the consequences that the facilitative branch may apply, especially the following:

(1) Provision of financial and technical assistance, including technology transfer and capacity building, taking into account Article 4, paragraphs 3, 4 and 5, of the Convention; and
(2) Formulation of recommendations to the Part concerned, taking into account Article 4, paragraph 7, of the Convention.[158]

The paragraphs of Article 3 mentioned above speak of the obligation of developing country Parties and other Parties included in Annex II to the Convention to provide financial resources, and technology transfer to developing country Parties. Article 3, paragraph 7 links the extent to which developing country Parties effectively implement their commitments under the Convention to, among others, the effective implementation by developed country Parties of their financial and technology-transfer related obligations. These references to developing country commitments in the Convention, which are mirrored in Article 10 of the Kyoto Protocol, could be used as a ground for this argument, especially in relation to the facilitative branch's overall mandate for "providing advice and facilitation to Parties in implementing the Protocol, and for promoting compliance by Parties with their commitments under the Kyoto Protocol."[159]

While it is theoretically possible to bring a question of implementation relating to a non-Annex Party before the Compliance Committee (or at least its facilitative branch), there will be some practical difficulties in doing so. As explained in previous sections, Compliance Committee action would, in most cases, be based on a previous review by an ERT. Non-Annex I implementation of the Kyoto Protocol is currently not subject to ERT review. An important question would therefore be the factual basis on which the facilitative branch could begin consideration of any request for facilitation. Another important

[158] Section XVI, subparagraphs (c) and (d), Procedures and mechanisms.
[159] Section IV, paragraph 4, Procedures and mechanisms.

question, given the restriction against providing advice and facilitation from sources established for developing countries, is what additional sources of financial and technical assistance the facilitative branch could direct a developing country to, should it ask for facilitation.

Another view held is that the consequences listed above are a placeholder for the future, when non-Annex I Parties may take on some form of binding commitments under a successor agreement to the Kyoto Protocol.

On the matter of the mechanisms, failure of an Annex I Party to obtain initial eligibility or suspension of its eligibility would affect the marketability of CDM project activities and/or the CERs that result from these project activities. The exact nature and extent of these effects have yet to be determined.

5.2. *The future*

The elaborate procedure set out in the preceding sections gives Parties a glimpse of the basket of commitments imposed on a Party that takes on GHG emission reduction commitments and point to the extent and nature of preparedness that Parties contemplating future commitments are expected to have. While the shape of any climate change related agreement beyond the first commitment period remains to be seen, it is reasonable to assume that many elements of reporting and review will remain components of such an agreement. Even the voluntary actions being discussed may entail a system of verification that would require reporting and review. Assuming that the enforcement function of the Compliance Committee or any successor institution falls away, it reasonable to believe that the facilitative function would continue to be carried out, especially if less developed Parties begin to take on commitments.

Compliance Under the Kyoto Protocol and Its Implications for the Asian Region: Postscript

Since the submission of the paper for publication in late 2007, several developments relating to compliance have occurred. This postscript highlights some of those developments.

In 2008, the enforcement branch of the Compliance Committee had the opportunity to consider its first two questions of implementation and the facilitative branch focused on substantive discussions with respect to its mandate.

On 29 April 2008, the date when Greece was supposed to have become eligible to participate in the mechanisms, Greece became eligible to participate only in JI track 2.[160] Partial eligibility resulted from a finding by the enforcement branch that Greece was not in compliance with the guidelines for national systems under Article 5, paragraph 1, of the Kyoto Protocol (decision 19/CMP.1) and the guidelines for the preparation of information required under Article 7 of the Kyoto Protocol (decision 15/CMP.1).

As a result of this finding of non-compliance, Greece was required to submit a plan under Section XV, paragraph 2, of the procedures and mechanisms.[161] The submission of that plan provided the enforcement branch with the opportunity to discuss how to carry out the review and assessment called for by the procedures and mechanisms. After carrying out such review and assessment, the enforcement branch requested Greece to submit a revised plan.

Following an in-country review by a new ERT and the submission by Greece of the revised plan requested by the enforcement branch, the branch decided, on 13 November 2008, that there no longer continues to be a question of implementation with respect to Greece's eligibility and that Greece was now fully eligible to participate in the flexibility mechanisms.

In the case of Canada, the enforcement branch decided to proceed with the question of implementation, which it identified as relating to compliance with the guidelines for the preparation of the information required under Article 7 of the Kyoto Protocol (decision 15/CMP.1), the modalities for the accounting of assigned amounts under Article 7, paragraph 4, of the Kyoto Protocol (decision 13/CMP.1), as well as the annex to decision 5/CMP.1 and the requirements of the technical standards for data exchange between registry systems.

---

[160] See discussion in on JI track 2 in p. 30 of the paper.

[161] Please see pp. 10 and 12 of the paper for a more detailed description of such a plan.

On 15 June 2008, the enforcement branch adopted a decision not to proceed further based on the information submitted and presented by Canada at the hearing of 14 June 2008. In the same decision, the branch concluded that "the status of Canada's national registry resulted in non-compliance with the guidelines and modalities on the publication date of the [ERT's] review report."

While welcoming the decision not to proceed further, Canada, in a "further written submission," took exception to the conclusion regarding its non-compliance, arguing that such a declaration of non-compliance was not consistent with the authority conferred upon the enforcement branch by decision 27/CMP.1. Since the decision not to proceed further concluded the proceedings with respect to the related question of implementation, the substantive arguments in the "further written submission" were not considered by the branch. Instead, the document was treated as a comment to the final decision and was appended to the third annual report of the Compliance Committee to the CMP, as requested by Canada.

Documents relating to these two questions of implementation are available on the UNFCCC website.[162]

Both questions of implementation were triggered by reports of expert review teams. In considering these questions of implementation, the branch drew on the expert advice of members of the relevant ERTs and other experts on the question of implementation (i.e., experts on national systems and national registries).

The consideration of these two questions of implementation allowed the enforcement branch to explore key questions relating to non-compliance (e.g., at what point non-compliance must be ascertained) and eligibility to participate in the mechanisms. The experience resulted in recommendations to amend the rules of procedure of the Committee. At its fourth session, the CMP adopted amendments to the Committee's rules of procedure that, among other things,

---

[162] See http://unfccc.int/kyoto_protocol/compliance/enforcement_branch/items/3785.php. These documents are also annexed to the third annual report of the Compliance Committee to the Conference of the Parties serving as the meeting of the Parties to the Kyoto Protocol, FCCC/KP/CMP/2008/5.

clarifies what a plan that a Party which is found in non-compliance should contain and how the enforcement branch should review and assess such a plan.[163]

In the same year, the facilitative branch began, and has decided to continue, discussions on how it can carry out its responsibility to provide advice and facilitation "with the aim of promoting compliance and providing for early warning of potential non-compliance" under Section IV, paragraph 6 (a), of the procedures and mechanisms.

Recognizing the critical importance of the work of expert review teams under Article 8 of the Kyoto Protocol, the plenary of the Compliance Committee has expressed concern about issues of consistency in the review process and resource limitations facing expert review teams. The issue of consistency will continue to be discussed in subsequent meetings of the plenary.

[163] See annex to decision 4/CMP.4, Compliance Committee.

CHAPTER 13

CLIMATE CHANGE AS A THREAT TO PEACE & SECURITY: GLACIAL MELTING & HUMAN SECURITY IN THE HIMALAYAS

Kishan Khoday[1]

Legal Research Fellow, Center for International Sustainable Development Law (CISDL), Canada.

1. Introduction

On June 16, 2007 UN Secretary General Ban Ki-moon published an article in the Washington Post titled "*A Climate Culprit in Darfur*". In it, the Secretary General made the argument that "(a)mid the diverse social and political causes, the Darfur conflict began as an ecological crisis, arising at least in part from climate change.... It is no accident that the violence in Darfur erupted during the drought. Until then, Arab nomadic herders had lived amicably with settled farmers... For the first time in memory, there was no longer enough food and water for all. Fighting broke out. By 2003, it evolved into the full-fledged tragedy we witness today ... There are many other parts of the world where such problems will arise."[2] With growing

[1] Kishan Khoday (JD, MSc, BSc) has served with the United Nations in Asia since 1997, as UNDP Deputy Coordinator for Environment in Indonesia (2001–05) and UNDP Assistant Resident Representative and Team Leader for Energy & Environment in China (2001–09). He currently serves as UNDP Deputy Resident Representative in Saudi Arabia and as Research Fellow with the Center for International Sustainable Development Law (CISDL) in Canada. The opinions expressed herein are solely the personal views of the author and do not represent the views of the United Nations, UNDP or CISDL.

[2] *See* www.washingtonpost.com/wp-dyn/content/article/2007/06/15/AR2007061501857.html.

levels of resource scarcity and rapidly emerging impacts from climate change, Asia stands at risk in this regard.

As further noted in a June 1, 2007 article in *Scientific American* by leading climate expert and Director of the Earth Institute, Professor Jeffrey Sachs, "(o)ver the course of a few decades, if not sooner, hundreds of millions of people may be compelled to relocate because of environmental pressures... Impacts will vary widely across the world. It will be important to keep our eye on at least four zones: low-lying coastal settlements which are especially vulnerable to rising sea levels; farm regions which are dependent on rivers fed by glacier melt and snowmelt; subhumid and arid regions which are likely to experience greater drought frequency; and humid areas in Southeast Asia vulnerable to changes in monsoon patterns."[3] Climate change is now seen as a major risk for security and conflict within and between countries. In Asia, a major threat takes the form of melting glaciers in the Himalayas, which provide water and food security to hundreds of millions of people in downstream countries like China and India. Low per capita water security and increasing resource scarcity is combining with rising temperature and glacial melting trends, further exacerbating risks to regional security and human security.

China and India have faced historical tensions over various issues. However recent trends towards greater cooperation and bilateral treaties for investment and development have caused optimism that any regional risks arising from climate change may be dealt with in a cooperative manner. Against the backdrop of improved China-India relations and bilateral dialogue arises a new threat — trends of glacial melting in the Himalayas which provide 40% of the world's population its water needs. Communities living alongside seven of Asia's great rivers derive at least half their water from the Himalayan glaciers. Hundreds of millions of people in China and India rely directly on rivers flowing from Himalayan glaciers.

As noted by Professor Sachs, "vast numbers of farmers in the Indo-Gangetic Plain and in China's Yellow River Basin will most likely face severe disruptions in water availability. Yet those regions are already experiencing profound water stress due to unsustainable rates

---

[3] *See* http://www.sciam.com/article.cfm?id=climate-change-refugees-extended.

of groundwater pumping performed to irrigate large expanses of Northern China and Northern India. Surface water bodies in these regions are already over-appropriated and degraded ... Until now, the climate debate has focused on the basic science and the costs and benefits of reducing greenhouse gas emissions. That stage is now ending, with a resounding consensus on the risk of climate change and the need for action."[4] With improved relations and cooperation between China and India, the scope exists to take joint action to address the causes and impacts climate change, and prevent the future risks that both countries face from melting glaciers.

1.1. *Climate change as a threat to peace and security*

From the halls of the United Nations to the World Economic Forum (WEF) and the G8 Summit, leaders are now debating climate change beyond the confines of global environmental agreements, with climate now seen as a key element of the general social and economic security of our planet. UN Secretary General Ban Ki-moon has hailed climate change as the "defining issue of our era", and in late 2007 he presided over two major events to seek action by UN member states. Firstly in late September 2007, the UN Secretary General convened world leaders on the side of the UN General Assembly for the largest ever high-level climate change summit in history. This meeting served as a prequel to the second major event, the UN Climate Change Convention in Bali in December, which focused on the new urgency for mitigation and adaptation measures across the world.

What we witnessed in 2007 was a marked shift in the position of climate change to the center of global economic and political discourse. On April 17, 2007, climate change also entered the realm of global security when the UN Security Council, for the first time, held a special session on climate change. The focus of the debate was on the security implications of a changing climate, including impact on the primary drivers of conflict (food and water security, migrations, etc). This was the highest level forum to date to address climate as a security matter.

---

[4] Ibid.

As noted in the Secretary General's remarks to the Security Council, "when resources are scarce — whether energy, water or arable land — our fragile ecosystems become strained, as do the coping mechanisms of groups and individuals. This can lead to a breakdown of established codes of conduct, and even outright conflict."[5] The April debate built on the 2005 special Security Council meeting at the level of Heads of State and Government (Security Council Resolution S/RES/1625 (2005)) which called for a broad strategy for conflict prevention to address the causes of conflict, including those related to sustainable development.[6]

The April 2007 Security Council debate also built on a series of reports on conflict prevention by former UN Secretary-General Kofi Annan, pointing to the threats emanating from environmental degradation and resource scarcity: "Environmental degradation has the potential to destabilize already conflict-prone regions, especially when compounded by inequitable access or politicization of access to scarce resources."[7]

In his report to the April 2007 Security Council session on Climate Change, UN Secretary General Ban Ki-moon pointed out several plausible scenarios resulting from rapid climate change:

> The adverse effects of changing weather patterns, such as floods and droughts, and related economic costs, including compensation for lost land, could risk polarizing society and marginalizing communities ... Extreme weather events and natural disasters, such as floods and drought, increase the risk of humanitarian emergencies, and thus the risk of instability and dislocation. Migration driven by factors, such as climate change could deepen tensions and conflicts, particularly in regions with large numbers of internally displaced persons and refugees. Scarce resources, especially water and food, could help transform peaceful competition

---

[5] *See* http://www.un.org/apps/sg/sgstats.asp?nid=2524.

[6] *See further* CK Penny, "Greening the Security Council: Climate Change as an Emerging 'Threat to International Peace and Security'" in *International Environmental Agreements: Politics, Law and Economics* (2007) 35, 40.

[7] Ibid 45.

> into violence ... The economic costs and losses of all these scenarios would impede the ability of countries to reach the Millennium Development Goals.[8]

These issues are of concern in Asia. Existing tensions over water and land resources may become exacerbated owing to climate change. Asia hosts the world's highest number of people in extreme poverty, and climate change brings new risks to development goals and human security. As noted by the UN Secretary General, climate change poses major obstacles to progress in meeting the Millennium Development Goals (MDGs), particularly for rural communities that continue to rely on natural resources for their livelihood and income. While countries like China and India have made major progress towards the MDGs over the past decade, there is now a clear risk that climate change could start to roll back some development gains during the twenty-first century, undermining local human security in areas on the front line of glacial melting, like the Tibet plateau and Kashmir, and creating broader national and regional security risks in the process.

Thus, apart from concerns over regional tensions that may arise from changing water regimes and increased migrations, a primary concern in Asia relates to the risks to development goals and human security. The concept of human security was first introduced by the UN Development Programme (UNDP) in its 1994 Global Human Development Report (GHDR). Following from this, in 1999, during Canada's presidency of the Security Council Canada first put human security on the Security Council agenda as an emergent element of peace and security.[9] More than a decade after the initial GHDR on human security, UNDP's GHDR 2007/08 has explored the emerging link between human development and climate change, with emphasis on how the science of climate change has added new complexities to the development and human security landscape.[10]

---

[8] *See* http://www.un.org/apps/sg/sgstats.asp?nid=2524.

[9] *See* Ben Wisner *et al*, *Climate Change and Human Security* (15 April 2007) 5 at www.afes-press.de/pdf/ClimateChange_and_HumanSecurity.

[10] *See* http://hdr.undp.org/en/reports/global.

1.2. *Current and future impacts of climate change*

Ten of the warmest years in recorded history have occurred since 1990, leading to more frequent and extreme storm events and higher rates of glacial melting on the world's mountains and polar icecaps. According to the Fourth Assessment Report (AR4) of the Inter-Governmental Panel on Climate Change (IPCC), we now have clear and convincing evidence (90% certainty) that our use of fossil fuels and emission of greenhouse gases are warming the Earth well beyond normal background levels, bringing far reaching consequences for the state of global ecosystems and the future of human development.[11]

The AR4 also shows that even if drastic measures were started today to greatly reduce global emissions, current climate change trends would continue to pose serious risks to humanity for decades to come. If current trends continue, by 2100 the planet will experience global temperatures not seen since 3 million years ago, while half of all species could be faced with extinction. With the impacts of climate change felt more frequently in various parts of the world in the form of severe droughts, water shortages and extreme weather events, the time for speculation over the causes and consequences of climate change is over. While stronger mitigation measures are critical, it is also clear that we are now entering a period of hard consequences, requiring urgent adaptation to secure development achievements and prevent a reversal of development goals.

Developing countries like China and India have made many hard-won development gains over the past decade, and will make many more in the lead up to the 2015 targets under the Millennium Development Goals (MDGs). However, the shifting nature of the world's climate and trends of glacial melting pose serious risks to sustaining any such gains in the decades to come. Unless global action is taken to greatly reduce emissions, and enact local adaptation measures, many of these hard won gains could start to reverse in decades

---

[11] *See* UN Framework Convention on Climate Change (UNFCCC), "IPCC Fourth Assessment Report" (2007).

to come. As with other global environmental trends, the impacts of climate change will be felt the hardest by the poor because poor communities often lack the capacities to adapt to an uncertain future.

The Asian region holds special challenges in this regard. With the largest concentration of people surviving on less than US$2 per day, and with the carrying capacities of land and water systems already being stressed owing to rapid growth and development, the future of development in Asia is closely tied to the ability of governments and communities to adapt to the emerging consequences of climate change. The frequency and severity of storms and floods is increasing across Asia. Examples include the historic typhoons in 2007 in Bangladesh, the 2006 floods in China, India and Indonesia where tens of millions were affected, and historic droughts in 2005–2006 in semi-arid areas of China and India that experienced severe water-stress conditions. Urbanization is a major challenge with Asia expected to host most of the world's largest cities by 2020, most of which are currently located in low-lying coastal zones at risk from sea-level rise. Cities like Dhaka, Jakarta, Mumbai and Shanghai are at risk from both sea-level rises and increased frequency of storm events.

The consequences of climate change will be critical for countries' ability to sustain progress towards, and achievement, of the MDGs. Critical to achieving the overarching goal of poverty reduction are countries' ability to achieve levels of food and water security. However, the water and agriculture sectors are the components of Asia's economies most sensitive to climate impacts, particularly those lying downstream of melting glaciers in the Himalayas. Food output and water supplies in parts of Asia are projected to decrease because of rising temperatures, more frequent droughts, floods, and continued land degradation. This is particularly important for communities in high-altitude mountain systems in Asia where fragile ecosystems already face severe pressures from rapid development.[12]

---

[12] *See* F Pearce, "Flooded Out: Retreating Glaciers Spell Disaster for Valley Communities", (1999) 2189 *New Scientist* 16, 18.

2. Glacial Melting in the Himalayas

2.1. *Trends and scenarios*

One of the most pressing concerns in Asia relates to the rapid melting of the Himalayan glaciers.[13] "Himalaya" is a Sanskrit word meaning abode of snow, known in the region as "Asia's water tower" or the "third pole". The Himalayas host tens of thousands of glaciers covering an area of 33,000 km^2, the largest freshwater reserve outside the polar ice caps.[14] These glaciers feed seven of the great Asian rivers – the Ganges, Indus, Brahmaputra, Salween, Mekong, Yangtze and Yellow rivers – the source of food and water security for approximately 2 billion people.[15]

While the Himalayas have supported development along these rivers for millennia, their ability to sustain future development towards and beyond the 2015 targets of the MDGs is now in question owing to rapid melting from climate change.[16] The IPCC has found that up to 25% of the total global mountain glacier mass could disappear by 2050 and up to half by 2100.[17] Data published by UNEP and the World Glacier Monitoring Service (WGMS) shows that mountain glaciers around the world are melting fast as global temperatures rise. The average rate of ice loss from all glaciers around the world since 2000 is over half a meter per year, three times the annual rate during the 1980s.[18]

In the Himalayas, as much as two thirds of glaciers are retreating at an alarming rate. The International Commission for Snow and Ice

---

[13] *See* J Bandyopadhyay, JC Rodda, R Kettelmann, ZW Kundzewicz and D Kraemer. "Highland Waters — A Resource of Global Significance" in B Messerli and JD Ives (eds), *Mountains of the World — A Global Priority* (1997) 131.

[14] *See* World Wildlife Fund, Nepal Program, *An Overview of Glaciers, Glacier Retreat, and Subsequent Impacts in Nepal, India and China* (2005).

[15] Ibid 3.

[16] *See* P Singh and L Bengtsson, "Impact of Warmer Climate on Melt and Evaporation for the Rainfed, Snowfed and Glacier-fed Basins in the Himalayan region" (2005) 300 *Journal of Hydrology* 140.

[17] *See* RT Watson, MC Zinyowera and RH Moss (eds), *Climate Change 1995: Impacts, Adaptation and Mitigation of Climate Change: Scientific and Technical Analyses* (1996).

[18] *See* www.wgms.org/mbb/mbb9/sum05.html.

(ICSI) in a recent study reported that Himalayan glaciers are shrinking quicker than anywhere else on the planet, and that if current trends continue they could disappear altogether by 2035.[19] WWF reports that Himalayan glaciers are receding at 10–15 meters per year, and that climate change is accelerating this.[20] Glacial melt will result in increased water flows for several years, followed by water shortages as glaciers gradually disappear.[21]

At the center of this challenge lays the Qinghai-Tibet Plateau in western China, known as the "Roof of the World". The plateau covers more than 1.6 million km^2, about a quarter of China's landmass. About the size of Western Europe, this plateau forms the largest high-altitude landmass on earth, stretching across China's provinces of Tibet and Qinghai with more than 45,000 individual glaciers, covering about 90,000 km^2 and extensive permafrost.

Many of Asia's great rivers originate in the Qinghai-Tibet Plateau. China's Yangtze and Yellow Rivers commence their flow on the plateau, flowing eastward to China's heavily populated rural and urban centers. The Mekong River also originates on the Tibetan Plateau as does the Yarlung Tsang Po River that travels eastwards from Tibet for 2,000 km before becoming the Brahmaputra River flowing south into India and Bangladesh.[22] Most of Nepal's rivers also originate in Himalayan melt waters, subsequently flowing south to contribute to India's Ganges River, while Pakistan's Indus River also derives part of its original flows in Tibetan glaciers.[23]

In the decades to come, rapid melting trends could result in decreased water flows and shortages for over 500 hundred million of

---

[19] *See* www.icsi.org. *See also* Meghna Tare, *Himalayan Glaciers* at www.csi.org.

[20] World Wildlife Fund, above n 14, 5.

[21] *See* P Singh and N Kumar, "Impact Assessment of Climate Change on the Hydrological Response of a Snow and Glacier Melt Runoff Dominated Himalayan River" (1997) 193 *Journal of Hydrology* 316.

[22] World Wildlife Fund, above n 14.

[23] *See* M Lal, "Climatic Change — Implications for India's Water Resources" (2001) 21 *Journal of Indian Water Resources Society* 101; GJ Young and K Hewitt, "Hydrology Research in the Upper Indus Basin, Karakoram Himalaya, Pakistan" in L Moln´ar (ed), *Hydrology of Mountainous Areas* (1998) 139.

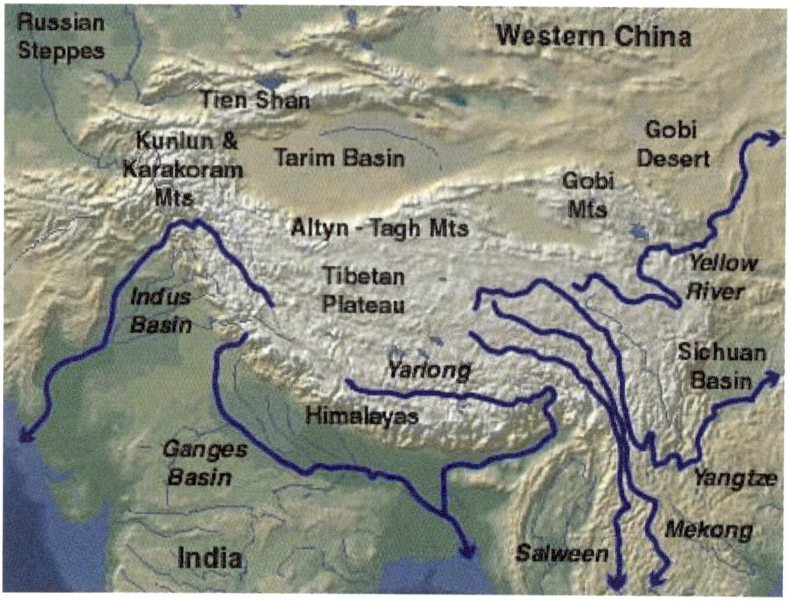

people directly reliant on these waters, unless actions are taken now to reduce the risk of rapid melting and water supply disruptions.[24] If glaciers continue to retreat at current rates, many of the afore-mentioned rivers systems will be altered, threatening food and water supplies, human security, and later in the twenty-first century the possible reversal of hard won development gains in this part of the world.[25]

The impacts of global climate change on Himalayan glacial melting is far more than a scientific concern, it can have profound political, economic and social repercussions in the region unless urgent action is taken.[26] As one sign of the importance of the issue to

---

[24] *See* HG Rees and DN Collins, "*Regional Differences in Response of Flow in Glacier-fed Himalayan Rivers to Climatic Warming*" (2006) 20 *Hydrological Processes,* 2157.

[25] *See* Wang Guo-qing, Zhang Jian-yun, Zhang Si-long, *et al*, "Impacts of Climate Change on Water Resources and its Vulnerability in China" (2005) 16 *Journal of Water Resources & Water Engineering* 8.

[26] *See* International Centre for Integrated Mountain Development, Kathmandu, Nepal (ICIMOD), *Inventory of Glaciers, Glacial Lakes and Glacial Lake Outburst Floods — Monitoring and Early Warning Systems in the Hindu Kush–Himalayan Region–Nepal* (2000).

countries in the region, institutions from China and India — China's Institute of Geology and Geophysics and the Indian Mountaineering Foundation — are undertaking the first joint Indian-Chinese research survey to the Qinghai-Tibet plateau to explore the process of glacial melting and jointly monitor the changing nature of these water sources.[27] This evidences the good will and potential for regional cooperation and joint action to prevent risks to human security and regional security.

2.2. *Vulnerability in the Qinghai-Tibet plateau*

In early 2007, the Ministry of Science and Technology and six other national ministries jointly issued China's first National Climate Change Assessment Report.[28] The report lays out the likely impacts of climate change on social and economic development over the upcoming decades, showing that few aspects of development will escape from emerging trends.[29] According to the Government report, China's average annual temperature may increase between 1.3 and 2.1 degrees Celsius by 2020, by as much as 3.3 degrees Celsius by 2050 and by as much as 6 degrees Celsius by 2100. The report also predicts rainfall to increase in some regions of China as much as 17% by 2100.[30] However, regional differences in precipitation patterns will become more extreme and higher temperatures could result in increased evaporation, resulting in net water reductions around some inland rivers.[31]

---

[27] G Rabinowitz, "Indian, Chinese Team to Map Glacier Melt: Two Expeditions Will Take Scientists into Remote Areas of Tibet", *Associated Press*, 21 December 2006.

[28] Chinese Ministry of Science and Technology (MOST), *National Climate Assessment Report* (2007).

[29] *See also* Qin Dahe, Ding Yihui, Su Jilan *et al*, *Climate and Environment Changes in China, Volume I: Climate and Environment Changes in China and their Projections* (2005).

[30] Chinese Ministry of Science and Technology, above n 28, 12.

[31] Wang Shun, Wang Yan-guo, Wang Jing *et al*, "Change of Climate and Hydrology in the Tarim River Basin during Past 40 Years and Their Impact" (2004) 12 *Journal of Water Resources & Water Engineering* 23.

According to China's Initial National Communication to the UN Climate Change Convention, glaciers in Western China could decrease overall by 27% by 2050.[32] With one quarter of the world average per capita water availability, China already faces serious water pressures. The findings suggest that this could get worse in some regions owing to reduced glacial water flows and rising temperatures, with the risk of per capita water availability in certain regions decreasing by a further 20 to 40% by 2080.[33] By 2050, China's grain output could fall by as much as 10% unless crop varieties adapt to new temperature and water regimes, while by the latter half of the century, production of wheat and rice could drop by as much as 37% according the Government report.

China is also expected to have more extreme weather. Typhoons, floods and droughts killed 2,704 people and caused a loss of $27.32 billion in 2006, China's warmest year on record since 1951. Some coastal zones of China have already seen sea level rise of about 5cm per year, among highest in the world. Major urban areas such as Shanghai are expected to face serious challenges from rising sea levels, including increased salinity of groundwater supplies.[34]

For western China, recent UNEP/GEF projections show a major warming very likely to develop over the region during the twenty-first century, based on projections using IPCC assessment data and the results run by China's National Climate Center and the Institute of Atmospheric Physics.[35] Western China could see temperature rise of

[32] National Climate Change Coordination Committee, *Initial National Communication to the UNFCCC* (2002) 39.

[33] Ibid.

[34] Chinese Ministry of Science and Technology, above n 28, 15.

[35] *See* Yongyuan Yin, "Vulnerability and Adaptation to Climate Variability and Change in Western China", in UNEP/GEF Assessments of Impacts and Adaptations to Climate Change (AIACC) Project, Report Project No AS 25 (2005) 15. *See also* Weihong Qian and Xiang Lin, "Regional trends in recent temperature indices in China" (2004) 27 *Climate Research* 119; ZC Zhao, Y Ding, Y Xu and Zhang Jin, "Detection and projection of climate change over Northwest China for the 20th and 21st centuries due to the human activity" (2003) 8 *Climate and Environment Research* 26.

1.0–2.5 degrees Celsius by 2050 (relative to 1961-1990 figures), with the region likely to see higher increases than the national average trends, and large increases seen in the northern reaches of the Tibetan Plateau.[36] At the same time, rainfall over western China could increase by 5-25% by 2050 relative to 1961–1990 levels according to the UN study.

Recent research by experts from the National Development and Reform Commission, the Ministry of Water, the Chinese Academy of Social Science, the National Natural Science Foundation and various local partners has found that by 2050, glaciers in western China will experience a strong retreat, and the areas of marine-type, sub-continental-type, and continental-type glaciers will reduce by 30%, 12% and 5% by 2030, and by 50%, 24% and 15% by 2050, respectively.[37] The study predicts sudden flood and glacial outbursts will increase in the Qinghai-Tibet Plateau and surrounding areas. Glacial runoff would increase 5%–30% across the plateau and increase to as high as 50% in specific areas. Areas of unstable snow cover will increase, causing more severe winter and spring droughts.

Meanwhile frozen soils overlaying the plateau, the permafrost, are expected to decrease by 5%–15% in the Qinghai-Tibet Plateau, while the permafrost boundary on mountain slopes could rise by 300 to 400 meters. Permafrost melting in the plateau will also increase through higher soil temperatures over the next 30 to 50 years.[38] With increased temperatures of approximately 2 degrees Celsius over the next 50 years in the area, permafrost with temperature above –0.42 degrees Celsius will only exist seasonally. Permafrost melting would reduce water runoff into streams and groundwater supplies, with estimated water loss as much as 5–11 billion m^3 per year in the Qinghai-Tibet Plateau region.[39] Permafrost degradation is expected to lead to a decline of

---

[36] Ibid.

[37] Chen Yiyu, Ding Yongjian, She Zhixiang *et al*, *Assessment of Climate and Environment Changes in China, Volume Two: Impacts, Adaptation and Mitigation of Climate and Environment Changes* (2005) 3.

[38] Ibid.

[39] Ibid.

natural forest areas, shifting habitat of vegetation species, and a disappearance of grasslands.

In the Qinghai-Tibet Plateau, temperatures hit record highs in early 2007. January temperatures in the Qamdo area of the Tibet Autonomous Region hit 21.8 degrees Celsius, 1.7 degrees higher than the previous record set in 1996.[40] And in Dengqen County in Tibet, temperatures reached 16.6 degrees Celsius, 2.5 degrees higher than the previous record in 2001. Eight other spots across the plateau also recorded record-breaking temperatures in 2007. According to the recent National Climate Change Assessment Report, warming trends are shrinking the glaciers by 7% a year, and can threaten the Yangtze and Yellow Rivers, and trigger more frequent droughts, sandstorms and desertification in other regions of China. According to recent studies, ice core records from the Dasuopu Glacier in Tibet indicate that the last fifty years have been the warmest in the past 1,000 years.[41]

The report finds that China's glaciers declined by 21% since the 1950s, an average of 131.4 km^2 annually over the past thirty years. If climate change continues at its current rate, two thirds of the glaciers could disappear by 2050, and almost all could be gone by 2100.[42] According to research on the Tanggula Mountain Glacier in Tibet, 50% of the 2213 m^2 glacier could melt by 2050 if the temperature rises by just 2.4 degrees Celsius, and 90% will disappear if it rises by 5.8 degrees Celsius.[43] The report predicts that the shrinking of the glaciers would seriously impact livelihoods of Tibetan mountain communities and exacerbate water supply and land degradation challenges further downstream. Analysis of data by the Chinese Academy of Social Science (CASS) from 680 Chinese weather stations shows that

---

[40] "Tibet`s Record Temperatures Spark Climate Change Fears", *Zee News* (Beijing) 8 January 2007 at http://www.zeenews.com/znnew/articles.asp?aid=346544&ssid=26&sid=ENV.

[41] *See* www.climatehotmap.org/fingerprints.

[42] Chinese Ministry of Science and Technology, above n 28, 27.

[43] "Glacier on Tibet-Qinghai Plateau Melting Fast", *Xinhua News Agency* (Beijing) 21 May 2005.

average temperatures in the plateau region have risen 0.9 degrees Celsius since the 1980s resulting in an annual 7% reduction in glacier extent and the widespread melting of permafrost.

In Lhaze County, in the western reaches of Tibet, increased levels of development has resulted in gradual deterioration of land and water resources over the past several years.[44] Given the fragile nature of the mountain ecosystems in the area, and the rapid loss of forest cover over the past decades and immature soils, the ability of the area to preserve moisture and soil fertility is already low. Trends of environmental degradation will become even more challenging with the emerging impacts of glacial melting and climate change in Tibet.[45]

Traditionally, summer precipitation accounted for more than 80% of the area's annual water and irrigation supply. Farmlands, grasslands, and other vegetation along the edges of local streams are now threatened. Climate change is already starting to alter this regime.[46] Increasing temperatures in the region are also changing the nature of wind patterns, bringing increased erosion and sandstorms. Data from local weather stations shows that sandstorms have increased in recent years.

In summary, three general consequences are foreseen from continued glacial melt in various regions in the plateau. First, continued melting trends will have serious implications for upstream and downstream water supply. As noted in the 2006 UNDP Global Human Development Report "Beyond Scarcity: Power, Poverty and the Global Water Crisis", recession of the Himalayan glaciers threatens to disrupt the water supply of over 100 cities and 300 million people across China.

---

[44] Yan-Zhi Zhao, Xue-Yong Zou *et al*, "Assessing the Ecological Security of the Tibetan Plateau: Methodology and a Case Study for Lhaze County" (2006) 80 *Journal of Environmental Management* 120.

[45] *See* Y Xu, ZC Zhao and D Li, "Simulations of Climate Change for the next 50 Years over Tibetan Plateau and along the Line of Qing-Zang Railway" (2005) 24 *Plateau Meteorology* 698. *See also* KP Sharma, CJ Vorosmarty and B Moore, "Sensitivity of the Himalayan Hydrology to Land-use and Climatic Changes" (2000) 47 *Climatic Change* 117.

[46] Zhao, Zou *et al*, above n 45, 128.

Second, glacial melting will exacerbate land degradation and desertification trends. Climate change is exacerbating unsustainable land-use practices in western China, increasing the frequency and severity of dust storms and catalyzing conversion of grasslands into semi-desert conditions in areas that have been the habitat of local herders for centuries, relying on livestock management for their livelihood.[47] Official estimates show 2,330 km^2 of land going to desert each year, while a much greater area will see reductions in productivity, as land faces the dual challenge of overuse and a changing climate.

Third, increasing temperatures in the Plateau are also associated with melting of the permafrost, frozen soils that underlay much of the region. Climate change is exacerbating the melting of these soils, threatening to further exacerbate the scale of potential desertification in area.[48] Also of concern is the likely release of greenhouse gases such as methane and carbon dioxide from such soil as it thaws. With estimates of the amount of carbon locked up in permafrost ranging between 60 and 190 billion tons, thawing of the soil of the Qinghai-Tibet Plateau will produce a large-scale release of greenhouse gases, further adding to the climate change process.

As highlighted in recent studies, local adaptation measures must consider factors unique to this area — infrastrcucture solutions are unlikely to be stable due to permafrost melting and resultant land instability; afforestation to support soil stability will be difficult given challenges of increasing water scarcity; and the area lacks local capacities to take aggressive measures. As development activities move forward, adaptation measures should be incorporated into investment and development plans to take into account the changing nature of the environment and implications for local human security and development goals.

Adaptation measures in such cases also need to be integrated into national actions. Land degradation and water supply are now top

---

[47] *See* Gao Qingzhu, Li Yu'e, Lin Erda *et al.* "Temporal and Spatial Distribution of Grassland Degradation in Northern Tibet" (2005) 60 *Acta Geographica Sinica* 965.

[48] *See* Cheng and Guodong (eds), *Assessing Climate Change Impacts on Snow Pack, Glaciers and Permafrost in China* (1997).

national priorities. Climate change risks need to be integrated into environmental assessments for domestic and foreign investments directly or indirectly linked to the glacial melting issues in the region. The capacity of local agencies need to be built in the source areas in Qinghai and Tibet to improve indicator and monitoring system to track the emerging vulnerabilities from rising temperatures, and permafrost and glacial melting.

2.3. *Risks in the Yangtze River Headwaters*

The Yangtze River Headwaters are situated in the interior of the Qinghai-Tibet plateau, lying between the Kunlun and Tanggula Mountains in southern reaches of Qinghai Province and northern reaches of Tibet.[49] The Yangtze River is China's largest river and provides crucial water supply to industry, agriculture and 500 million domestic users along its course. In its downstream reaches, the Yangtze is home to the Three Gorges Dam, now the world's largest hydroelectric dam, slated to commence electricity provision in 2009.

The year 2006 was China's warmest year in almost half a century and the upper reaches of the Yangtze saw their lowest levels since the 1920s. Many now worry about the consequences for the Yangtze from long-term decreases in water flows owing to glacial melting. According to recent studies by experts from the National Development and Reform Commission and the Ministry of Water, climate change will also cause sudden melting in the Qinghai-Tibet Plateau that could cause an increase of flash floods in the middle reaches of the Yangtze.[50]

The source of the Yangtze is in the semi-arid alpine zone of the plateau in Tibet and Qinghai Provinces. At about 4,000 meters above sea level, the area has low population, extensive permafrost and annual average temperatures between 2 to 4 degrees Celsius. However,

---

[49] Wang Xin, Xie Zi-chu, Feng Qing-hua *et al*, "Response of Glaciers to Climate Change in the Source Region of the Yangtze River" (2005) 27 *Journal of Glaciology and Geocryology* 498.

[50] *See* Chen Yiyu, Ding Yongjian, She Zhixiang *et al*, *Assessment of Climate and Environment Changes in China, Volume Two: Impacts, Adaptation and Mitigation of Climate and Environment Changes* (2005) 3.

according to recent studies, climate change has brought increases in average annual temperature by as much as 2 degrees in the last 50 years, bringing about accelerated permafrost melting.[51] Over the next 50 years, the temperature is expected to increase even faster, altering the ecological nature of the area and eventually reducing glacial melt flows into the Yangtze.

With a naturally dry and cold climate with sparse vegetation, the area is already characterized as a fragile mountain ecosystem. The current and future impacts of climate change are adding to this fragility with land degradation and reduction of permafrost already seen throughout the area.[52] Indeed the area is now officially designated as "desertified land".[53] Recent field research and remote sensing data has confirmed the area of land degradation and soil erosion to have passed the 100,000 km^2 mark by 2006: over two-thirds of the total area.[54]

Such trends pose serious risks not only for local communities in western China dependent on the waters for agriculture, livestock and domestic use, but also for the hundreds of millions of downstream users. Sediment deposition in past years has already raised the riverbed and reduced the capacity for flood discharge. Rapid changes owing to climate change are expected to exacerbate this process unless preventive and adaptive measures are taken.

Changing climates and glacial melting will make future watershed conservation efforts increasingly difficult and will aggravate sand deposits into the Yangtze silting up its middle and downstream reaches and further reducing its flood discharge capacity.[55] Such

---

[51] *See* Fan Qi-shun, Sha Zhan-jiang, Cao Guang-chao *et al*, "Assessment of the Influence of Glacial Resources on Climatic Changes in Qinghai Plateau" (2005) 19 *Journal of Arid Land Resources and Environment* 56.

[52] *See* Gao Qingzhu, Li Yu'e, Lin Erda *et al*, "Temporal and Spatial Distribution of Grassland Degradation in Northern Tibet" (2005) 60 *Acta Geographica Sinica* 965.

[53] P Yan, GR Dong *et al*, "Desertification Problems in the Yangtze River Source Area, China" (2004) 15 *Land Degradation Development* 177, 177.

[54] Ibid 179.

[55] Wang Xin, Xie Zi-chu, Feng Qing-hua *et al*, "Response of Glaciers to Climate Change in the Source Region of the Yangtze River" (2005) 27 *Journal of Glaciology and Geocryology* 498.

changes bring risks related to water supply and disaster prevalence in the entire Yangtze River basin. Future impacts will come from glacial melting and reduced water flows, permafrost-melting exacerbating land degradation, and increased wind erosion resulting from changing local climates. Permafrost melting is of particular concern given the local and national threats from rapid desertification. Reduced water flows resulting from glacial melting will also exacerbate the process of dryland expansion and siltation into the Yangtze.[56]

3. Adaptation and Human Security

3.1. *Sustainable development and human security*

Climate change poses risks not only to physical infrastructure and ecosystem change, but also to the sustainability of human development in the Himalayan region.[57] Climate change is increasingly seen as a fundamental threat to human security with risks for reversal of hard won development gains over the next decades.[58] While China and India are experiencing rapid reductions of human poverty, human development at the global level could face a major "U-turn" by mid-century owing to climate change impacts.[59]

If current trends continue, climate change could jeopardize achievement and sustainability of the Millennium Development Goals in the Himalayan region unless urgent measures are taken to "climate proof" development. There has been growing awareness within the region itself of the impact of climate factors on local livelihoods — particularly for rural communities whose livelihoods are dependent

[56] Y Yin, Q Miao, and G Tian (eds), *Climate Change Impact Assessment and Sustainable Regional Development in the Yangtze Delta* (1999) 234.

[57] *See* MC Goldstein, B Jiao *et al*, "Development and Change in Rural Tibet: Problems and Adaptations" (2003) 43 *Asian Survey* 758.

[58] *See* JC Ribot, AR Magalhães, and S Panagides, *Climate Variability, Climate Change, and Social Vulnerability in the Semi-arid Tropics* (1996).

[59] *See* the Stern Report on "The Economics of Climate Change" at www.sternreview.org.uk.

on natural resources but lack the capacities to respond to changing land and water dynamics.[60]

Local communities have traditionally been effective in adapting to changing environments and conditions, but the projected magnitude of climate change from rising temperatures and glacial melting in the plateau could stretch this adaptive ability to its outer limits, with serious consequences for local development. Given the unprecedented scale and consequences of projected impacts in the Himalayan region, environmental changes can render the customary knowledge and traditions in the area ineffective.

Current strategies for increasing investments, trade and growth are no longer a guarantee for improved social development, once the implications of climate change are taken into consideration. Indeed rapid developments that fail to take climate change into account will likely exacerbate risks to human security if they add to ecological challenges in the fragile Himalayan environment.

Climate change calls for a shift in the development paradigm in which development policy decisions fully integrate the vulnerabilities local communities will face from climate change. Using a Human Development Approach to climate change adaptation would go beyond the reorientation of policies to climate proof infrastructure and investments projects. A focus on human capabilities in the Himalayan region ensures communities are able to act as agents of change, taking proactive measures to adapt to climate risks to ensure their own security, local development, ways of life and livelihoods.

In the 1990s, Nobel economist Amartya Sen articulated the foundation of the Human Development Approach.[61] In *Development as Freedom* Professor Sen argued that development is "not the mere

---

[60] *See* Yongyuan Yin, *Evaluation Tools to Identify Implications of Climate Change and Economic Development for Sustainability in Lijiang and Yulong Mountain Region of China*, Environment Canada, (2003). *See* S Brogaard and J Seaquist, *An Assessment of Rural Livelihood Vulnerability in Relation to Climate — A Case Study in Agro-pastoral Northern China*, paper presented at International Workshop on Human Security and Climate Change, Oslo (2005).

[61] Amartya Sen, *Development as Freedom* (2000).

accumulation of goods but the enhanced freedom to choose, to lead the kind of life one values".[62] Beyond being beneficiaries of economic and social development, climate-affected communities need to be active agents of change to ensure that local development gains can be sustained by future generations. Under Sen's human capability approach, the focus of climate impacts in the region shifts from climate proofing infrastructure and investments to focusing on aspects of community life beyond income, so as to better understand the aspects of local life against which climate change poses its greatest risks.[63]

Two principles have historically been applied under the capability approach — (i) that the chosen capability be universally valued by people across the world and (ii) that the capability be so basic that, without it, many other capabilities would be foreclosed.[64] A capability approach allows measures to adapt and accommodate to new challenges in a world of rapid change. Climate change now stands at the center of this process. Beyond being a mere "market externality", climate change is shifting the foundations of life on earth, with implications for our concept of human development itself. The ability to achieve and sustain a certain quality of life for current and future generations is now under threat for communities across the Himalayan region. Indeed one could say that the capacity to deal with climate change is quickly becoming a universally valued capacity, and that without this capacity many other human development goals are in jeopardy.

The purpose of adaptation measures under a human security approach is to improve human capabilities and lives, to ensure

---

[62] SP Marks, *The Human Rights Framework for Development: Five Approaches*, Submission to the 2nd Global Forum on World Development, UNDP, Rio de Janeiro (2000) 4, *See also* P Alston, "Making Space for New Human Rights: The Case of the Right to Development" (1988) 1 *Harvard Human Rights Yearly Bulletin* 3, 20.

[63] *See* T Cantrell, E Holthaus and A King, "China and Its Climate: The Impacts on Culture" in *Using the Science of Weather in Business and Public Policy* (2005). *See also* Cheng-Bang An, Lingyu Tang *et al*, "Climate Change and Cultural Response around 4000 cal yr B.P. in the Western Part of Chinese Loess Plateau" (2005) 63 *Quaternary Research* 347.

[64] Fukuda-Parr, *Operationalising Amartya Sen's Ideas on Capabilities*, UNDP Working Paper 6–7 (2002).

communities' abilities to achieve and sustain their development goals in a rapidly changing environment. From this point of view, adaptation should be about removing vulnerabilities to climate change as obstacles to achievement of future development goals. A human development approach would focus measures on future communities affected by glacial melting in areas of China and India being able to achieve a standard of living as good as those of the present generation; the right of future generations to live in a healthy and productive environment.

3.2. A rights-based approach

For communities standing in the path of glacial melting impacts in China and India, climate change poses fundamental threats to their security — food and water security and the ability to sustain standards of living, local ways of life and the exercise of their basic right to development. The UN Declaration on the Right to Development was adopted by the UN General Assembly in 1986 by an overwhelming majority. The first article in the Declaration puts forward the concept of the right to development as "an inalienable human right by virtue of which every human person and all peoples are entitled to participate in, contribute to and enjoy economic, social, cultural, and political development in which all human rights and fundamental freedoms can be fully realized."[65] This resonates with national policies in the region, with policy frameworks in both China and India both placing heavy emphasis on the fundamental right of communities to achieve social and economic development goals.

A stable environment is critical to achieving basic social and economic rights.[66] The process of integrating environmental concerns

---

[65] A Sengupta, *The Right to Development as a Human Right,* Working Paper, Harvard School of Public Health (1999) 2.

[66] *See* UNEP, *Human Rights and the Environment*, Proceedings of UNEP Geneva Environment Network Roundtable (2004). *See also* S Watt-Cloutier, *Climate Change and Human Rights*, Carnegie Council Human Rights Dialogue, Series 2, No 11, (2004) 10.

into broader rights frameworks started in 1972 with the Stockholm Declaration on the Human Environment stating: "man has the fundamental right to freedom, equality and adequate conditions of life, in an environment of a quality that permits a life of dignity and well-being". This was taken forward in Principle 1 of the 1992 Rio Declaration on Environment and Development that stated "human beings are at the center of concern for sustainable development" and that they are "entitled to a healthy and productive life in harmony with nature".[67]

Since then, climate change has accelerated and has now become a much larger global concern. Climate change will transform the landscape of development and the ability of communities' to achieve a minimum level of human security and socio-economic rights. Many would argue that urgent problems of survival are more crucial for these communities than climate-proofing development activities. This would reflect a traditional "need-based approach" to development, which the capability approach includes but goes well beyond. Sustainable human development is "development which meets the needs of the present generation without compromising the ability of future generations to meet their own needs".[68]

The UN Special Rapporteur on Human Rights and the Environment further elaborated the link between the right to development and global environmental change, arguing that the right to development would be violated if trends of climate change and other environmental harms continue.[69] The UN Special Rapporteur encapsulated this in the idea of a "right to prevention of ecological harm", proposed as being part and parcel of the right to development, particularly given the clear impacts of global environmental change on poor and vulnerable communities.

Given the stark changes expected in the future, a stable climate may now evolve to become an essential underlying element of the right to development, as part of the "enabling environment" in which

---

67 A Dias, *Human Rights, Environment and Development,* UNDP Human Development Report 2000 Background Paper (2000) 3.

68 Ibid.

69 Ibid.

communities — current and future — can lead secure and healthy lives. Making this connection lifts ecosystem services from being a subset of other rights and endows the complex elements of ecosystem change, such as climate change, to a higher moral, policy and legal plane. Applying a rights-based approach in the Himalayan context construes adaptation as a way of protecting and achieving socio-economic rights and human security in vulnerable communities, emphasizing equity between upstream and downstream communities, between urban and rural communities and between current and future generations.

The human security aspects of climate change are becoming central to the basic concept of human development, and as such a new paradigm is needed, one which takes us from Sen's original "Development as Freedom" model to a new "Sustainable Development as Freedom" model with ecological security as a core freedom within broader human development and right to development frameworks. In addition to the five freedoms outlined in Sen's Development as Freedom framework, one may argue that ecological security has become a sixth freedom. As noted by Professor Sen himself in 2004,

> To use a medieval distinction, we are not only patients, whose needs demand attention, but also agents, whose freedom to decide what to value and how to pursue it can extend far beyond the fulfillment of our needs. The question can thus be asked whether environmental priorities should be seen in terms also of sustaining our freedoms. Should we not be concerned with preserving — and when possible expanding — the substantive freedoms of people today without compromising the ability of future generations to have similar, or more, freedoms? Focusing on "sustainable freedoms" may not only be conceptually important (as a part of a general approach of "development as freedom"), it can also have tangible implications of immediate relevance.[70]

[70] Amartya Sen, "Why We Should Preserve the Spotted Owl" (2004) 26 *London Review of Books* 2.

Analyzing response measures in this context recognizes the fundamental value of a stable climate as a basic condition of life, indispensable to human welfare and to the fulfillment of human security. Given the gravity of the situation resulting from glacial melting in the Himalayas, actions to address climate change are more than an academic exercise. They are quickly becoming issues of 'common concern for humanity', a growing ethical concern for people across the Himalayan region.

In the Himalayas, concerns over ecological change align with cultural norms and traditions regarding not just rights but also the duty to respect nature and protect the environment. In the sacred *Sutta Nipata*, the Buddha argued that since humanity is far more powerful than other species, we have special duties to the environment linked to this asymmetry of influence. A common sense of urgency among China, India and the broader Himalayan community to the growing crisis of climate change brings forth an opportunity to engage this responsibility to support regional actions and community-based capacities for adaptation, based on a sense of the right to a healthy environment and an emerging sense of moral duty to address future risks of climate change as a common concern of humanity.

3.3. *Conclusion*

By focusing on the connections between climate risks and human security, response measures can bring synergies between natural and socio-economic policies to reduce vulnerabilities to glacial melting in upstream and downstream communities along Asia's major rivers. Responses should focus on marginalized communities, those within mountain ecosystems and poor communities along downstream riverways, who have the fewest resources and the least capacity to manage change. Those at the forefront of glacial melting — rural communities in the Himalayas — require new adaptive capacities if future risks to human development and security are to be managed.

Governments in the region, such as China and India, also recognize that upstream dynamics can have serious consequences on communities and industries far downstream. Urbanization and infrastructure plans in both countries should be reviewed for future water and food security impacts, as well as for adequate flood and disaster risk management systems. In areas where impacts of glacial melting are likely to be felt, there is a need to develop incremental adaptation strategies to manage climate risks. In stressing the importance of climate in planning, designing and implementing development strategies and investment activities, two overarching strategies should be developed.

The first is a *macro strategy* and the need for integration of climate risks into regional and national development strategies, foreign direct investment policies, water and food security policies, and disaster preparedness systems so that future generations along the Ganges and Yangtze river basins can sustain the hard won development gains made by the current generation. Stronger regional cooperation to address transboundary issues is needed. Support should go to supporting "South-South cooperation" between China, India and other Himalayan nations, helping to upscale lessons to address policy issues and to coordinate strategies for response to glacial melting trends. This should also focus on using new Strategic Environmental Assessment (SEA) policies and tools, meant to integrate environmental concerns into upstream investment, trade and development policies, including new bilateral China-India trade and investment agreements.

In designing responses to the climate crisis, countries in the Himalayan region can use best practices for designing regional and local measures which can build on successful UN models such as the UNEP/GEF project on *Global Assessments of Impacts of and Adaptation to Climate Change*. Its Integrated Assessment (IA) approach provides a robust framework to integrate climate change scenarios, socio-economic scenarios, climate vulnerability identification, climate change impact assessment, sustainability indicator specification, adaptation option evaluation, and multi-stakeholder participation.[71] The IA approach provides an effective means for the synthetic assessment of climate vulnerabilities and evaluation of the general performance levels of a set of adaptation options through a multi-criteria and multi-stakeholder decision making process.

This should be supported by in-depth research on the local effects of climate change and glacial melting around key glaciers in the Himalayas and particularly the Tibetan Plateau through collation of new data. This should build on new initiatives such as India-China joint climate research. The results can be used to develop models to predict the behavior of the glaciers under future climate scenarios and map out possible scenarios of local, national and regional development pressures and responses.

The second is *a micro strategy* and involves a focus on key sectors and geographic areas within the Himalayas most sensitive to the climate change. The scale of future impacts, livelihood risks and vulnerabilities all vary both between and within countries in the region. If measures are to have real effect in coming years, an urgent need exists to develop the regional and local policies, partnerships and implementation capacities to take action. In particular, response measures should focus on ways to strengthen capacity of countries to integrate adaptation measures into local action, using best practices such as UNDP's Adaptation Policy Frameworks (APFs). The impact on the achievement and sustainability of MDGs in the glacial melt

---

[71] *See* Yongyuan Yin, *Vulnerability and Adaptation to Climate Variability and Change in Western China*, UNEP/GEF Assessments of Impacts and Adaptations to Climate Change (AIACC) Project, Project Report No 25 (2005). *See also* R Benioff, S Guill, and J Lee (eds), *Vulnerability and Adaptation Assessments: An International Guidebook* (1996); TR Carter, ML Parry, H Harasawa and S Nishioka (eds), *IPCC Technical Guidelines for Assessing Climate Change Impacts and Adaptations* (1994); J Feenstra, I Burton, J Smith and R Tol (eds), *Handbook on Methods for Climate Change Impact Assessment and Adaptation Strategies, Version 2.0*, UNEP (1998); S Fankhauser, *The Costs of Adapting to Climate Change*, GEF Working Paper No 13 (1997); JM Callaway, "The Benefits and Costs of Adapting to Climate Variability and Change" in *The Benefits and Costs of Climate Change Policies: Analytical and Framework Issues*, OECD (2004).

impact zone must be explored and the changing vulnerabilities faced by poor and vulnerable communities addressed. A participatory process should identify climate risks facing specific communities in the glacial melting hotspots and vulnerable downstream communities along Ganges and Yangtze Rivers, detailing how risks will intensify and change, assess changing vulnerabilities at the community level, and how these impact on human security and social and economic rights.

A rights-based approach would bring attention the impacts that climate change will have on the ability of local communities in places like the Tibetan plateau and Kashmir to achieve MDGs, human security and socio-economic rights for current and future generations, preventing risks that climate change and degrading ecosystem can pose to local poverty and social discontent. This involves developing new institutional and individual capacities to promote links between climate and human security, and improving community access to information, participation and remedy related to climate impacts. This would support community-based actions with participation of relevant local stakeholders in mountain and downstream communities.

It will be critical to support local capacities for making informed decisions on ways to adapt development to emerging climate shifts. This includes increasing access of communities to information on the scientific trends of glacial melting and climate change, and to information on development and investment projects with a high likelihood of being impacted by climate risks. Increased participation in designing and implementing response measures can ensure greater correlation with local adaptive capacities and local needs. Finally, creating mechanisms to address remedy and compensation for impacts will become increasing important as communities deal with primary impacts from climate change trends, and secondary impacts from inadequately adapted development projects.

CHAPTER 14

THE KYOTO PROTOCOL AND BEYOND: A SOUTH ASIAN PERSPECTIVE

Asanga Gunawansa

School of Design and Environment, National University of Singapore

There is overwhelming scientific consensus that the climate is changing, the earth is warming up and it is primarily human-induced. Most climate change models predict increasing storm severity with greater precipitation levels and higher wind velocities in the years to come. It is claimed that climate change is at least partially responsible for many catastrophes in the world including the Asia tsunami, hurricanes Katrina, Rita and Wilma in 2005, and the four Florida hurricanes in 2004. Thus, there is no doubt that climate change is going to be to be one of the biggest environmental problems the world will face in the coming decades.

Since the Earth Summit in 1992, several nations of the world have been working towards reducing the emission of Green House Gases (GHGs). However, to date, Kyoto Protocol (KP) is the only multilateral framework we have to address climate change. Many opponents and critics of KP argue that it is fatally flawed as it actually imposes obligations of GHG reductions only on a few developed countries. Further, the critics point out that, as far as developing countries are concerned, the Clean Development Mechanism (CDM) that was introduced under the KP to enable developing countries to achieve sustainable development instead of meeting emission reduction levels imposed by the KP has allowed countries such as India and China, which are amongst the highest

GHG emitters in the developing world, to continue with business as usual and not take effective measures to reduce the current levels of GHG emissions. That said, it cannot be ignored that, there is growing concern both in developed countries and developing countries about climate change. Thus, various measures to mitigate climate change as well as measures to adapt to climate change are being taken and/or are under consideration at national and regional levels.

The aim of this paper is to present a brief summary of the key objectives of the KP, identify its shortcomings and summarize the initiatives taken by and the challenges faced by three developing countries in South Asia, namely, India, Pakistan and Sri Lanka in implementing the goals of the KP. The paper argues that climate change will spare no country and, thus, in order to develop an internationally acceptable approach to addressing climate change, a post-2012 climate regime will need to balance the diverse needs of all countries while striving to prevent the potentially serious economic and social consequences of the impacts of climate change.

1. Introduction

There is an overwhelming scientific consensus that the climate is changing, the Earth is warming up and it is primarily human-induced. Climate change and global warming are at least partially responsible for many catastrophes, including the Asia tsunami, hurricanes Katrina, Rita and Wilma in 2005, and the four Florida hurricanes in 2004. Further, most climate change models predict increasing storm severity with greater precipitation levels and higher wind velocities in the years to come.

Thus, there is no doubt that a changing climate will disrupt complex environmental, social and economic systems that have built up over centuries, and which cannot withstand rapid fundamental change. The Intergovernmental Panel on Climate Change (IPCC) in its Third Assessment Report in 2001 warned that global warming, even at existing levels, had already impacted a number of important physical and biological systems. It predicted significant further impacts, including:

- Increased risk of flooding for tens of millions of coastal dwellers worldwide;
- Increased incidence of extreme weather events;
- Reduced yields of the world's food crops; and
- Decreased water availability in many water-scarce regions.[1]

Thus, climate change is going to be one of the biggest environmental problems the world will face in the coming decades.

The causes of climate change can be divided into two categories, namely, natural causes and those that are man-made. The natural factors responsible for climate change include the continental drift, volcanoes, ocean currents, the earth's tilt, and comets and meteorites. The human factors include industrial waste, use of natural resources for human consumption and habitation, population increase and the extreme reliance on the use of fossil fuels and natural gases.

Since the Earth Summit in 1992,[2] several nations of the world have been working towards reducing the emission of Green House Gases (GHGs). The Kyoto Protocol (KP),[3] which was agreed on 11 December 1997 at the 3rd Conference of the Parties to the United Nations Framework Convention on Climate Change (UNFCCC),[4]

---

[1] The full report is available at: http://www.ipcc.ch/ipccreports/assessments-reports.htm. Accessed on 23 June 2007.

[2] United Nations Conference on Environment and Development (UNCED), Rio de Janeiro, 3–14 June 1992.

[3] KP is a protocol to the United Nations Framework Convention on Climate Change. The UN Framework Convention on Climate Change (UNFCCC) of 1992 calls for the avoidance of "dangerous anthropogenic interference with the climate system".

[4] UNFCCC (1992) calls for the avoidance of "dangerous anthropogenic interference with the climate system".

was envisaged to be a first step on the road to reducing greenhouse gas emissions among developed nations.[5] To date, the KP is the only multilateral framework we have to address climate change.

Many opponents and critics of the KP argue that it is fatally flawed as it actually imposes obligations of GHG reductions only on a few developed countries. Further, it is argued that the force of the KP as an effective mechanism to control GHGs is modest, particularly as its coverage is limited, and as developed countries such as US and Australia,[6] which amongst the main GHG emitters in the worlds, are not parties to it. Further, the critics point out that, as far as developing countries are concerned, the Clean Development Mechanism (CDM)[7] has allowed countries such as India and China, which are amongst the highest GHG emitters in the developing world, to avoid meeting GHG reduction targets imposed under the protocol.

2. The Kyoto Protocol

The KP which was agreed upon on 11 December 1997 was created as an effort to force action by the international community. This was done since, despite the establishment of the UNFCCC in March 1994, very little action had been taken at that time at the international level to reduce GHGs as a global response to climate change.[8]

---

[5] The Kyoto Protocol entered into force on 16 February 2005. To date, approximately 175 countries have ratified it.

[6] It should be noted that Australia subsequently ratified the Kyoto Protocol on 3 December 2007.

[7] The Clean Development Mechanism (CDM) was introduced under the KP to enable developing countries to achieve sustainable development instead of meeting emission reduction levels imposed by it. See, http://unfccc.int/kyoto_protocol/background/items/2881.php. Accessed on 23 June 2007.

[8] *See*, Environment Canada, "A Discussion Paper on Canada's Contribution to Addressing Climate Change", Government of Canada (Ottawa, 2001).

Under the KP, industrialized nations agreed to cut GHG emissions to a certain percentage below 1990 levels.[9] According to a press release from the United Nations Environment Programme (UNEP):

> *"The agreement aims to lower overall emissions from a group of six greenhouse gases by 2008–2012, calculated as an average over these five years. Cuts in the three most important gases — carbon dioxide (CO_2), methane (CH_4), and nitrous oxide (N_20) — will be measured against a base year of 1990. Cuts in three long lived industrial gases — hydrofluorocarbons (HFCs), perfluorocarbons (PFCs), and sulphur hexafluoride (SF_6) — can be measured against either a 1990 or 1995 baseline."*[10]

The year 1990 was chosen as a baseline because that was the year in which the United Nations initially launched negotiations on climate change. The KP requires that the agreed cuts in emissions would have to be accomplished by the target period of 2008–2012.

When UNFCCC was ratified by many countries (including the United States who later backed out of the subsequent KP), the principle of *common but differentiated responsibilities* to reduce "dangerous anthropogenic interference with the climate system" was generously acknowledged. This principle recognized that:

- The largest share of historical and current global emissions of greenhouse gases has originated in developed countries;
- Per capita emissions in developing countries are still relatively low;
- The share of global emissions originating in developing countries will grow to meet their social and development needs.

Thus, developing countries such as China, India and others were not included in any numerical limitation of the KP because they were

[9] The Kyoto Protocol requires the developed countries to reduce their emissions to an average of 5.2 percent below the 1990 levels by 2012.

[10] UNFCC press release dated 11 December 1997, "Industrialized countries to cut greenhouse gas emissions by 5.2%". See http://unfccc.int/cop3/fccc/info/indust.htm. Accessed on 19 June 2007.

not considered to be amongst the main contributors to the GHG emissions during the pre-treaty industrialization period. Further, their per capita GHG emissions were considered to be much lower than those of developed nations. This decision also took into account the fact that the poorer economies of the developing countries would be unable to absorb the costs of switching from a fossil fuel based system to cleaner fuels. The aim was to bring poorer countries into future climate change agreements as cleaner technologies develop and become less expensive.[11]

As far as developing countries were concerned, the KP did not require them to reduce the emissions of GHGs. However, it provided that the developing countries do share the common responsibility that all countries have in reducing GHG emissions. Thus, they were encouraged under CDM to benefit from transfer of technology, and foreign investments into sectors such as renewable energy, energy generation and afforestation projects, which would contribute to mitigate GHG emissions and to achieve the ultimate objective of the UNFCCC. CDM was also seen as a tool that would provide Certified Emissions Reduction Units (CERs) to developed countries, which they can use to meet their GHG reduction commitments under the KP.[12]

In brief, a CDM project activity generates CERs. CERs are created through CDM investments and project activities in Non-Annex 1 countries[13] that lead to the reduction of emissions, which otherwise would occur in the absence of the project. CDM projects could be developed by a private or public sector project developer or by a state. If the development of a project results in the reduction of one or more out of the six GHGs identified in the KP against a validated

[11] Jaimet, Kate, "A primer on Kyoto", The Ottawa Citizen, 30 August 2002.

[12] Under the CDM mechanism, a developed country could take up GHG reduction project activities in developing countries, where the costs of GHG reduction project activities are usually much lower, with the purpose to assist developing country parties in achieving sustainable development, and to assist developed country parties in achieving compliance with their quantified emission limitations.

[13] Non-Annex I refer to the developing countries without any commitments to reduce GHG emissions.

baseline, the credits when produced can be marketed and transferred to an Annex-I party, which can be either an Annex-I government or an entity operating within an Annex I country.[14] All CDM projects must be validated by a Designated Operational Entity (DOE) and approved by the Designated National Authority (DNA). Once validated successfully and approved, the project activity will have to be submitted for registration by the Executive Board of the CDM.[15]

As to targets set by the KP to be meaningful, the countries have to ratify it and put the appropriate measures in place to achieve the required reductions. Further, in order for the KP itself to come into effect, a minimum of 55 countries, together producing at least 55% of the world's CO_2 emissions (as per 1990 level) were required to ratify the Protocol.[16]

The KP suffered a major setback in March, 2001, when the United States, which produces approximately 36% of the carbon dioxide emissions of KP's Annex I countries, decided not to ratify it. As a result, the effective date of the KP was delayed until 16 February 2005,[17] approximately eight years.

---

[14] Annex I of the UNFCCC lists all the countries in the Organization of Economic Cooperation and Development (OECD), plus countries with economies in transition, Central and Eastern Europe (excluding the former Yugoslavia and Albania). By default the other countries are referred to as Non-Annex I countries. Under Article 4.2 (a and b) of the Convention, Annex I countries commit themselves specifically to the aim of returning individually or jointly to their 1990 levels by the year 2000. Ratification of the Kyoto Protocol would mean their acceptance of emission targets for the period 2008–2012 as per Article 3 and Annex B.

[15] *See* Article 12 of the Kyoto Protocol.

[16] This is known as the 55/55 target. See, Minns, Asher, "An Introduction to the Kyoto Protocol", changingclimate.org, 15 March 2002. *See* www.changingclimate.org/content/articles/article/data/section_4/article_85/part_433/. Accessed on 18 June 2007.

[17] The Protocol entered into force on 16 February 2005, the ninetieth day after at least 55 Parties to the Convention, incorporating Annex I Parties which accounted in total for at least 55% of the total carbon dioxide emissions for 1990 from that group deposited their instruments of ratification, acceptance, approval or accession.

3. Shortcomings of the Kyoto Protocol

The fact that no penalties exist for countries that ratify it but fail to meet its reduction targets can be listed as one of the major shortcomings of the KP. In effect, the absence of such penalties makes it a soft instrument which could be easily ignored by the parties. Another shortcoming is the fact that, although the ratifying countries of the KP agreed to GHG "reductions", there was no general acceptance on what exactly was to be considered "reductions".[18] Yet another shortcoming of the KP is the alleged ineffectiveness of the three mechanisms introduced under it to allow countries and companies to buy, generate, or trade CERs, which in turn assist them to meet the required GHG reduction targets. These mechanisms are known as:

- International Emissions Trading, which allows the purchase of credits from other industrialized countries that have exceeded their reduction targets;
- Joint Implementation, which allows investing in emissions reduction projects in other industrialized countries; and
- CDM, which allows investing in clean energy and other emission reduction projects in developing countries.

The logic behind the above mechanisms is that the planet as a whole does not care where the reductions in emissions are achieved, so long as they are achieved somewhere. As such, by investing in a reduction project on any side of the globe, a country could still contribute to its own reduction quota. The main criticism levelled at the above mechanisms is that the methods for their actual use have not been specifically finalized. Another criticism is that the emergence of a global market for energy credits would lead to such credits being traded in the same way as commodities like oil, tea or coffee are traded, allowing prices to fluctuate with supply and demand, with ample opportunities for profits and losses. In the absence of a substantive

---

[18] Ibid Note 8.

mechanism for regulating, the creation of such a market may detract from the KP.[19]

As far as CDM is concerned, it has been burdened by potentially unrealistic expectations about what it can deliver, in terms of the potential size of the market and the ability to bring about major changes in developing countries. Project-based mechanisms are, by their nature, more administratively demanding and costly than cap-and-trade. These limitations, in turn, restrict the ability of project based mechanisms to effect the types of infrastructure change and technology shifts that many non-Annex I countries, in particular, had hoped to achieve through the CDM.

While many agree with the principle of *common but differentiated responsibilities* enshrined in the KP, on the premise that it would not be viable to require developing nations to meet reduction targets as it might slow down their economic progress, the complete exclusion of the developing countries also raises serious questions about the overall effectiveness of the agreement.

It is a fact that cannot be denied that many developing countries make use of out-of-date and environmentally unfriendly technologies in their activities, mainly due to the lack of infrastructure, and policies to develop environmentally-friendly alternatives. In these circumstances, the non imposition of GHG reduction targets may result in such countries continuing to rely on the older technologies as their economies and populations grow. Thus, their emissions will continue to grow without being limited by the KP. This might result in any gains made by the developed nations under the KP being easily offset by the growth of emissions in the developing countries. The fact that rapidly growing economies such as China and India, which together represent one-third of the world's population, have been left without any emission reduction targets is a good example to stress the gravity of this shortcoming.

4. Implementing the Kyoto Protocol in South Asia

South Asia is a southern geopolitical region of the Asian continent, consisting of India, Pakistan, Sri Lanka, Bangladesh, Nepal, Bhutan

---

[19] Ibid Note 6.

and the Maldives. In South Asia, during the 10th Summit of the South Asian Association for Regional Cooperation (SAARC)[20] held in 1998, the Heads of States expressed their satisfaction over the adoption of a common position by Member States prior to the Third Session of the Conference of the Parties to the UNFCCC, held in Kyoto. They welcomed the adoption of the KP in December 1997. However, since that time SAARC has failed to take any specific and effective regional initiatives towards implementing the KP. Issues such as the indo-Pakistan conflict and the volatile political situation in Sri Lanka may have contributed to this.

It should be noted however that, although SAARC has not taken any meaningful regional initiatives, some of its members have taken individual initiatives to promote the objectives of UNFCCC and the KP, and to benefit from CDM.

4.1. *India*

India ratified the KP in August 2002. Being a developing country, India is not required to reduce emission of GHG under KP. However, it is expected to gain from it in terms of transfer of technology and related foreign investments.

With the aim of achieving its key objectives in acceding to the protocol, i.e. to fulfil prerequisites for implementation of CDM projects in accordance with national sustainable priorities, the central government of India has established the National Clean Development Mechanism Authority (NCDMA).[21] It is in charge *inter alia* of receiving projects for evaluation and granting of approval as per the guidelines and general criteria laid down in the relevant rules and modalities pertaining to CDM in addition to the guidelines issued by

[20] South Asian Association for Regional Cooperation (SAARC) is an economic and political organization of the countries in Southern Asia. It was established in 1985 by seven countries namely, India, Pakistan, Sri Lanka, Bangladesh, Nepal, Bhutan and the Maldives. Afghanistan joined SAARC at its 14th Summit in April 2007.

[21] *See*, http://cdmindia.nic.in/cdm_india.htm. Accessed on 18 June 2007.

the CDM Executive Board and Conference of Parties serving as Meeting of Parties to the UNFCCC.[22]

India has also taken several bilateral initiatives to implement the objectives of KP. One significant initiative is the establishment in August 2003 of the GTZ CDM-India, through an agreement between German Technical Cooperation (GTZ) and the Bureau of Energy Efficiency (Ministry of Power, Government of India) under the Indo-German Energy Programme (IGEN). This is as a capacity building facility that could help reduce transaction costs in the early market development process involving CDM projects. According to the information available on the NCDMA website, as of September 2007, 60 CDM project proposals with the potential to generate over 40 million CERs have been received by GTZ. It has already supported the development of 10 projects of which six projects have received host country approval and three have signed validators' contracts.

It is interesting to note that, although at the early stages of negotiations concerning the KP, India was opposed to it on the premise that CDM would help the developed countries to escape direct responsibility for emission reductions, along with Brazil and China, India has benefited the most from the CDM mechanism. In fact, India is the largest beneficiary under the CDM mechanism. As of September 2007, it has 268 registered CDM projects, more than any other country in the world.

India, together with China, is often criticized as being irresponsible towards mitigating climate change despite being two of the highest pollutant countries in the developing world. It should be pointed out however that, according to the Climate Change Performance Index (CCPI)[23] of 2008, of the 56 countries that are collectively responsible for more than 90 percent of global energy-related CO_2 emissions, India (ranked 5th) is the only Asian country

---

[22] *Ibid.*

[23] CCPI is prepared by Germanwatch, an independent, non-profit and non-governmental North-South Initiative.

Registered CDM Projects as of 2007

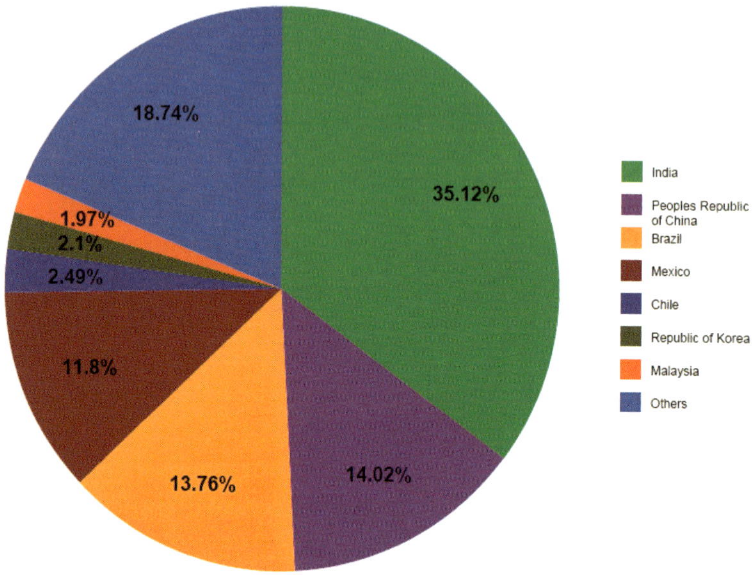

Source: UNFCC

that has made it to the top 10 climate protection performers.[24] The other countries that have made it to the top 10 performers list include: Sweden, Germany, Iceland, Mexico, Hungary, United Kingdom, Brazil, Switzerland, and Argentina. It should be noted that after India, the next best Asian performer is China (ranked 40th) with, Japan, Singapore, Malaysia, the Republic of Korea, and the other Asian countries who have made it to the list of 56 countries, lagging behind. The presence of developing countries such as Mexico, Argentina, Brazil and India in the list of top performers in CCPI,

[24] The Climate Change Performance Index (CCPI) is an innovative instrument that enhances transparency in international climate politics. On the basis of standardized criteria the index evaluates and compares the climate protection performances of the 56 countries that, together, are responsible for more than 90 percent of global energy-related CO_2 emissions. The objective of the index is to increase the political and societal pressure on those countries which up to now have failed to take initiatives in climate protection and which still neglect the importance of the issue. *See*, http://www.germanwatch.org/klima/ccpi2008.pdf. Accessed on 18 June 2007.

evidences the fact that developing countries are not ignorant of their responsibility towards mitigating climate change. Further, the impressive ranking of India in the CCPI evidences the fact that, although India is under no obligation to meet any GHG reduction targets under the KP, it seems to have taken some effective steps towards implementing the objectives of the KP.

4.2. *Pakistan*

Pakistan ratified the KP in January 2005. In 2006, Pakistan announced an energy policy which highlighted the importance of renewable energy.[25] The Government granted legal status to the Alternate Energy Development Board (AEDB) established in 2003, enabling it to take the lead in exploring renewable energy sources. AEDB has signed an agreement with a manufacturer in Germany (Fuhrlaender) and another with a manufacturer based in the United States (Access Energy Inc.), to manufacture wind turbines in Pakistan to help establish wind based power projects. In addition, AEDB has recently (October 2006), signed a framework financing agreement with the Asian Development Bank (ADB) to finance renewable energy investment projects in Pakistan over the next 10 years.[26]

Pakistan has appointed a Designated National Authority (DNA) within the Ministry of Environment to manage the CDM process in Pakistan efficiently and transparently, and in line with national sustainable development goals. The DNA is guided by the Prime Minister's Committee on Climate Change and consists of a National CDM Steering Committee, several Technical Committees, and a CDM Secretariat.[27] Thus, the DNA in Pakistan appears to be on

---

[25] *See*, http://www.aedb.org/PakistanREDevelopmentPolicy_FinalFormatte.pdf. Accessed on 23 June 2007.

[26] The agreement is available online. See http://www.adb.org/Documents/Framework-Financing-Agreements/PAK/34339/34339-PAK-FFA.pdf. Accessed on 23 June 2007.

[27] *See* http://www.environment.gov.pk/act-rules/PakCDM-NatOpelStrgy.pdf Accessed on 23 June 2007.

a good footing, with direct access to, and direction from, the relevant ministry in the Government. However, as far as CDM projects are concerned, as of September 2007, Pakistan has been successful in registering only a single project.[28]

4.3. *Sri Lanka*

Sri Lanka started considering the KP in 1999. After several rounds of public consultation meetings, the country finally ratified the protocol in 2002. A DNA to manage CDM projects and to promote the government's initiatives in implementing the KP has been established within the Ministry of Environment.

As of September 2007, Sri Lanka has successfully completed four hydropower projects eligible for carbon trading. According to the Central Environmental Authority of Sri Lanka, these four projects marketed 50 000 tonnes of carbon credits in 2006. According to a statement by Sri Lanka's Minister of Information in June 2007, the country plans to set up a Carbon Fund to promote carbon trading and provide financial as well as technical support for potential project developers.[29] The fund is expected to undertake carbon asset management activities such as financing and investment of CDM projects, buying and selling of carbon credits and providing consultancy services. According to the same source, a total of 45 CDM projects are currently being developed in Sri Lanka.[30] Thus, it is likely that, if the projects in development come through smoothly, Sri Lanka's share of registered CDM projects would substantially increase, making it a prominent host country for CDM projects.[31]

---

[28] Catalytic N_2O Abatement Project in the Tail Gas of the Nitric Acid Plant of the Pakarab Fertilizer Ltd (PVT) in Multan, Pakistan.

[29] Gayan Rathnayake, "Smoke Sales", Lanka Business Online, 15 June 2007. www.lankabusinessonline.com. Accessed on 23 June 2007.

[30] *Ibid.*

[31] Currently, only India, China, Brazil and Mexico have more than 20 registered CDM projects.

In addition to the initiative to establish the Carbon Fund, Sri Lanka has taken several other initiatives to promote the objectives of UNFCCC and the KP. Some of the key initiatives are listed in the following table:

Sector	Initiatives
Energy	• A study funded by USAID SARI (South Asian Regional Initiative) Energy program to consider natural gas options and alternative energy has been completed. • A national energy policy with renewable energy targets has been formulated in 2006. • An energy labeling scheme for energy consuming appliances has been introduced. • A scheme has been established to provide concessionary funding for demand side management (DSM) and renewable energy projects. • Alternative energy studies (wind energy in particular) have been undertaken. • Initiatives have been taken to harness the maximum potential of hydropower (a hydro project with 70 MW capacity has been already completed). • Government energy policy has set a target of replacing 10% of fossil fuels that supply grid electricity with rentable energy by 2020. • A program called "20/20 formulation" has been set forth to replace 20% of transport fuel with bio-fuel by the year 2020.
Industrial and Transport Sectors	• Stationary source emission standards for industries and power plants have been drafted but have not yet been gazetted. • Standards for mobile sources have been gazetted, but not yet enforced due to delays in building vehicular emission testing centres. • An "Energy Efficiency Building Code" has been introduced. • The private sector has been encouraged to recover methane from solid waste. Private waste management companies and local authorities have already begun working on promoting recycling programs and some waste-to-energy projects.

(*Continued*)

(*Continued*)

Sector	Initiatives
Others	• Initiatives have been taken to promote fast-growing fuel wood in place of coal or oil to preserve the carbon reservoir contained in fossil fuels. • Steps have been taken to incorporate climate change concerns into national wetland conservation programmes, recognizing that, in addition to being a national heritage that should be conserved, restored, and sustainably managed, the wetlands provide important ecosystem services including flood retention and detention areas; sediment traps; sources of water; aquatic habitats; and could function as the green lungs of the nation.

Source: Draft Second National Communication under the United Nations Framework Convention on Climate Change, 2006.

5. Impediments faced by South Asian Countries in Implementing the Kyoto Protocol

Perhaps excluding India, the other members of SAARC faces the problem of lack of investor interest in financing CDM projects in their jurisdictions, a problem that is common to many developing countries. The main reasons for this lack of interest among investors in financing CDM projects in developing countries have been stated as financial and institutional risks, uncertainty associated with delays in approving CDM project activities, and the lack of sufficient capacity building to train validators.[32] The political risk associated with some developing countries in South Asia could also be added to this list. It would not be difficult to argue that the

[32] Ellis, J., Corfee-Morlot, J. and H. Winkler (2004), "Taking Stock of the Progress under the Clean Development Mechanism (CDM)", Paris: OECD/International Energy Agency, COM/ENV/EPOC/IEA/SLT(2004)4/FINAL. *See*, www.oecd.org/dataoecd/58/58/32141417.pdf. Accessed on 18 June 2007.

prevailing political situation in countries like Sri Lanka, Pakistan, and Nepal would have contributed to lack of interest on the part of some investors in financing and developing CDM projects in these countries.

The following table provides a vivid picture of the imbalance in technology and investment flows into developing countries in South Asia under the CDM mechanism:

CDM Projects in SAARC Countries as of September 2007.

SAARC Member	Registered CDM Projects
Bangladesh	2
Bhutan	0
India	268
Maldives	0
Nepal	2
Pakistan	1
Sri Lanka	4

Although, some argue that the investment flows into CDM projects in developing countries are slowly increasing, it should be pointed out that the main foreign direct investment (FDI) flows do not go to CDM projects nor to developing countries. As can be seen from the following chart, in the year 2003, the greatest recipients of FDI were the developed countries (69%), followed by developing countries excluding the People's Republic of China (20%), People's Republic of China (10%), and lastly the least developed countries (1%).[33]

[33] Raymond Saner, "Hype or Reality: Can the CDM Trigger FDI", European Climate Platform Background Paper 2, September 2005.

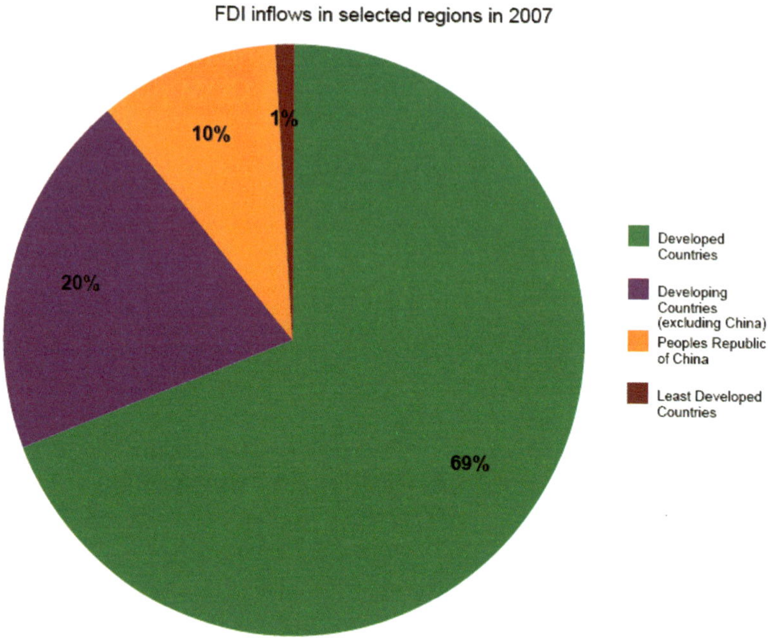

Source: Taffere Tesfachew & Karl P. Sauvant (2005).[34]

As can be seen from the following table which shows the FDI flows during the years 2004–2006, this dominance in attracting FDI inflows by the developed countries continues:

Table. FDI inflows, by host region and major host economy, 2004–2006 (Billions of dollars).

Host region/economy	2004	2005	2006[a]	Growth rate (%)
World	**710.8**	**916.3**	**1230.4**	**34.3**
Developed economies	**396.1**	**542.3**	**800.7**	**47.7**
Europe	217.7	433.6	589.8	36.0

(*Continued*)

[34] Tafere Tesfachew and Karl P. Sauvant, "Making FDI work for the poor", Development and Cooperation, No. 32 (2005).

Table. (*Continued*)

Host region/economy	2004	2005	2006[a]	Growth rate (%)
European Union	213.7	421.9	549.0	30.1
EU-15	*185.2*	*387.9*	*510.7*	*31.7*
France	31.4	63.6	88.4	39.0
Germany	–15.1	32.7	8.1	–75.1
Italy	16.8	20.0	30.0	50.2
United Kingdom	56.2	164.5	169.8	3.2
New 10 EU member states	*28.5*	*34.0*	*38.4*	*12.8*
Czech Republic	5.0	11.0	5.4	–50.8
Hungary	4.7	6.7	6.2	–7.3
Poland	12.9	7.7	16.2	109.9
United States	122.4	99.4	177.3	78.2
Japan	7.8	2.8	–8.2	–395.5
Developing economies	**275.0**	**334.3**	**367.7**	**10.0**
Africa	17.2	30.7	38.8	26.5
Egypt	2.2	5.4	5.3	–1.9
Morocco	1.1	2.9	2.3	–20.9
Nigeria	2.1	3.4	5.4	60.0
South Africa	0.8	6.4	3.7	–42.7
Latin America and the Caribbean	100.5	103.7	99.0	–4.5
Argentina	4.3	4.7	3.3	–29.5
Brazil	18.1	15.1	16.0	5.9
Chile	7.2	6.7	9.9	48.4
Colombia	3.1	10.2	4.9	–52.0
Mexico	22.3	18.9	18.9	0.0
Asia and Oceania	157.3	200.0	229.9	15.0
West Asia	18.6	34.5	43.3	25.5
Turkey	2.8	9.7	17.1	76.3
South, East and South-East Asia	138.0	165.1	186.7	13.1
China	60.6	72.4	70.0	-3.3
Hong Kong, China	34.0	35.9	41.4	15.4
India	5.5	6.6	9.5	44.4
Indonesia	1.9	5.3	2.0	–62.9
Korea, Republic of	7.7	7.2	0.5	–92.6
Malaysia	4.6	4.0	3.9	–1.6

(*Continued*)

Table. (*Continued*)

Host region/economy	2004	2005	2006[a]	Growth rate (%)
Singapore	14.8	20.1	31.9	58.8
Thailand	1.4	3.7	7.9	114.7
South-East Europe and CIS	**39.6**	**39.7**	**62.0**	**56.2**
Russian Federation	15.4	14.6	28.4	96.4
Romania	6.5	6.4	8.6	34.1
Kazakhstan	4.1	1.7	6.5	275.5

Source: United Nations Conference on Trade and Development (UNCTAD)[35]

The table shows that although the FDI inflows into developing countries have grown by 10% in 2006, the growth of the flow to developed counties is almost 48%.

In South Asian countries like developing countries from any other region, social and behavioral preferences for existing technologies and lifestyles at the household level are a major barrier to introducing new and environmental technologies. Further, having been pampered for decades by government subsidies and welfare state systems, some may resist politically-driven technological change if there are insufficient incentives. In these countries, in the near term, patterns of capital consumption are driven by factors largely unrelated to climate change, and will likely only use climate change as an investment decision factor when obligated to do by regulations.[36]

Another impediment faced by the developing countries in South Asia in the effective implementation of the objectives of KP is the foreign ownership restrictions or the absence of state policy that encourages investment inflows into CDM type investment projects. For example, China requires that all power investments over US$30 million be approved by the central government.

[35] UNCTAD Press Release: UNCTAD/PRESS/PR/2007/00109/01/07.

[36] Lempert, R.J., S.W. Popper, S.A. Resetar and S.L. Hart, "Capital Cycles and the Timing of Climate Change Policy", Washington, D.C., Pew Center on Global Climate Change (2002).

In South Asia, India caps foreign ownership on energy investments at US$350 million.[37]

Further, many developing countries face additional challenges when trying to attract investment into the energy and infrastructure technologies needed for social development, due to the perception of high risk and inadequate returns, alongside substantial transaction costs.[38] In addition, IEA's World Energy Investment Outlook notes that:

> *"Investing in energy projects in developing countries and the transition economies is generally riskier than in the OECD because of less well-developed institutional and organizational structures, lack of clear and transparent energy, legal and regulatory frameworks, and poorer political and economic management."*[39]

In an ideal world, it could be easily argued that under the CDM mechanism, developed-countries could meet their GHG reduction targets by funding emission reduction projects in the developing world and in doing so, contribute to sustainable development of the developing countries, a perfect win-win-win scenario for all.[40] However, it is clear from what has been explained above that the CDM mechanism does not ensure that every developing country would equally benefit under it. As there is no control over the

---

[37] Purvis, Nigel, "Climate Change and the L20: Options for Non-Emission Target Commitments. Presented at the "Post-Kyoto Architecture: Toward an L20?", New York City, September 20–21, 2004.

[38] Van Aalst, Paul, "Innovative Financing Options for Climate Change-related Transfer and Development of Technologies. Presentation at the UNFCCC Workshop on Innovative Options for Financing the Development and Transfer of Technologies, Montreal, Canada, September 27–29, 2004. http://unfccc.int/meetings/workshops/other_meetings/items/3142.php. Accessed on 23 June 2007.

[39] International Energy Agency (IEA), 2004b, World Energy Outlook 2004, Paris: IEA. http://www.iea.org/techno/index.htm. Accessed on 23 June 2007.

[40] *See*, Anne Arquit Niederberger and Raymond Saner, "Exploring the relationship between FDI flows and CDM potential", Transnational Corporations, Vol. 14, No. 1 (April 2005).

distribution of projects and investment inflows into developing countries, the investors from developed countries are likely to continue their current practice of choosing safe and profitable host countries to invest. This is clearly the case in South Asia. In the circumstances, if the South Asian region is to benefit from CDM and be active in the effective implementation of the KP, a lot needs to be done at national, regional and global levels.

6. The Way Forward

It is known that around 1.3 billion people live on less than a dollar per day and have no access to modern forms of energy or energy services. The Asia Pacific region is the home to the majority of the poor as it includes the two most populous developing countries in the world, namely, India and China.[41] As far as South Asia is concerned, out of a population of 1.37 billion, nearly 400 million are poor.[42] Further, it is said that, worldwide, 1.6 billion people are without access to electricity and an equal number continue to use traditional solid fuels for cooking. According to the Asian Development Bank (ADB) a good number of such people, approximately 50% (in 2003) are living in South Asia with 35% of them in India.[43] Thus, it could be argued that the current energy systems that are fuelling the economic growth have not been successful in addressing the basic needs of all people in South Asia.

As noted above, the CDM is expected not only to assist Annex 1 countries to achieve their emission reduction targets stipulated under the KP, but also to be conducive to the sustainable development of host countries listed as Non-Annex 1. Thus, even in the realm of CDM anomalies noted above, the mechanism could still be

---

[41] United Nations, "Implementation of the Clean Development Mechanism in Asia and the Pacific — Issues, Challenges and Opportunities, Economic and Social Commission for Asia and the Pacific", United Nations, New York, 2003.

[42] Shantayanan Devarajan and Ijaz Nabi, "Economic Growth in South Asia: Promising, Un-equalizing ... Sustainable?", World Bank, South Asia Region, June 2006.

[43] ADB, "Access to Electricity", January 8, 2003.

effectively used for addressing poverty reduction issues in regions such as South Asia.

Energy is a major contributor to the GHG induced global warming and associated climate change. Due to the dependence on imported energy, there are issues of energy security that concerns most countries in the South Asia. According to the World Bank, a range of social issues, including poverty alleviation, population growth, urbanization and a lack of opportunities for women, is closely linked to energy consumption patterns in South Asia. Further, according to the same source, there is a logical nexus between poverty-energy and climate change-CDM as shown in the following table:[44]

Poverty-Energy Interface	Poverty-Climate Change Linkages	CDM-Poverty Nexus
• Poor people have limited energy supply options (depend on fuel wood for cooking and carry water/ physically uplift water from natural water bodies — lakes, rivers, streams, wells, etc.) • Poor spend much of their income on energy • Poor in rural as well as urban areas do not have access to basic social amenities • Poor infrastructure access limit livelihood options	• Poor rely on inefficient devices resulting in increased emissions • Poor are most vulnerable to the impacts of climate change (agriculture, health, etc.) • Poor have limited adaptive capacity	• Promote sustainable energy technologies for the poor (e.g. biogas, biomass energy systems, solar PV systems) • Improve incomes through community type energy projects • Develop infrastructure that benefits poor (efficient use of material in housing, roads, alternate fuels for transportation, etc.)

(*Continued*)

[44] See ibid note 41 for a more comprehensive table.

(*Continued*)

Poverty-Energy Interface	Poverty-Climate Change Linkages	CDM-Poverty Nexus
• Low incomes and lack of purchasing power lead to slow rate of investments • Limited institutional capacity to deal with energy delivery issues and increasing energy prices		• Create enabling environment for inducing CDM investments for poor (sectoral CDM projects; partnerships with communities) • Widen energy choices for long-term stability (reduced energy price shocks) and improved income sources • Reduce inequities through greater access to modern infrastructure for the poor

Thus, CDM could be effectively used as tool for poverty reduction and sustainable development in South Asia. However, in order to fully exploit the full potential of the mechanism, two sets of objectives: those of the investors and those of the developing countries, should be carefully integrated. These should be done at a national, regional and global level.

6.1. *National initiatives*

Risk sharing can increase financial investment and stimulate local private-sector participation. As the political and financial risks associated with South Asian countries contribute to the lack of investor interest in investing in CDM projects in the region, the governments in the region have an essential role in risk management. This includes establishing priority investment areas, setting out framework conditions for transactions and shaping financial flows, and providing a stable legislative and regulatory environment for investment.

The countries in South Asia should realize that putting in place such a legislative and regulatory environment alone might not be sufficient to attract investor interest, given that many developing countries in the world are competing for investments. Thus, certain policies will have to be put in place, and programs at the national level will have to be implemented which enhance technology development, and provide local capital for developing environmentally friendly development projects that contribute to sustainable development. The resolving of ongoing civil wars in countries such as Sri Lanka and Nepal would help such countries to divert current budgetary allocations for equipping the armed forces and for servicing the war to be directed towards environmentally friendly development projects.[45]

The interface between FDI inflows and sustainable development is neither simple nor automatic. Furthermore, not all FDI brings the desired benefits. Therefore, the developing countries in South Asia should understand that maximizing the positive effects of FDI investments in CDM projects requires careful and intelligent planning. Furthermore, in designing investment promotion strategies to attract FDI for CDM projects, the developing countries in South Asia should take into consideration their national development priorities and objectives. The inflow of investments into CDM projects should not be viewed merely as a source of capital. It should be understood that investment inflows into CDM projects can contribute to the transfer of other tangible and intangible assets, including technology and skills, and can create new employment opportunities and new markets. Further, it is important that each country assess and promote the

---

[45] Lampert (supra, note 35) argues that government funded research plays an essential role in developing new technologies and advancing new technologies to market readiness, through such policies as tax credits, appreciated deceleration of investments reducing GHG emissions and government procurement of low-emitting technologies.

types of CDM projects that will enable it to maximize benefits and contribute to sustainable development. If not, the true purpose of the KP in introducing the mechanism would be lost.

6.2. *Regional initiatives*

Agreements to cooperate on climate technology could take place outside of the UNFCCC and the KP. In fact, a number of countries have signed bilateral agreements on technology and scientific cooperation. For example, the European Union cooperates on international science policy with more than 30 countries and is engaged in a number of technology agreements aimed at the improvement of energy technologies and climate-friendly production processes. Further, although it has not ratified the KP, the United States has signed various bilateral climate technology agreements with countries such as Australia, Japan, Russia, Italy, India and China. Japan has strengthened its role in climate cooperation within Asia, including a joint research initiative with seven developing nations aimed at providing technological assistance to these countries to reduce their GHGs in exchange for carbon dioxide emission credits.[46] The Asia-Pacific Partnership on Clean Development and Climate (signed by Australia, China, India, Japan, Korea and the USA in July 2005) aims to promote the development, diffusion, deployment and transfer of existing and emerging cost-effective cleaner technology and practices. This is a good example of a regional initiative.

Thus, the countries in South Asia should be more effective in promoting regional initiatives towards attracting more investment to the region for CDM projects and taking collective efforts towards implementing the goals of the KP. South Asia remains the least integrated region in the world. Starting from such a low base, greater integration can generate enormous benefits along several dimensions. For example, considerable expertise in the area of bioenergy and other alternative sources of energy exists in developing countries such as India and Pakistan. It would be much easier and less costly to transfer

---

[46] *See*, Buchner, B. and C. Carraro, "Economic and environmental effectiveness of a technology-based climate protocol." Climate Policy 4 (2005), pp. 229–248.

and share such knowledge with other countries in the region than seeking Northern solutions.

Further, as a result of the ongoing cross-border conflict, both India and Pakistan incur large military expenditures at the expense of much needed development. It is important that measures are taken to manage such conflicts in the region in a way that social investments are protected and intra-regional trade, investment and people's mobility is restored to levels enjoyed by other regional groupings.

The development of regional infrastructure would also contribute to attracting investor interest to the region. Connectivity in general in the region is poor. Bangkok and Dubai, the two international airline hubs on either side of the region are better connected to capitals outside the region than in any regional capital. This adds several hours of additional flying time and substantial cost to mobility within the region. Further, road/rail connections in the region are sub-standard. Visa restrictions and inadequate intra-regional support services, such as banking, also create a discouraging investment climate in the region. Thus, in addition to improving the legislative and regulatory environment at the national level, steps should be taken to make the region as a unit more attractive for investors.[47]

6.3. *International initiatives*

As FDI accounts for the major portion (approximately 60–80 per cent) of global financial flows in recent years, private sector investment should be viewed as one of the main vehicles for developing CDM projects despite the fact that private investment remains low in many developing countries.[48] In this context, there is a need to ensure that the financial markets and public finance institutions are mutually reinforcing efforts to more effectively support sustainable development

[47] Shantayanan Devarajan and Ijaz Nabi, "Economic Growth in South Asia: Promising, Unequalizing, ... Sustainable?" World Bank, Washington D.C., June 2006.
[48] Violetti, Daniele, "Trends in Financial Flows and Technology Transfer", Presentation at the UNFCCC Workshop on Innovative Options for Financing the Development and Transfer of Technologies, Montreal, Canada, September 27–29, 2004.

if the world is to stabilize GHG emission reductions over the long term.

Public-private partnerships are increasingly being used as an innovative means to promote private sector investment in climate-friendly technologies. The following initiatives could be cited as two examples of steps taken in the right direction:

- The UN Foundation's (UNF) support for emerging climate change policies that promotes public-private partnerships with UN agencies that advance innovative sustainable energy programs in developing countries; and
- The UNF and the Shell Foundation investing with UNEP in an initiative to accelerate the market for financing solar power in India by helping two commercial banks develop lending portfolios.

More initiatives of such nature are required.

It is important that any post-Kyoto mechanism that will be put in place after 2012 addresses the above issue if the CDM mechanism to be meaningful. A fair distribution of investment flows into all developing countries will be difficult to ensure unless the free choice of selecting host destinations by the developed countries is not in some way regulated. A regional quota system which would ensure that the investment inflows into CDM projects are not made in a country specific manner but are allocated according to regional needs and then fairly distributed among the developing countries in the region is a method that could be considered. Offering incentives to investors who would invest in the neediest developing countries is also something that could be considered. However, for any of these methods to be effective, their should be concensus amongst both developed and developing nations to accept climate change and global warming as a threat that will spare no country, and is thus a universal issue which should be dealt with by both the developed and developing countries accepting responsibility and adopting necessary mitigatory measures.

The point that global negotiations and consensus is essential for tackling the issue of climate change cannot be contradicted.

However, it has to be pointed out that whilst the global efforts are ongoing to reach an understanding on the effective role that should be played by each nation, more specific and result oriented mechanisms should be put in place that would ensure that the national efforts to mitigate the dangers of climate change and control the GHG emissions could be achieved. In order to put in place such mechanisms, a new focus is needed. The following suggestions should be considered:

- Changed priorities for the international funding agencies such as the World Bank.
- Creation of a global renewable energy agency to facilitate technology transfer and reciprocal environmental quality requirements on imports and domestic production of energy.
- A ban on subsidized energy exports.
- The establishments of an International Court for Environmental Justice or an environmental chamber for the International Court, with power to hear.

As the famous German Parliamentarian at the Deutscher Bundestag and Alternative Nobel Prize Winner Dr Herman Scheer argues, the result of such an effort would be "a dynamic, goal-oriented climate change policy, free of bureaucratic impediments, and a step forward from simply prolonging and refining the current series of international conferences."[49]

7. Conclusion

Climate change will spare no country. In order to develop an internationally acceptable approach to addressing climate change, a post-2012 climate regime will need to balance the diverse needs of all countries while striving to prevent the potentially serious economic and social consequences of the impacts of climate change. However

[49] Asymmetric Threats Contingency Alliance (ATCA) Briefing, London, UK, 3 September 2006.

much certain developed countries would argue that an effective climate change control mechanism should impose equal obligations to reduce GHG emissions on both developed and developing countries, it is unlikely that most developing countries would agree to any mechanism that would substantially slow down their economic development. Thus, achieving a balance in any post Kyoto agreement will require the reaching of a common understanding of the issues associated with the following key elements:

- The need to develop clean technologies;
- The need to ensure sustainable economic development for developing countries;
- The need to establish and ensure the continuity of an effective international carbon market with adequate regulatory measures; and
- The need to ensure that the international agreements reached to mitigate climate change are not in disharmony with national development goals of countries.

As far as the developing countries in South Asia are concerned, further research should be carried out to determine how they could attract CDM investment and improve their ability to profit from investment inflows, technology transfer, and revenue from export of CERs. Further, measures should be found to encourage fair distribution of CDM benefits among all developing countries in the region by encouraging the investors to consider countries which have not yet benefited from the CDM mechanism for investment. For this to be achieved, in addition to the initiatives that should be taken by the developing countries (for example, by making their jurisdictions investor friendly by enacting appropriate legislation and establishing a market friendly investment and regulatory environment), other players such as international organizations and NGOs will have to play a key role in working with the developing countries, assisting them with their reforms. Further, national efforts need to be complemented at the international level by efforts from developed countries to provide capacity building, technology and financing to developing countries.

The South Asian countries should not delay any further to establish a regional mechanism for coordination of activities to reduce the impact of GHGs and to exchange information on policies, initiatives and technologies. They should begin to work together in respect of the CDM to promote GHG reduction technologies and improve their capacity in sustainable practices.

CHAPTER 15

BEYOND KYOTO: CLIMATE CHANGE INCLUDING A DISCUSSION OF THE AP6 INITIATIVE FROM THE AUSTRALIAN PERSPECTIVE*

Michael I. Jeffery, QC**

Professor of Law and Director, Centre for Environmental Law Macquarie University (MU-CEL), Sydney, Australia; former Deputy Chair of the International Union for the Conservation of Nature and Natural Resources (IUCN)

This paper presents an overview of the Australia Commonwealth government's response to the challenges of climate change, that has been very much tempered by concerns over the adverse impact that any such response may have on the Australian economy. It focuses in particular on the establishment of the Asia Pacific Partnership on Clean Development and Climate (commonly referred to as the AP6 initiative) and discusses Australia's options within a regional framework.

Included are appendices containing the "Communiqué" from the AP6 Inaugural Ministerial Meeting held in Sydney in January 2006; the AP6 Charter; the AP6 Action Plan Guidelines and the AP6 Taskforce Guidelines.

*Based on a presentation to the Macquarie Forum, 3 April 2007, Sydney.

**The author is indebted to Wendy Howe, MU-CEL PhD candidate for her dedicated and enthusiastic assistance in the preparation of this paper.

Brief Background on Climate Change in Australia

While a variety of environmental issues have been at the forefront of Australian politics since the 1960s,[1] climate change only emerged as an issue in 1988[2] in response to a number of scientific conferences[3] and events that alerted the world to increasing concentrations of greenhouses gases (GHGs) in the atmosphere.

In 1989 the Federal Government established a National Climate Change Program which included the National Greenhouse Advisory Committee of scientific advisers and a Prime Ministerial Working Group to assess achievable targets. However, right from the start, the greenhouse debate became polarized between the environment ministers and departments and the energy ministers and departments.

The environment ministers felt that climate change is primarily an environmental issue and therefore, requires an appropriate environmental response. However, the energy ministers and departments believed that as their departments are held accountable for the bulk of greenhouse gas emissions, their departments' responses on ways of reducing those emissions should be given priority. They argued that, as the Australian economy is based on major exports of coal, natural gas and uranium, any realistic strategy to reduce GHGs must focus on the energy supply system.

In the emerging climate change debate in Australia that followed, it was clearly understood that any strategy to contain GHG emissions must focus primarily on the energy system.

In October 1990 the Federal Government announced an interim planning target for stabilising Australia's greenhouse gas emissions,[4]

---

[1] For example, Lake Pedder, Kakadu and Fraser Island campaigns, Green Bans in Sydney, and forest conservation and urban pollution issues detailed in Hutton, D & Connors, L., 1999, *A History of the Australian Environment Movement* (Cambridge University Press, 1999).

[2] Hamilton, Clive, *Running from the Storm — The Development of Climate Change Policy in Australia* (New South Wales Univ. Pr. Ltd., 2001) at p. 31.

[3] In November 1988 the Intergovernmental Panel on Climate Change (IPCC) was established with the purpose of reporting on the science and impacts of global warming and strategies to deal with the possibility of climate change.

[4] The targets were for stabilization at 1988 levels by the year 2000, and reducing them by 20% by 2005 — in Hamilton, Clive, p. 33.

but made it very clear that "the Government will not proceed with measures which have net adverse economic impacts nationally or on Australia's trade competitiveness in the absence of similar action by major greenhouse gas producing countries".[5]

This statement by a Labour government then in power was designed to address industry concerns about GHG emissions and the Howard government has steadfastly adhered to this fundamental policy position ever since.

Australian Business Interests in Climate Change

In November 2004, the Australian Business Roundtable on Climate Change group was formed and publicly acknowledged that climate change represented a significant core risk to businesses and corporations.

The Roundtable comprised representatives from six of Australia's largest businesses and from a cross section of Australian industries. The group includes BP Australia, Insurance Australia Group, Origin Energy, Swiss Re, Visy Industries, Westpac, and the Australian Conservation Foundation.

The group's aim is to increase the understanding of business risks and opportunities associated with climate change and to help develop effective policy frameworks and market conditions. The Business Roundtable has clearly recognised that climate change poses serious threats to Australia's economy and that uncertainty about future climate policy heightens the risks associated with investment.

In April 2006, the Business Roundtable on Climate Change released a report, entitled "*The Business Case for Early Action*". This report was based on preliminary studies made by the CSIRO (scientific aspects of climate change) and the Allen Consulting Group

---

[5] Attributed to Labor Government proposing stablilization of greenhouse gas emission targets to appease industry and to presage climate change policy to come, in Hamilton, Clive, p. 33.

(economic aspects of climate change) and made the following key recommendations to the Australian government:

- The need to design a "long, loud and legal" framework to establish a carbon price signal.
- The need to encourage innovation and investment in emerging and break-though technologies.
- The need to build national resilience to the impacts of climate change.

The Australian Business Roundtable's report, which included the Australian Government's GHG emission projections, the impacts of climate change on industry such as tourism, water and primary industries, and infrastructure and insurance issues, recommended that business and government work together to frame policies to create necessary investment conditions to reduce GHG emissions while maintaining economic growth.

In November 2006, a new coalition of Australian blue-chip companies[6] was formed and began evolving a business policy response to climate change. The new group, which includes members of the Australian Business Roundtable on Climate Change, has held a number of preliminary discussions "to look at which policy tools might accelerate and deepen the commercialisation of low-emission technologies such as clean coal and energy efficiency in Australia".

Global Roundtable on Climate Change[7]

In May 2005 the Global Roundtable on Climate Change was formed to bring together senior executives from the private sector and leaders

[6] These companies include Rio Tinto, ANZ, Mirvac and Anglo Coal. See George Lekakis & John McCarthy, "Blue Chips Go Green" (Herald Sun, 7 November 2006), online: http://www.news.com.au/heraldsun/story/0,21985,20712644-664,00.html.
[7] Online: http://www.earthinstitute.columbia.edu/grocc/ (Accessed on 21 February 2007).

of international governmental and non-governmental organisations to discuss and explore areas of common scientific and technical interest critical to shaping sound public policies to climate change.

The Global Roundtable has five objectives[8]

- To explore the potential for developing an improved global consensus on core scientific, technological, economic and policy issues related to climate change.
- To explore technological and policy options for mitigating climate change while meeting global energy needs.
- To champion demonstration projects that test and scale sustainable energy technologies and other activities and policies that address climate change.
- To provide a unique forum for discussion, analysis and exchange of ideas among businesses from all economic sectors and all parts of the world, international institutions, non-governmental organisations, and leading academic experts.
- To help catalyse new initiatives and interactions among Roundtable participants that address climate change mitigation and adaptation.

The Roundtable has held several meetings[9] on climate science, technology and economics, and practical proposals for adaptation to climate change. Recently, in February 2007, the Roundtable released their post-Kyoto framework for affecting change at the levels of policy and industry. This new release "The Path to Climate Sustainability: A Joint Statement by the Global Roundtable on Climate Change" makes six recommendations[10] aimed at meeting the challenge of climate change while recognizing the business opportunities.

---

[8] *Ibid.*

[9] http://www.earthinstitute.columbia.edu/grocc/conferences/ (Accessed on 10 March 2007).

[10] Full details of the recommendations can be found at http://www.earth.colombia.edu/grocc/grocc4_statement.html (Accessed on 8 March 2007).

The recommendations include

- The world's governments should set scientifically informed targets, including an ambitious but achievable interim, mid-century target for global CO_2 concentrations, in accordance with IPCC objectives.
- All countries should be party to this accord — commitments for actions by individual countries should reflect differences in levels of economic development and greenhouse gas emissions.
- Clear, efficient carbon emission market mechanisms should be established — consistent worldwide and across all sectors to reward efficiency and emission avoidance, encourage innovation, and maintain equitable technology options.
- Government policy initiatives should address energy efficiency in all sectors — allowing businesses to choose among a range of options and encourage the development and deployment of low emission technologies.
- Governments, the private sector and non-governmental organizations should cooperate with climate change mitigation strategies.
- Signatories to this statement will continue to support the IPCC processes, public awareness campaigns and emissions trading schemes, champion demonstration projects and support public policy efforts to mitigate the effects of climate change.

Roundtable participants are drawn from all regions of the world and every major economic sector, including industry manufacturing, transportation, non-renewable resources, renewable resources, energy, finance, insurance, health, food and agriculture.

Australian Government Response to Climate Change — AP6

Meanwhile the Australian government has responded to these calls from business with the establishment, in July 2005, of the Asia Pacific Partnership on Clean Development and Climate (or AP6 as it is known). The AP6 is a multilateral multi-stakeholder alliance and was established by developed and developing countries to focus on technology transfer addressing global climate change issues. The partnership aims to be consistent with, and contribute to, the United Nations Framework Convention on Climate Change (UNFCCC) and will complement, but not replace, the Kyoto Protocol.

The nations involved in the partnership are Australia, China, India, Japan, Republic of South Korea and the United States of America. And the partners have agreed to collaborate on the three core areas of energy security, climate change and air pollution issues in a way that encourages economic development and reduces poverty.

The six partner countries represent about half of the world's economy, population and energy use, and they produce about 65 percent of the world's coal, 48 percent of the world's steel, 37 percent of world's aluminium, and 61 percent of the world's cement.[11]

AP6 intends to build on existing multilateral climate initiatives[12] including the Renewable Energy and Energy Efficiency Partnership,[13] the Carbon Sequestration Leadership Forum,[14] the International Partnership for a Hydrogen Economy,[15] and Methane to Markets.[16]

First AP6 Partnership Meeting

The AP6 partnership held its inaugural meeting in Sydney, Australia on 11–12 January 2006[17] and involved Government Ministers, diplomats, researchers and business representatives. This first meeting issued a Communiqué, Charter and Work Plan.

---

[11] Online: http://www.asiapacificpartnership.org/ (Accessed on 26 October 2006).

[12] For example, the Australia-US Climate Action Partnership; Australia-China Climate Change Partnership; Japan-Australia Practical Collaboration on Climate Change. Information is available on http:///www.greenhouse.gov.au/international/partnerships.cap.html (Accessed on 27 February 2006).

[13] Online: http://www.reeep.org/ (Accessed on 26 February 2007).

[14] Online: http://www.cslforum.org/ (Accessed on February 2007).

[15] Online: http://www.iphe.net (Accessed on 26 February 2007).

[16] Online: http://www.methanetomarkets.org/ (Accessed on 26 February 2007).

[17] Online: http://www.asiapacificpartnership.org/default.htm (Accessed on 9 May 2006).

These have been summarized in Box 1. The full text of the AP6's Communiqué and Charter are set out in Appendix 1 and 2.

Box 1

Summary of inaugural AP6 meeting outcomes

Communiqué

The Communiqué acknowledges that the private sector is critical to AP6 efforts and recognises the need to direct domestic and foreign investment into low emission technologies.

The Communiqué also introduces the ideas of a Partnership Work Plan and Task Forces.

Charter

The Charter makes it clear that the partnership is a voluntary, non-binding framework for international cooperation on developing and deploying cleaner, more efficient technologies, to meet national pollution reduction, energy security and climate change objectives.

The preamble declares that AP6 is consistent with the principles of the U.N. Framework Convention on Climate Change and specifies that AP6 will pursue "development, energy, environment and climate change objectives" through technology cooperation.

AP6's purpose is to integrate all four objectives rather than an isolationist approach to climate change.[18]

(Continued)

[18] As Alexander Downer, Australian Minister for Foreign Affairs described it when the AP6 was announced in 2005, "The core element of the Partnership is policy integration. The Partnership sensibly doesn't seek to address issues in isolation. It is a coming together to find ways to meet our energy, climate change and air pollution issues in a practical way that make economic sense. Technology cooperation will be a core means of delivering outcomes" (Press Conference, Vientiane, Laos, 28 July 2005). See Joint Press Release at http://www.foreignminister.gov.au/releases/2005/index.html (Accessed on 26 February 2006).

(*Continued*)

Work Plans

The Work Plans focus on power generation in key industry sectors of partnership economies and establish eight public-private sector Task Forces.

The Task Forces include 1. cleaner fossil energy; 2. renewable energy and distributed generation; 3. power generation and transmission; 4. steel; 5. aluminium; 6. cement; 7. coal mining; and 8. buildings and appliances.

The Task Forces are to assess current standards, drive best practice and identify any relevant existing and emerging technologies and to develop Action Plans that identify specific commercial opportunities.

The Partnership Work Plan[19] sets out the innovative approach of using government/industry task forces to develop sustainable solutions to shared challenges through bottom-up practical action. Under the Work Plan, the Partnership has "agreed to work together with private sector partners to meet goals for energy security, national air pollution reduction, and climate change in ways that promote sustainable economic growth and poverty reduction. The Partnership will focus on expanding investment and trade in cleaner technologies, goods and services in key market sectors".[20]

The Work Plans also establish the framework for the eight public-private sector Task Forces with each Task Force expected to include representatives from government, research and industry.

[19] Online: http://www.dfat.gov.au/environment/climate/ap6/work_plan.html (Accessed on 26 February 2007).

[20] Online: http://www.asiapacificpartnership.org/default.htm (Accessed on 9 May 2006).

AP6 Organization Under the Charter

The AP6 partnership is to be governed by the non-legally binding Charter with membership terminable upon written notice 90 days prior to the anticipated date of termination.

Participation in the AP6 is voluntary, and funds, personnel and other resources are to be provided at the discretion of the partners themselves. This means the cost for any activities contemplated by the partnership are to be borne by the incurring country, subject to any special arrangements that may be made on an *ad hoc* basis.

The AP6 operates through two main bodies: The Policy and Implementation Committee and the Administrative Support Group.

Box 2

Summary of the role of AP6 Policy and Implementation Committee[21]

The Policy and Implementation Committee is the main steering body made up of three representatives from each partner country. Decisions are made by consensus and the Committee may meet whenever it is deemed appropriate. It has broad responsibilities over both technical and policy issues. This may include managing the implementation of partnership activities, engaging the private sector, multilateral financial institutions, research and governmental organisations, directing the Administrative Support Group and periodic review of the partnership.

The main activity however is the supervision of Task Forces and other subgroups which it can create whenever appropriate to assist in implementing its objectives.

---

[21] Summarized from Global Clean Energy and Climate Change Newsletter — January 2006 at http://www.bakernet.com/ (Accessed on 13 February 2007).

Box 3

Summary of the role of the AP6 Administrative Support Group[22]

The Administrative Support Group facilitates communication between the partner countries and is responsible for all administrative functions. The United States is initially responsible for this group, although each country will have a designated administrative liaison. This structure is subject to review.

Accommodation of National Policies under the Charter

The AP6 Charter acknowledges the supremacy of the national policies and national circumstances of the partners. For example, in China[23] development, population pressures and economic concerns are the main drivers for government response. These different development concerns are also acknowledged in India.[24]

---

[22] *Ibid.*

[23] "Climate change is not only an environmental issue but also a development issue. The essence of addressing climate change is to achieve sustainable development. The key lies in innovation and the transfer of technology information to concrete international cooperation ... We believe this new Partnership will further facilitate international cooperation on energy technology and with the framework of the Climate Change Convention we are ready to enhance scientific and technological cooperation with other countries in the world and work together to address energy issues to pursue sustainable development and address challenges posed by climate change", Mr. Liu Yongxing, Chinese Ambassador to Laos (Press Conference, Vientiane, Laos, 28 July 2005). *See*, Joint Press Release at http://www.foreignminister.gov.au/releases/2005/index.html (Accessed on 26 February 2007).

[24] "We are developing countries, we have our own agendas for our development activities, so we cannot give any promise, any commitment to reduce further our emissions", Mr. A Raja, Indian Minister for Environment, Press Conference, Sydney, 12/01/06. See Joint Press Release at http://www.dfat.gov.au/environment/climate/ap6/ (Accessed on 26 February 2007).

In the United States,[25] AP6 is viewed as a framework for addressing climate change objectives in the broader context of sustainable development and energy security.

While in South Korea[26] the government intends to engage local businesses and develop investment opportunities. The Japanese[27] perspective on innovation and improvements in energy efficiency and international relationships also receive major attention.

Second AP6 Partnership Meeting

The second AP6 meeting was held in Berkeley, California on 18–21 April 2006 and formally approved the eight public-private sector task forces which will deal with implementing and monitoring agreed work plans. About 300 representatives from government, industry, and research institutions from the six partner countries attended the meeting.

The Policy and Implementation Committee also presented their initial Action Plan Guidelines and Task Force Guidelines (See Appendix 3 and 4).

---

[25] Condoleeza Rice, US Secretary of State, "Asia-Pacific Partnership on Clean Development and Climate" Press Conference, Washington, DC, 9 February 2006. See http://www.state.gov/secretary/rm/2006/60858.htm (Accessed on 27 February 2007).

[26] Press release referring to the Ministry of Commerce, Industry and Energy plans to customise a strategy to expand into natural resource technologies and markets "The Ministry intends to capture the interest of local financial businesses and industrial sectors by explaining the climate change treaty as a new investment opportunity" — of Kyoto Protocol and "the Roh Moo-hyun administration has conducted natural resources diplomacy toward 17 nations" — "Gov't turns to market place to spur greenhouse gas reductions, resource procurement", Korea.net, 22 February 2007, online: http://www.korea.net/News/News/NewsView.asp?serial_no=20070221026 (Accessed on 23 February 2007).

[27] "Climate Change, Asian Economy, Energy and Policy" (Chapter 1) in Taishi Sugiyama & Stephanie Ohshita (eds.) *Cooperative Climate — Energy Efficiency Action in East Asia* (CRIEPI, University of San Francisco and International Institute of Sustainable Development (IISD), 2006).

The purpose of the Action Plan Guidelines is to promote consistency between the various Action Plans while recognizing the specific circumstances relevant to each task force theme and to each partner country.

As Task Forces are not permanent Partnership structures, existing only as long as is needed to achieve their objectives, the Task Force Guidelines are intended to assist in the development, implementation and monitoring of operating procedures within the Action Plans.

Third AP6 Partnership Meeting

From 11–13 October 2006, the Policy and Implementation Committee of the AP6 held its third meeting in Jeju, Korea during which it reviewed the draft Action Plans developed by the Partnership's eight Task Forces.

Details of the Action Plans were released on 31 October 2006, and these plans emphasised the involvement of business enterprises in the AP6 projects and focusing on cost competitive technologies.[28]

Table 1. Summary of Task Force Action Plans[29]

Task Force	Technology Focus	Objectives
Cleaner Fossil Energy Chair: Australia Co-chair: China	Advanced coal and gas technologies Air borne pollutants and advanced power generation systems e.g., Integrated Gasification Combined Cycle (IGCC),	Improve efficiency and environmental performance of fossil fuel use. Improved gas processing and

(*Continued*)

[28] "*Virtually all of the actions identified will involve business, and a number of the activities will be undertaken primarily or exclusively by companies and associations representing commercial enterprises*" (*See*, Asia-Pacific Partnership on Clean Development and Climate Executive Summary of Task Force Action Plans (Accessed on 31 October 2006).

[29] Summarized from Executive Summary of Task Force Action Plans — Third Policy and Implementation Committee meeting of the AP6 held in Jeju, Korea 12–13 October 2006 http://www.asiapacificpartnership.org/ActionPlans.htm (Accessed on 31 October 2006).

Table 1. (*Continued*)

Task Force	Technology Focus	Objectives
	producing hydrogen from coal, and Ultra Supercritical Pulverized Coal, post combustion capture and storage, CO_2 capture and storage Potential barriers to delivery of cross border liquefied natural gas Coal bed methane recovery	transportation technologies. Improved methane gas, coal bed, liquefied natural gas marketing opportunities in Asia-Pacific region. Encourage uptake of geosequestration technologies
Renewable Energy and Distributed Generation Task Force Chair: Republic of Korea Co-chair: Australia	Alternative fuels such as hydro, solar, geothermal and wind Distributed generation for customised rural power solutions based on local fuel sources; renewable energies for independent electricity supply system; solar enhanced fossil fuels; and petroleum coke oven gas to produce electricity	Address cost competitiveness, technology dissemination, intermittency, electricity storage. Increased energy security Rural-remote area requirements addressed
Power Generation and Transmission Chair: US Co-chair: China	Best Practice management for Power Generation, Transmission and Distribution, Demand Side Management and Information Sharing including thermal efficiency of power plants; electricity market reform	Increased understanding of investment opportunities in efficient power systems. Development of transmission and demand side management technologies
Steel Chair: Japan Co-chair: India	Best Practice Clean Technologies Handbook; survey of state of industry in partner countries; waste product recycling increased Initial focus on China and India	Improved benchmarking and reporting. Increased energy and material efficiencies. Technology development and deployment

(*Continued*)

Table 1. (*Continued*)

Task Force	Technology Focus	Objectives
Aluminium Chair: Australia Co-chair: USA	Best Practice use of existing equipment focussing on perfluorocarbon emissions management; energy efficiency; process emissions management Several MOUs signed	Adopt benchmarking and reporting energy and material efficiencies. Technology development and deployment
Cement Chair: Japan Co-chair: tbc	Best Practice and benchmarking in use of clean energy technologies, alternative fuels and raw materials, and power processing from plant waste	Encourage and deployment of new technologies. Information sharing and reporting improvements
Coal Mining Chair: US Co-chair: India	Improvements in coal mining and beneficiation efficiency; best practice and technologies in coal preparation, coalmine methane capture and mine health and safety	Facilitating technology transfer, methane capture and recovery techniques. Information sharing and adoption of best practice standards
Building and Appliances Chair: Republic of Korea Co-chair: US	Energy efficiency reductions in residential and commercial sectors; best practice standards and codes; financing and leasing options	Support best practice recommendations. Respond to energy efficient technologies

Funding for AP6 Projects

The overall financial commitment made to the AP6 by the partners is small, with pledges and in-kind contributions from companies and research organisations making up the bulk of the arrangements.

However, on 1 November 2006, the Prime Minister of Australia announced the first tranche of Australian Government in-principle support for AP6 projects, representing spending of $60 million for

42 projects.[30] These were selected from over 90 cooperative projects developed in the AP6 Action Plans to date. The publication "Partnership for Action 2006" provides a summary of Action Plans developed by the AP6 and the projects being supported by Australia.

Supporting Domestic Regimes

At the Annual Dinner of the Business Council of Australia in Sydney on 13 November 2006, the Prime Minister announced[31] that the Federal Government would consider a new emissions trading scheme which would protect the natural advantages that Australia has, not only as major coal and gas supplier, but of uranium as well. He also announced the idea of "a new Kyoto" that operates differently to the present model, to try and find a new global solution that must include all the major emitters and operates on a basis that is appropriate to Australia's interests and needs. The Government revealed plans to use the AP6 group as "a bridge" to reduce global greenhouse gas emissions[32] and that a new government-business taskforce would be set up to assess the best options for Australia.

At the same dinner, the President of the Business Council of Australia[33] supported and encouraged the Federal Government's decision on an emissions trading scheme and encouraged the government to pursue this through the AP6.

---

[30] "PM blows his climate trumpet". The Age, 1 November 2006, online: http://www.theage.com.au/news/national/pm-blows-his-climate-trumpet/2006/11/01/1162339901697.html (Accessed on 1 November 2006).

[31] Transcript of the Prime Minister's Address at the Business Council of Australia Annual Dinner, online: http://www.pm.gov.au/News/speeches/speech2246.html (Accessed on 15 November 2006).

[32] Steve Lewis & Dennis Shanahan, "Howard pushes for 'new' Kyoto agreement", The Age, 1 November 2006, online: http://www.news.com.au/story/0,23599,20679728-2,00.html (Accessed on 1 November 2006).

[33] The BCA President's Address to the BCA 2006 Dinner 2006, online: http://www.bca.com.au/ (Accessed on 15 November 2006).

In December 2006, the joint government-business task group[34] was established to look at a national emissions trading scheme using the AP6 as a bridge to international schemes. A discussion paper, "The Task Group on Emissions Trading Issues Paper" was released for comment on 7 February 2007. After taking any comments into consideration, the Task Group was to report back to the Prime Minister by 31 May 2007. In its final report, the Task Force concluded that Australia should not wait until a truly global agreement has been negotiated but rather it should adopt an appropriate emissions constraint. The Task Force also concluded that the most efficient and effective way to manage risk is through market mechanisms, in particular, an emissions trading scheme with the Australian Government setting a national framework for reducing GHGs and the market being allowed to set the carbon price.[35] The Government responded to the Report by pledging to design an emissions trading scheme by 2011 with trading to begin in 2012.

Australian Government Climate Change Risk Management

Over the same period as the establishment of the AP6 and the formation of the Australian Business Roundtable on Climate Change, the Australian Government was preparing the "Climate Change Impacts and Risk Management Guidelines for Businesses and Government".[36] The publication was released in May 2006 and was accompanied by a series of workshops held around the country aimed at both the public and private sector. The workshops were aimed at informing executives of climate change risks and suggested appropriate ways to deal with those risks. Presenters at the workshops included

[34] The Task Group on Emissions Trading comprises representatives from government departments and businesses. See http://www.pmc.gov.au/emissionstrading/index.cfm for details (Accessed on 10 March 2007).

[35] *See*, Report of the Task Group on Emissions Trading at http://www.pmc.gov.au/publications/emissions/docs/introduction.pdf (Accessed on 16 July 2007).

[36] *Climate Change Impacts and Risk Management — A Guide for Business and Government*, 2006, Australian Greenhouse Office.

Australian Greenhouse Office representatives, leading Australian climate scientists and specialists in economics, risk management and public policy.

The climate change risk management guidelines are based on the Australian and New Zealand Standard for Risk Management (AS/NZS 4360:2004) and the workshops demonstrated how this methodology could be extended to include climate change risks. The guidelines focus on the assessment and prioritisation of climate risks and establish a process for managing these identified risks.

Box 4 contains a summary of the climate change risk management process.

Box 4

Climate change risk management process[37]

1. Establish the context — this includes defining business objectives, identifying stakeholders, establishing criteria against which risks can be evaluated, identifying and/or developing areas of responsibility within the organisation and determining the relevant climate change risks to the organization.
2. Identify the risks — describing and listing how climate change will affect key elements of the organization.
3. Analysing the risks — which covers reviewing the controls, management and regimes already in place to deal with each specific risk, and then assessing the consequences and determining the level of climate risk to the organization.
4. Evaluate the risks — this is reaffirming the judgements and estimates of the climate risks, ranking them in terms of their severity and prioritizing those risks that need addressing.
5. Treat — or respond to climate change risks — by identifying relevant options available to manage, adapting or mitigating climate risks and selecting the best options and incorporating these into company plans.

[37] *Ibid*, p. 19.

The Australian Government's climate change risk management guidelines provide a framework for managing the increased risks to organisations due to climate change impacts. And while they establish an initial strategic assessment and prioritisation process, the guidelines are primarily concerned with the physical risks of climate change.

Business Climate Change Risk Management

From a business perspective, climate change risk has been increasingly identified as a legitimate corporate risk.[38] A clear understanding of not only physical climate change risks, but legal liabilities and opportunities that exist, or are likely to exist in the future, must also be considered by businesses and included in corporate planning and investment decisions.

While business leaders routinely include risk management into strategic planning, climate change risk management is unusual because of the uncertainty of regulatory risk at all levels of government, domestic and international, and the unpredictability of climate change's impact on weather, markets and trade issues. Businesses need to understand the impacts of climate change on their own operations, and recognize that managing climate change risk is also about identifying climate change opportunities.

Even though Australia has not ratified the Kyoto Protocol, Australian businesses are preparing for a regulatory environment which could affect their output and profitability. While current climate change legislation is mostly voluntary in Australia,[39] businesses are anticipating and preparing for a more stringent regulatory environment. The ramifications for Australian businesses of possible

---

[38] "In Control of Carbon Climate Law and Policy; managing obligations, liabilities, commercial risks and opportunities" Baker and McKenzie; "A Guide for Company Analysts — Climate Change and Company Value", Baker and McKenzie; "Managing the Risks and Opportunities of Climate Change: A Practical Toolkit for Corporate Leaders" Ceres and the Investor Network on Climate Risk; "*Climate Change: Risk and Opportunity — Key issues for Businesses and Risk Managers*" Marsh.

[39] Each State and Territory has Acts governing climate change. For example, Victoria's Renewable Energy Bill, South Australia's Climate Change Bill, New South Wales' Greenhouse Gas Abatement Scheme and Queensland's 13% Gas Scheme.

restrictions to international trade opportunities[40] are also being considered.

Generally, business climate change risks can be grouped into four categories. These are physical risks, regulatory risks, competitive risks and reputational risks. Box 5 provides an explanation of these risk categories.

Box 5

Business climate change risks

Physical risks include damage to property and assets and project delays resulting from extreme weather events.

Regulatory risks result from unknown national and international government regulations designed to reduce greenhouse gas emissions. Regulatory risks also include the possibility of climate change litigation arising from corporate action, or inaction.

Competitive risks could result from a decrease in consumer demand for energy intensive products and a rise in costs for energy intensive processes.

Reputational risks to businesses come from the perception of inaction, or little action, on climate change issues. This risk includes not only the perceptions of the general community but also includes shareholders' and investors' concerns for competitiveness and investment returns.

Businesses evaluate these climate change risks as part of their environmental risk registers and consider which solution is the most viable for their operations.

[40] Future trade opportunities with countries that have imposed climate change regulations as part of their Kyoto Protocol requirements. A newspaper article on the French proposal to tax the imports of countries that refuse to ratify the Kyoto Protocol is a recent example — see Peter Hartcher, "US pours scorn on international greenhouse tax proposal", The Sdyney Morning Herald, 20 November 2006, online: http://www.smh.com.au/news/world/us-pours-scorn-on-international-greenhouse-tax-proposal (Accessed on 20 November 2006).

Company directors and officers also need to consider the extent to which they are responsible for managing these climate change risks as shareholders and investors may hold them liable if their company is devalued because it failed to plan for the effects of climate change. It was recently reported in Australia[41] that insurers may not insure company directors that have not satisfactorily developed a climate change risk management plan for their company.

From the insurance industry's perspective,[42] climate change is recognised as a reality that must be addressed urgently. At a basic level, climate change threatens the insurance industry's core business of underwriting weather and climate related catastrophes. Insurers underwrite these catastrophes by calculating, pricing and spreading the risk and then meeting claims as they arise. Climate change, characterized by less predictable weather related events, has the potential to reduce the insurance industry's capacity to calculate, price and spread the risk.

The role of the insurance industry in underwriting climate change and weather related events is also an important component of the national economy. In Australia, the increasing population in marginal and more vulnerable areas such as along coastlines[43] also increases the number of communities exposed to extreme weather related events such as tropical cyclones, storm surges and flooding of coastal rivers. The resulting stress of increased claims on the insurance industry has the potential for decreasing insurance cover, leaving affected communities dealing with both the hardships of an extreme weather event and possible problems associated with rising or unavailable premiums.

---

[41] "The Business of Climate Change", Radio National 16/10/05 — interview with Federal Government Environment Minister and various business representatives — http://www.abc.net.au/rn/talks/bbing/stories/s1480714.htm (Accessed on 22 November 2006).

[42] Insurance Australia Group, 2005, *The Impact of Climate Change on Insurance against Catastrophes* — http://www.tec.org.au/dev.index.php (Accessed on 4 August 2006).

[43] "More than 80% of (Australia's) population resides within 50 km of the coast "*Ibid* — p. 8.

Government intervention is often required, and requested, through Emergency Management Australia[44] — a co-ordinated emergency government department response involving the government at the local, State, Territory[45] and Federal levels, non-governmental organizations (NGOs), the private sector, local communities and volunteers.

In response to increased climate change risks to the insurance industry, opportunities for new business strategies to mitigate against further climate change damage are being developed in Australia. Some examples of these strategies include promoting public awareness programs that identify vulnerable areas such as flood-prone land, lobbying governments to change or enforce building codes, introducing policies or products such as cheaper premiums to reduce car emissions, supporting public transport systems, as well as investment in research into the understanding of climate change and extreme events. One of the loudest calls from the insurance industry is to promote effective risk mitigation and help governments create market based mechanisms to address and resolve environmental challenges.[46]

Other climate change opportunities for both governments and businesses have also been identified[47] and are summarized in Box 6.

---

[44] See Emergency Management Australia, online: http://www.ema.gov.au (Accessed on 23 November 2006).

[45] Under the Australian Constitution, the primary responsibility for the protection of life and property rests with State and Territory governments. The Federal Government provides extra guidance and financial support in times of emergency, *ibid.*

[46] Insurance Council of Australia Strategic Blueprint 2007 — Environment and Climate Change — Core Statement and Objectives — online: http://www.insurancecouncil.com.au (Accessed on 27 November 2006).

[47] A most recent example is in the Stern Review, "The Economics of Climate Change", issued on 30 October 2006, in the Executive Summary recommending immediate action on climate change suggests "that there will be wide range of opportunity for growth and development along the way".

Box 6

Business climate change opportunities

Regulatory opportunities include compliance and emissions trading reductions for companies active in the development of these schemes, as well as the opportunity to help shape government regulations. Also, long-term market signals concerning growth potential of new technologies can assist in planning company portfolios.

Technical opportunities cover growth in developing new technologies or products which offer new financial opportunities. Businesses are also able to implement their own energy efficiency strategies to reduce emissions and save money.

Competitive opportunities could result from businesses investing in new technologies that may seize new markets or market share.

Reputational benefits may come from businesses' ability to capitalise on these green investment strategies to enhance corporate reputations.

In late March 2007 the Australian Prime Minister, partly in response to a the visit to Australia of Sir Nicholas Stern, the former World Bank economist and author of the review prepared for the UK government on the economics of climate change, published in October 2006, which is considered the most comprehensive review to date of the costs of inaction on climate change versus the costs of action,[48] announced an initial $200 million over five years commitment to reduce deforestation in Southeast Asia as a more practical and more timely way to reduce GHG emissions.

---

[48] Stern, Sir Nicholas, "The Economics of Climate Change — The Stern Review" (Cambridge University Press, January 2007).

Firmly rejecting calls from environmental groups, the Greens and Sir Nicholas Stern for the imposition of harsh carbon emission targets, the Government reiterated its long-held view that its response would be guided by protecting jobs and safeguarding the Australian economy. In contrast, the opposition Labour Party's leader Kevin Rudd announced Labour's commitment to a 60 percent reduction by the year 2050 just a few days earlier. Entirely lacking in specifics as to how that reduction would be achieved and at what cost to the Australian economy, Mr Rudd instead indicated that, when it comes to designing an emissions trading scheme, Labour is going to have to look at how that is engineered, with a series of staged targets between now and 2050.

It is likely, in this writer's view, that the Howard government will tackle the important issue of climate change that is emerging as one of the critical issues on which the forthcoming election will be fought, by relying on a series of measures such as clean coal technology, the establishment of some form of emissions trading scheme, a mix of renewable energy initiatives that include solar, wind, biofuels and at least a serious look at nuclear energy production.

It is of passing interest to note that, notwithstanding the Australian government's steadfast refusal to ratify the Kyoto Protocol, it is nevertheless projected to be one of the very few developed countries that will actually meet its Kyoto emission reduction target in the year 2012. In contrast, many countries which did ratify Kyoto in Europe, the former Soviet bloc, North America and elsewhere are projected to fall drastically short of meeting their respective defined reduction targets.[49] In response to the often-raised comment — "So what! Australia's assigned target at the conclusion of the Kyoto negotiations was 108% above 1990 emission levels and Australia was successful in being allowed to take into account land-clearing in the formula" — The fact remains that this was the target for Australia that was agreed to by both Australia and the other Contracting Parties to

[49] Canada, for example, has admitted that it has no chance of meeting its quantified reduction target of −6% by the year 2012 and some estimates place it as much as 65% above its targeted reduction agreed to at Kyoto.

the Protocol and that Australia's quantified emission reduction target in the first Commitment Period (2008–2012) *would not have changed* had Australia ratified the Convention.[50]

APEC Meeting, Sydney, September 2007

APEC (Asia-Pacific Economic Cooperation) is an inter-governmental forum facilitating economic growth and prosperity, cooperation, trade and investment in the Asia-Pacific region, and operates on the basis of non-binding commitments, open dialogue and equal respect for the views of all participants regardless of the size of their economy.[51] At its most recent meeting hosted by the Australian government in Sydney in early September 2007, climate change was placed at the top of the forum's agenda. To the surprise of many, a limited consensus was reached by APEC leaders including both China and the US calling for concerted international action to reduce GHG emissions. Whilst not setting an overall target for cutting greenhouse gas emissions the Sydney Declaration on Climate Change does commit APEC members to trying to improve energy efficiency by at least 25% by the year 2030. It also contains a commitment to increase forest cover in APEC countries by at least 20 million hectares by 2020 thus reducing 1.4 billion tonnes of carbon. In addition the APEC leaders also pledged to generate new forms of collaboration on clean energy technologies and renewable energy sources as well as to expand the trade in environmental goods and services, including possible carbon trading. The real breakthrough, however, is that for the first time the US and China, the world's two largest emitters, joined other APEC

[50] This in no way implies that this writer agrees with the extraordinarily generous reduction targets and conditions accorded to a developed country like Australia with one of the highest performing economies of all OECD countries. To put this in perspective, the only other country that won similar concessions during the negotiations was Iceland with a population of just over 300 000 people and an economy a fraction of the size of Australia's economy. Iceland was allowed a 10% increase over 1990 emission levels.

[51] As defined on the APEC website http//www.apec2007.org (Accessed on 29 October 2007).

nations including Australia and other large developing countries such as Malaysia, Mexico and Indonesia in a common effort.[52]

Both the United States under President George Bush and Australia under John Howard have clearly signaled that they will not be parties to a post-Koyoto agreement that does not place any obligation on developing countries such as China and India although it is clearly acknowledged in the Sydney Declaration that they are not opposed to the application of the principle of differentiated responsibility for some countries depending on their circumstances.

In the third week of the Australian federal election campaign, the Australian Labor Party's shadow environment minister, Peter Garrett, stated that, if Labor was elected, it would not be a "deal breaker" if developing nations such as China did not accept binding targets to cut greenhouse gas emissions. By late the following day, after initially supporting Garrett, Labor Leader Kevin Rudd had endorsed John Howard's long-held position and flatly contradicted his environment spokesman. Rudd said a Labor government would not ratify a post-Kyoto agreement unless developing nations accepted binding targets. He is quoted as saying "I have made it absolutely clear that we would need to see clear-cut commitments from the major emitters from the developing world for us to become a party to that agreement." In this remarkable turn of events, Labor seriously undermined its election promise to ratify Kyoto while at the same time categorically stating that, if elected, it would not ratify a post-Kyoto agreement unless developing countries were also bound by legally binding reduction commitments.[53]

Concluding Comments

It is apparent that no matter which of the two major political parties in Australia forms the government of the day on 24 November 2007,

[52] For an analysis of the Sydney Declaration see http//www.globalinsight.com/SDA/SDADetail10562.htm (Accessed on 29 October 2007).

[53] See Sid Marris, Dennis Shanahan & Patricia Karvelas, "Rudd sets 20% renewable target" (The Australian, 31 October 2007).

Australia's position at the forthcoming UNFCCC negotiations due to take place at Bali in December 2007 will be substantially consistent in demanding that some definitive action be undertaken by the major emitting countries. Given the stated positions of the United States, India, China and many other developing countries against binding emission reduction targets, a post-Kyoto agreement if one is achieved will likely be structured radically differently than the 1997 agreement and may well follow the approach taken by AP6 and APEC. This should not come as a total surprise as many of the same countries are members of both organizations.

Addendum

Since the delivery of this paper at the Singapore Conference, the Australian federal election took place and the Howard coalition government was replaced by a Labour government led by Prime Minister Kevin Rudd. Although the information in this paper was accurate at the time of delivery, subsequent events after the new Labour government took power necessitate a brief addendum to clarify the current situation.

The 2007 Australian federal election was described as the first to be fought over climate change. It is therefore unsurprising that the first official action of the new Labor government was to ratify the Kyoto Protocol. Given that Australia is expected to meet its obligations under the protocol,[54] ratification primarily paves the way for greater Australian involvement in post-Kyoto negotiations. This largely symbolic action on the part of the Rudd government took place in the context of his earlier humiliation of his newly appointed Environment Minister, Peter Garett who had indicated prior to the election that Australia would sign up to a post- Kyoto agreement *even if the major developing country emitters failed to agree to binding reduction targets.*

---

[54] For more information, see 'The Australian Government's Initial Report Under the Kyoto Protocol' (2008). Online: http://www.greenhouse.gov.au/inventory/publications/pubs/unfccc-report.pdf. Accessed 10 May 2008.

Rudd immediately reputed Garett's statement stating categorically that a Labour government would not ratify any post-Kyoto agreement unless developing countries accepted binding targets.[55] He further signified his displeasure over his apparently gaffe prone environment minister by appointing Penny Wong as the Minister specifically responsible for Climate Change issues.

Barely a week after taking office, Prime Minister Kevin Rudd joined the Australian delegation at the UN Climate Change Conference in Bali. The purpose of the Bali conference was to build a timetable for post-Kyoto negotiations within the UNFCCC framework. In a speech during the High Level Segment of the conference, Prime Minister Rudd called for all developed nations to 'embrace a further binding set of binding emissions targets' and for all developing countries 'to play their part- with specific commitments to action'.[56] This approach was further emphasised in a press conference the following day, where Rudd stated:

> 'We need to be working together so that all developed countries — and as I said yesterday, all developed countries — undertake binding targets for the future and that we have specific commitments from the major developing countries and the developing countries in general.'[57]

It would appear from this statement that the Labor Government will support a post-Kyoto agreement provided major emitters from the developing world commit to action; this commitment need not be in the form of legally binding emission reductions.

While the United States was reluctant to approve the framework for further negotiations (the Bali Action Plan) without a commitment

---

[55] See note 53.

[56] Prime Minister Kevin Rudd (Speech delivered at the High Level Segment of the Conference of the Parties; United Nations Framework Convention on Climate Change Bali Indonesia, 12th December 2007).

[57] Prime Minister of Australia, 'Press Conference, Bali' (Media Release, 13th December 2007. Online: http://www.pm.gov.au/media/Interview/2007/interview_0012.cfm. Accessed 10 May 2008).

from major developing countries to adopt binding emissions reductions, international consensus was ultimately reached. The Bali Action Plan recognises the importance of poverty reduction and development in developing countries and calls for 'nationally appropriate mitigation actions by developing country Parties', rather than quantified emissions reductions.[58] The Bali Action Plan contains no reference to an interim target for global emissions reductions by 2020. While an interim target of 25–40% reduction for developed countries was considered at the Bali conference, it proved a key hurdle in negotiations and was excluded from the decision.[59] The Australian government position is therefore in line with broad international consensus.

In addition to international negotiations, the Rudd Government has intensified efforts to establish a national emissions trading regime. Since taking office in December, the new government has remained committed to reducing emissions by 60% below 2000 levels by 2050. The emissions trading scheme is to be the centre-piece of Labor policy to reduce greenhouse output. A detailed schedule for implementation has been announced that would see the scheme commence in January 2010.[60] The design of the scheme is currently being debated and will depend upon the final results of the Garnaut Climate Change Review and economic modelling undertaken by the Department of Treasury.

In July 2007, the Australian Labor Party commissioned a major review of Australian climate change policy and the measures necessary

---

[58] UNFCCC (United Nations Framework Convention on Climate Change (2007) Draft decision -/CP.13: Fourth Assessment Report of the Intergovernmental Panel on Climate Change. Decisions adopted by COP 13 and CMP 3, United Nations Climate Change Conference 2007, Bali, Indonesia. Online: http://unfccc.int/files/meetings/cop_13/application/pdf/cp_arfour.pdf. Accessed 9 May 2008.

[59] Garnaut Climate Change Review 'Interim Report to the Commonwealth, State and Territory Governments of Australia' [41]. Online: http://www.garnautreview.org.au/CA25734E0016A131/pages/reports-and-papers. Accessed 10 May 2008.

[60] For more information see, Minister For Climate Change and Water 'Government Announces Detailed Timetable on Emissions Trading' (Media Release, 13th March 2008. Online: http://www.environment.gov.au/minister/wong/2008/pubs/mr20080317.pdf. Accessed 10 May 2008).

to mitigate the effects of climate change. The Garnaut Climate Change Review is chaired by Professor Ross Garnaut and is due to deliver the final report in September 2008. The report will help guide long term policy development and the adoption of short and medium term emissions reduction targets. Such guidance is critical to achieving the long term government goal of a 60% reduction in emissions by 2050 and will affect the position taken by the government in upcoming UNFCCC negotiations.

The interim report, released in February 2008 provides qualified support to the initiatives and long term goal of the government however it questions the wisdom of the international position taken in the Bali Action Plan. The report argues that binding emissions targets for major developing countries are necessary. If such targets are left until 2020 when a post-Kyoto agreement will expire, the scientific evidence suggests it will be too late to achieve climate stabilisation at moderate levels. Binding targets for all Parties will also encourage developed nations to commit to ambitious reductions in emissions.[61] The interim report also suggests that the government may need to aim higher than a 60% reduction in emissions by 2050 if a significant global agreement is to be reached. While recognising the complexities of international negotiations and therefore the value of bilateral agreements and financial incentives, the interim report concludes that 'there is no prospect within this framework of holding the risk of dangerous climate change to moderate levels'.[62]

Since December, the Labor government has also announced numerous projects to improve household energy efficiency, encourage small business involvement in the energy debate, develop clean coal technology, establish bi-lateral agreements with Australia's trade partners and to increase renewable energy generation.[63] These initiatives are all aimed at addressing the three 'pillars' of Labor's climate

---

[61] Garnaut Climate Change Review, Ibid. at [33].

[62] Ibid. at [36].

[63] See generally Ministerial Media Releases, Prime Minister's Department (http://www.pm.gov.au/media/index.cfm), Department of Environment, Heritage and the Arts (http://www.environment.gov.au/media/index.html).

change policy: reducing domestic emissions; adapting to unavoidable environmental impacts; and building an international solution. In addition to the long term emissions reduction of 60%, the government aims to increase renewable energy production to 20% of Australia's electricity supply by 2020 and is co-ordinating efforts between the state and territory governments.

The 2008–2009 Federal Budget, the first budget of the new government, commits $2.3 billion over four years to climate change initiatives. This amount, however, has been severely criticised by the environmental community as being woefully deficient and substantially undermining the new government's green credentials, particularly in the light of the $20 billion surplus amassed in the current budget year largely on the back of the commodity boom fuelled by the rapid industrialisation of both India and China.

When one compares the $9 billion of taxpayer subsidies directed to fossil fuels annually with the $232 million annually in federal subsidies going to renewable energy and energy efficient measures, the governments commitment to seriously addressing climate change issues must be questioned. The government has dedicated $700 million to improving energy and water efficiency in homes and includes funding for low-interest loans to install energy efficient hot water systems, an improved energy-labelling program for electrical appliances and rebates to landlords who insulate their rental properties. Grants will also be provided to schools to install energy efficient equipment.

One particular budget measure, however, has drawn the ire of government and non-government supporters alike. Under a scheme introduced by the Howard government rebates of up to $8000 were available to homeowners and community groups who installed accredited photovoltaic devices. While the rebate initiative has been maintained and funding increased, it will now be subject to means testing.[64] Households with a taxable income greater than $100 000 are excluded from the scheme, and many will be unable to afford the

[64] *Climate Change Budget Overview 2008–2009*, 2008, Department of Climate Change. Online: http://www.greenhouse.gov.au/budget/0809/pubs/ccbo-0809.pdf. (Accessed 16 May 2008).

average $13 000 – 15,000 capital outlay formerly offset by the rebate. Given current housing prices in all major capital cities, few families earning less than $100,000/annum can afford to own homes and would be unlikely to take up the rebate to convert to solar power. Using a perverse form of logic, the government sought to justify its action by stating that the reason for introducing the means test was that the program was "too successful" and could not be sustained in the long term. In the few days since the budget was released there have been more than 50,000 homeowners who cancelled their contracts for installation of solar systems on the basis that the rebates were no longer available. Those who recently entered the solar industry to supply was thought to be a growing market for these systems have seen their growth prospects severely curtailed. Such action under the guise of cutting costs to fight inflation sends the wrong message to the Australian public who have one of the largest per capita carbon footprints of the OECD countries.

Furthermore, $500 million has been promised over seven years under the National Clean Coal Fund to develop carbon capture technology and infrastructure. Funding will begin this financial year. A similar fund designed to accelerate the development and commercialisation of renewable energy technology, the Renewable Energy Fund, will also receive $500 million however funding will not begin until next financial year.[65] The prioritisation of coal initiatives over renewable energy technologies has not passed uncriticised.[66] The budding geothermal industry is also concerned that development of an important renewable energy source will be delayed by the postponement of funding.[67]

While it may be argued that the funding commitments set out by the Rudd government represent modest progress in addressing climate

---

[65] Ibid.

[66] See for example, Marian Wilkinson, 'Wrong Emphasis say Green Groups', *The Sydney Morning Herald*, 14 May 2008, Online: http://www.smh.com.au/news/national/wrong-emphasis-say-green-groups/2008/05/14/1210444530258.html (Accessed on 16 May 2008).

[67] See Matthew Warren and Christian Kerr, 'Repair Job to boost Geothermal development', *The Australian*, 15 May 2008, Online: http://www.theaustralian.news.com.au/story/0,25197,23700833-11949,00.html (Accessed on 16 May 2008).

change, they fall far short of the serious actions proposed by non-governmental organisations such as the Australian Conservation Fund.[68] Despite being considered 'the challenge of our generation', the $2.3 billion funding commitment of the government over four years appears increasingly pathetic alongside the $22 billion dedicated to Defence this financial year alone.

It remains to be seen how the post-Kyoto process will unfold and whether the Rudd Government will adopt the broader emission reduction commitments and greater leadership role that the final Garnaut Report is likely to recommend.

Given the current positions of major emitting countries and the importance of the coal industry to the Australian economy it will take a great deal more than empty long term emission reduction promises to be realised in the year 2050 or even mid-term 2020 reduction promises for which the present government will never (given Australia's three year election cycle) be held accountable by the electorate. It may be good 'politics', however it does little to combat the effects of global warming.

Appendix 1: Asia-Pacific Partnership on Clean Development and Climate Inaugural Ministerial Meeting — Sydney, January 2006

Communiqué

We met at Sydney for our first Ministerial meeting of the Asia-Pacific Partnership on Clean Development and Climate on 11–12 January 2006. We adopted a Charter that sets out a framework to implement the Vision Statement of the Partnership announced in Vientiane on 28 July 2005. At the core of this vision is our conviction of the urgent need to pursue development and poverty eradication. By working

---

[68] The Australian Conservation Fund's submission on the 2008–2009 provides insight into the measures they consider necessary to seriously address climate change. Online: http://www.acfonline.org.au/uploads/res/080118_-_ACF_2008_09_Federal_Budget_Submission_FINAL.pdf (Accessed on 17 May 2008).

together we will be better able to meet our increased energy needs and associated challenges, including those related to air pollution, energy security, and greenhouse gas intensity.

Our energy needs are growing rapidly, and will necessitate large-scale investments in the coming decades. We recognized that renewable energy and nuclear power will represent an increasing share of global energy supply. We recognized that fossil fuels underpin our economies, and will be an enduring reality for our lifetimes and beyond. It is therefore critical that we work together to develop, demonstrate and implement cleaner and lower emissions technologies that allow for the continued economic use of fossil fuels while addressing air pollution and greenhouse gas emissions. We undertook through this Partnership to cooperatively promote the deployment of promising technologies that offer greater energy efficiency and lower air pollution and greenhouse gas intensities.

Energy security is another major concern. Access to a diverse range of reliable and affordable energy sources underpins economic development and improved living standards and is a major determinant of energy security. Thus, our efforts to reduce the greenhouse gas intensity of a wide range of fossil fuels provide an important energy security benefit to us all.

We view climate change in particular as a serious problem that warrants a long-term commitment to substantive action. The Partnership will be consistent with and contribute to our efforts under the United Nations Framework Convention on Climate Change and will complement, but not replace, the Kyoto Protocol.

We reviewed the extensive range of existing national programmes and projects our Governments are pursuing with regard to clean development and climate. Each Partner will bring significant value to the Partnership and our Governments have pledged a serious commitment to Partnership projects and activities. We view the private sector as critical to this effort, and we will marshall considerable financial, human and other resources both from the public and private sectors. The Partnership aims to mobilize domestic and foreign investment into clean and low emission technology by fostering the best possible enabling environments.

We prepared the Partnership Work Plan, which explores a new approach for harnessing the power of our private sectors, our research communities and our government sectors to drive sustainable development. We will bring together the key experts from the public, private and research sectors of our economies to tackle these matters. We will also share experiences on related matters, such as workplace safety and technologies that help ensure the safety and well-being of our peoples. Our Work Plan focuses on power generation and key industry sectors of our economies. We established eight public-private sector Task Forces covering. (1) cleaner fossil energy; (2) renewable energy and distributed generation; (3) power generation and transmission; (4) steel; (5) aluminium; (6) cement; (7) coal mining; and (8) buildings and appliances.

We have directed the Task Forces to drive improvements with regard to best practices and ensure that a range of technologies is developed and repeatedly demonstrated so that scale is increased and costs are reduced.

In this regard we have asked each Task Force to:

- review the current status within their thematic area with regard to clean development and climate,
- share knowledge, experience and good practices of how efficiency can be improved,
- systematically roadmap, where appropriate, relevant existing and emerging technologies, and
- develop an action plan that identifies specific opportunities for co-operation, and wherever possible, ambitious and realistic goals.

The Partnership Work Plan will be dynamic, evolving as the Task Forces elaborate their work. Initially the Partnership chose to focus on a number of specific areas. The vision statement also detailed a rich array of other sectors, such as transport, where we will explore co-operation as the Partnership develops. There are also cross-cutting opportunities to advance clean development and climate beyond the current Task Forces, such as skills exchange. In this regard, we will positively consider the proposal to establish an "Asia-Pacific Energy Technology

Co-operation Centre", to focus on the development and implementation of an energy audit program and its follow-up projects. We envisage that future meetings will address these other sectors of interest and cross-cutting matters as well as provide a forum for sharing experiences in developing and implementing our sustainable development and energy strategies.

The Partnership brings together a grouping of key nations to address the serious and long-term challenges of climate change, energy security and air pollution in ways that support sustainable economic development. Working together, we can make a significant contribution to global clean development and climate. http://www.dfat.gov.au/environment/climate/ap6/communique.html — 09/05/06

Appendix 2: Charter for the Asia-Pacific Partnership on Clean Development and Climate

We, the representatives of the national governments of Australia, China, India, Japan, the Republic of Korea, and the United States of America (collectively referred to as the "Partners"), meeting in Sydney, Australia on 12 January 2006:

Guided by our Vision Statement for a New Asia-Pacific Partnership on Clean Development and Climate of 28 July 2005 (Annex I), which is an integral part of this Charter;

Bearing in mind that the purposes of the Partnership are consistent with the principles of the United Nations Framework Convention on Climate Change and other relevant international instruments, and are intended to complement but not replace the Kyoto Protocol;

Decide to create the Asia-Pacific Partnership on Clean Development and Climate (referred to as the "Partnership") and set forth the following non-legally binding Charter for the Partnership. This Partnership will serve as a framework for supporting agile, constructive, and productive international cooperation among the Partners to meet our development, energy, environment, and climate change objectives.

1. Shared Vision

The Partners have come together voluntarily to advance clean development and climate objectives, recognizing that development and poverty eradication are urgent and overriding goals internationally. By building on the foundation of existing bilateral and multilateral initiatives, the Partners will enhance cooperation to meet both our increased energy needs and associated challenges, including those related to air pollution, energy security, and greenhouse gas intensities, in accordance with national circumstances. The Partners recognize that national efforts will also be important in meeting the Partnership's shared vision.

2. Purposes

The purposes of the Partnership are to:

- Create a voluntary, non-legally binding framework for international cooperation to facilitate the development, diffusion, deployment, and transfer of existing, emerging and longer term cost-effective, cleaner, more efficient technologies and practices among the Partners through concrete and substantial cooperation so as to achieve practical results;
- Promote and create enabling environments to assist in such efforts;
- Facilitate attainment of our respective national pollution reduction, energy security and climate change objectives; and
- Provide a forum for exploring the Partners' respective policy approaches relevant to addressing interlinked development, energy, environment, and climate change issues within the context of clean development goals, and for sharing experiences in developing and implementing respective national development and energy strategies.

3. Functions

Through this Partnership, the Partners are to cooperate to:

- Exchange information on Partners' respective policy approaches relevant to addressing interlinked development, energy, environment, and climate change issues within the context of clean development, including any gaps and overlaps in national policy approaches, as well as other areas of mutual interest;

- Share experiences and exchange information about developing and implementing national clean development strategies and efforts to reduce greenhouse gas intensities;
- Identify, assess, and address barriers to the promotion and creation of an enabling environment for development, diffusion, deployment, and transfer of existing, emerging and longer term cost-effective, cleaner, more efficient, and transformational technologies and practices in accordance with the Partners' priorities;
- Identify and implement bilateral and multilateral cooperative activities among Partners for the development, deployment, diffusion, and transfer of existing, emerging and longer term cost-effective, cleaner, more efficient, and transformational technologies, in accordance with the Partners' priorities;
- Facilitate collaboration among existing bilateral and multilateral initiatives and promote information-sharing on climate-related technologies of respective Partners;
- Incorporate human and institutional capacity-building elements, as appropriate, into activities as a means to strengthen cooperative efforts;
- Engage the private sector as an integral part of the cooperative activities of the Partnership, as well as development banks, research institutions, and other relevant governmental, intergovernmental, and non-governmental organizations, as appropriate;
- Develop and implement work programs decided by the Partners; and
- Assess regularly the progress of the Partnership to ensure its effectiveness.

Each Partner will undertake activities contemplated by this Charter in accordance with the laws, regulations, and policies under which it operates and applicable international instruments to which it is a party.

4. Organization

A Policy and Implementation Committee and an Administrative Support Group will be formed to facilitate implementation of the Partnership.

The Policy and Implementation Committee will govern the overall framework, policies, and procedures of the Partnership, periodically review progress of collaboration, and provide direction to the Administrative Support Group. It will be responsible for management of the implementation of the cooperative activities of the Partnership, and for engaging representatives of the private sector, as well as representatives of development banks, research institutions, and other relevant governmental, intergovernmental, and non-governmental organizations, as appropriate. It will undertake activities in the promotion and creation of enabling environments within Partners and in support of Partners' efforts to meet relevant national-level clean development objectives. The Policy and Implementation Committee may form appropriate task forces and other subgroups to assist it in its work. The Policy and Implementation Committee should meet as often as is determined necessary by its members to accomplish its work, and may focus its agenda on policy issues or technical issues, or both, as appropriate. Policy and Implementation Committee decisions are to be made by consensus of the Partners on the Committee.

The Administrative Support Group, which serves as the principal coordinator of the Partnership's communications and activities, will be responsible for: (1) organizing meetings of the Partnership; (2) arranging special activities, such as teleconferences and workshops; (3) coordinating and communicating information regarding actions of the Partnership; (4) serving as a clearinghouse of information regarding the Partnership; (5) maintaining procedures and responsibilities for key functions that are approved by the Policy and Implementation Committee; and (6) performing such other tasks as the Policy and Implementation Committee directs. The Administrative Support Group's function will be administrative in nature, and will not include matters of substance except as specifically instructed by the Policy and Implementation Committee.

The Policy and Implementation Committee comprises representatives from Partners. Each Partner included in Annex II may designate up to three representatives to meetings of the Policy and Implementation Committee.

The Policy and Implementation Committee may, at its discretion, permit other experts to attend its meetings.

The United States Government is to serve initially as the Partnership's Administrative Support Group. This arrangement will be reviewed at two year intervals and may be changed by decision of the Policy and Implementation Committee. Each Partner will designate an administrative liaison to serve as its principal point of contact for the Administrative Support Group.

The Administrative Support Group may, as required, utilize the services of personnel employed by the Partners and made available to the Administrative Support Group. Unless otherwise determined by the Partners, such personnel are to be remunerated by their respective employers and remain subject to their employers' conditions of employment.

Each Partner will individually determine the nature of its participation in Partnership activities.

5. Funding

Participation in the Partnership is on a voluntary basis. Each Partner may, at its discretion, contribute funds, personnel, and other resources to the Partnership subject to the laws, regulations, and policies of the Partner. Any costs arising from the activities contemplated in this Charter are to be borne by the Partner that incurs them, unless other arrangements are made.

6. Intellectual Property

All matters related to intellectual property and the treatment thereof arising from cooperative activities of the Partnership are to be addressed on a case-by-case basis within the specific context in which they appear, bearing in mind the purposes of the Partnership.

7. Amendments

The Policy and Implementation Committee may amend this Charter and its Annex II at any time by consensus of the Partners on the Committee.

8. Term of Charter

Cooperation under this Charter will commence on 12 January 2006. Any Partner may terminate its membership upon written notice 90 days prior to the anticipated termination.

We will develop a non-binding compact in which the elements of this shared vision, as well as the ways and means to implement it, will be further defined. In particular, we will consider establishing a framework for the partnership, including institutional and financial arrangements and ways to include other interested and like-minded countries.

The partnership will also help the partners build human and institutional capacity to strengthen cooperative efforts, and will seek opportunities to engage the private sector. We will review the partnership on a regular basis to ensure its effectiveness.

The partnership will be consistent with and contribute to our efforts under the UNFCCC and will complement, but not replace, the Kyoto Protocol.

Annex I

Vision Statement of Australia, China, India, Japan, the Republic of Korea, and the United States of America for a New Asia-Pacific Partnership on Clean Development and Climate, 28 July 2005.

Development and poverty eradication are urgent and overriding goals internationally. The World Summit on Sustainable Development made clear the need for increased access to affordable, reliable and cleaner energy and the international community agreed in the Delhi Declaration on Climate Change and Sustainable Development on the importance of the development agenda in considering any climate change approach.

We each have different natural resource endowments, and sustainable development and energy strategies, but we are already working together and will continue to work to achieve common goals. By building on the foundation of existing bilateral and multilateral initiatives, we will enhance cooperation to meet both our increased energy needs and associated challenges, including those related to air pollution, energy security, and greenhouse gas intensities.

To this end, we will work together, in accordance with our respective national circumstances, to create a new partnership to develop, deploy and transfer cleaner, more efficient technologies and to meet national pollution reduction, energy security and climate change concerns, consistent with the principles of the UN Framework Convention on Climate Change (UNFCCC).

The partnership will collaborate to promote and create an enabling environment for the development, diffusion, deployment and transfer of existing and emerging cost-effective, cleaner technologies and practices, through concrete and substantial cooperation so as to achieve practical results. Areas for collaboration may include, but not be limited to: Energy efficiency, clean coal, integrated gasification combined cycle, liquefied natural gas, carbon capture and storage, combined heat and power, methane capture and use, civilian nuclear power, geothermal, rural/village energy systems, advanced transportation, building and home construction and operation, bioenergy, agriculture and forestry, hydropower, wind power, solar power, and other renewables.

The partnership will also cooperate on the development, diffusion, deployment and transfer of longer-term transformational energy technologies that will promote economic growth while enabling significant reductions in greenhouse gas intensities. Areas for mid- to long-term collaboration may include, but not be limited to: hydrogen, nanotechnologies, advanced biotechnologies, next-generation nuclear fission, and fusion energy.

The partnership will share experiences in developing and implementing our national sustainable development and energy strategies, and explore opportunities to reduce the greenhouse gas intensities of our economies.

Annex II

Australia
China
India
Japan
Republic of Korea
United States of America

From http://www.dfat.gov.au/environment/climate/ap6/charter.html — 09/05/06

Appendix 3: Asia — Pacific Partnership on Clean Development and Climate

Action Plan Guidelines

The purpose of these guidelines is to promote consistency between Action Plans, to the extent practicable. It is recognized that while the specific circumstances relevant to the themes addressed by each Task Force differ, and that each Task Force will need some latitude in determining the content of its Action Plan, there should also be some common elements that apply across the range of Task Forces. Action Plans will be public documents.

1. Introduction

1.1. These guidelines have been prepared by the Policy and Implementation Committee (the 'Committee') of the Asia-Pacific Partnership on Clean Development and Climate (the 'Partnership'), to provide guidance for all Partnership Task Forces in the preparation of their Action Plans.

1.2. These guidelines should be read in conjunction with the Partnership:

- Charter, January 2006.
- Work Plan, January 2006.

- Communiqué, Inaugural Ministerial Meeting, January 2006.
- Task Force Guidelines, April 2006.

2. Action Plan Scope

2.1. The Communiqué of the inaugural Partnership ministerial meeting and the Partnership Work Plan established a framework for the initial eight Task Forces, including the generic scope and nature of Task Force activities and specific objectives for each Task Force. Action Plans should reflect these ministerial decisions.

2.2. Task Force Action Plans are intended to serve as a blueprint for cooperation among the Partners within the thematic or sectoral area identified. For this purpose, each Action Plan should, to the extent practicable, provide a strategic framework for identifying opportunities and implementing priority actions to advance clean development and climate goals. Each Action Plan should seek to address the elements in 2.3 and 2.4 below, recognizing differences both in the specific circumstances of each theme and the national circumstances of each Partner.

2.3. For the purpose of providing strategic context, Action Plans should, to the extent practicable, include a review of the current status of the relevant sector or theme with regard to clean development and climate, including where applicable, information on current data (e.g. energy intensity and emissions data), information on the current state of the technology in terms of cost, performance, market share and barriers, as well as relevant knowledge relating to good practice. This review need not be exhaustive, but reflect readily accessible information and be sufficient to articulate the basis for priority setting within the Partnership over time. More rigorous reviews may be undertaken following approval of the Action Plan.

2.4. Where possible, Action Plans should set out realistic and ambitious goals relating to the thematic area. Goals may take a number of different forms, and can either be aggregated or distinct to specific projects and activities, as the Task Force considers appropriate. To the extent possible, goals should be results-oriented and measurement systems should be aligned, so that progress toward achieving the goals can be gauged. For this purpose, Task Forces should consider using relevant metrics to measure progress in energy efficiency, air pollution, greenhouse gas emissions, or other relevant criteria (e.g. recycling), either in the context of specific goals or as an aggregate measure of progress within the areas under its scope.

3. Projects and Activities

3.1. Each Task Force is expected to build on the wide range of actions already in place in Partner countries through national programs and other international cooperative arrangements and, where appropriate, seek to leverage existing initiatives to ensure maximum return on resources. Projects and activities advancing technology and improving best practices in each Partner country may also be linked, where useful, with others in the region, allowing the sharing of lessons across the Partnership.

3.2. The Action Plan should identify specific projects and activities that will further Task Force goals, and identify or propose the entities that could undertake these projects and activities. Projects and activities may include technology-based research, pilot projects, demonstration and deployment activities, skills enhancement and exchange, commercial and information exchanges (workshops and dialogues), and measures to disseminate best practices. It is expected that there will be a range of government and industry entities involved in implementing specific projects and activities.

3.3. With a view to advancing the goals of the Partnership on a significant scale, Action Plans should seek to include both discrete projects and activities, and portfolios of investment opportunities to advance specific clean development and climate goals where possible. The goal of the Action Plans should be to facilitate and engender self-perpetuating action.

3.4. We foresee a significant role for private sector and financial entities, as well as Partner governments, in implementation. Action Plans will be most successful where they catalyse actions by the private sector and leverage resources available to the Partnership. In this regard, focus should be given to identifying opportunities to enhance the enabling environment for technology research, development, deployment, dissemination and transfer, including identifying policy opportunities to foster or remove barriers to investment.

3.5. The Action Plans should identify, and promote ways and means to deploy, the best of current commercially viable technologies and practices, as well as the uptake of emerging cleaner and more efficient technologies.

3.6. Action Plans should address both immediate and medium-term specific actions. To the extent possible, early milestones and 'next steps' should be identified for each action.

3.7. Task Forces should focus on those projects and activities that will most efficiently and realistically advance the goals of the Partnership, within available resources.

3.8. Task Forces should identify who is to be responsible for project and activity implementation. This may involve entities associated with individual Partners, or it may involve the Task Force as a whole. Task Forces may form sub-structures such as working groups or project teams to engage in implementation activities. Task Forces should consider and identify resource needs and avenues for implementing projects and activities.

3.9. At their discretion and consistent with guidance issued by the Committee, Task Forces may propose 'flagship' projects and activities as part of their Action Plans. Proposed flagship projects or activities will only be designated as such with the approval of the Committee.

4. Action Plan Documentation

4.1. Task Forces should seek to finalise their Action Plans by mid-2006, and submit them for approval by the Committee. The Committee may ask the Task Force to revise the Action Plan if it so decides.

4.2. The Committee and Partnership Administrative Support Group will arrange for draft Action Plans to be shared among Task Forces, and for any meetings required to coordinate the initial planning work and to avoid unnecessary duplication. Task Force Chairs should communicate with each other to address any boundary issues during development of the Action Plans.

4.3. A sample Action Plan outline is provided at Attachment A. This may be used as a template, or be modified, where appropriate.

4.4. As there is a relatively short time frame for drafting the Action Plans, it is understood that in some cases proposed projects and activities will require further investigation and development. In these cases, the steps for amending the Action Plan should be identified.

4.5. The Committee expects that more detailed planning, project/activity management and risk management documentation will be prepared by the responsible entities to assist in implementation. The initial task, however, is for Task Forces to produce a robust and cohesive set of Action Plan.

4.6. Project and activity plans may evolve and change over time, particularly where the initial stages involve some form of review or research activity where the outcome is unpredictable.

For some larger medium-term projects and activities, an initial milestone may be the development of a more detailed plan following reviews or workshops. Version 1.0 18 April 2006.

Attachment A

Action Plan Outline

The Action Plan Outline should be reasonably brief, and comprise three parts:

1. Sector Review

This is a brief review of the current status, opportunities and key barriers for the sector. The text under each Task Force in the January

2006 Work Plan should be used as the starting point for drafting this section.

2. Objectives

This is a high-level statement of the objectives for the Task Force including 'ambitious and realistic' goals. More specific objectives and goals may be included for major initiatives, projects or activities. This section should also identify the priority areas of focus (types of projects and activities) for the Task Force. The objectives (dot points) listed under each Task Force in the January 2006 Work Plan should be used as the starting point for drafting this section.

3. Partnership Actions

This is a summary of the specific projects and activities that have been identified to contribute to the Action Plan goals. As the time frame for conclusion is mid-2006, it is understood that in many cases projects and activities will comprise proposed ideas for follow-up and investigation, recognising resource considerations.

The Action Plan should include a paragraph or two on each action or set of actions and how they will contribute to the objectives of the thematic area. For each project and activity, to the extent practicable, the Action Plan should briefly identify the entities involved, any specific goals, next steps, specific milestones, the individuals responsible for keeping the Partnership informed of progress, and possible resources.

Where appropriate, milestones may be presented in a timetable format, as illustrated below, or other suitable format. Major milestones may be outputs such as technical reports, training programs, or the opening of new facilities.

Action plan milestones

2006 2007 2008 etc
Project/Activity Major milestones
Project/Activity Major milestones Major milestones
Project/Activity Major milestones Major milestones Major milestones etc.

Where appropriate, one-page project/activity plan summaries may be included, using the headings in the table below. Where the detail has not yet been determined, the timeframe and process for finalisation should be identified.

Project/Activity plan

(1) *Project*

- Short project/activity title (such as 'Next generation widget technology', 'Capacity building in widget sector').
- Short (one paragraph) description of the project/activity (what is this project/activity about, and why are we doing it?).

(2) *Participation*

- Management (who is the Partner, business or agency leading or managing this project/activity?).
- Participation (which Partners, businesses or agencies will be involved?).

(3) *Objectives*

- Goals or outcomes sought (what will happen or change as a result of this project/activity?).
- Performance indicators (what are the ambitious/realistic indicators of success?).

(4) *Milestones*

- List of key intermediate outputs or deliverables, and their target date. What will this project/activity produce, and when (reports, conferences, openings, training programs, etc; month, quarter or year, as appropriate?).

(5) *Location*

- List of the project/activity locations, if relevant.

(6) *Resources*

- What are expected or proposed resource needs and avenues, including financial and in-kind? http://www.dfat.gov.au/environment/climate/ap6/ — accessed 09/05/06.

Appendix 4: Asia-Pacific Partnership on Clean Development and Climate Task Force Guidelines

This document is intended for guidance only. Task Forces may develop and agree their own operating procedures, as appropriate to their circumstances, except where a statement is specifically identified as a ***policy requirement***.

1. Introduction

1.1. These guidelines have been prepared by the Policy and Implementation Committee (the 'Committee') of the Asia-Pacific Partnership on Clean Development and Climate (the 'Partnership') to provide guidance for all Partnership Task Forces.

1.2. These guidelines should be read in conjunction with the Partnership:

- Charter, January 2006.
- Work Plan, January 2006.
- Communiqué, Inaugural Ministerial Meeting, January 2006.
- Action Plan Guidelines, April 2006.

2. Establishment and Purpose

2.1. The Partnership Charter established the Committee to govern the overall framework, policies, and procedures of the Partnership, and to periodically review progress. The Charter authorises the Committee to form appropriate Task Forces and other subgroups to assist it in its work.

2.2. It was decided to establish the initial eight Task Forces at the inaugural Partnership Ministerial Meeting in Sydney, Australia, in January 2006.

2.3. Task Forces are not permanent Partnership structures, existing only as long as is needed to achieve their objectives. They are created to deal with issues identified in the Work Plan and develop agreed Action Plans. Task Forces will propose modes for implementation and monitoring of the Action Plans. Task Forces should monitor progress made towards implementation of the agreed Action Plans, and may recommend further roles to be approved by the Committee.

2.4. The Committee will decide when to establish, disband, or hold in abeyance, a Task Force, taking into account any recommendation of existing Task Forces.

3. Leadership and Membership

3.1. Each Task Force will have a Chair and may have a Co-Chair. The initial Partner country responsibilities for Task Force Chairs and Co-Chairs were agreed at the inaugural ministerial meeting. Each Chair will be occupied by a senior government official named by the relevant Partner country. Co-Chairs may be a senior government or industry representative, named by the relevant Partner country.

3.2 Task Force Chairs are responsible for:

- organizing and chairing Task Force meetings;
- managing the development of Task Force Action Plans (by mid 2006);
- managing any Task Force role in the implementation of Action Plans agreed by the Committee, and monitoring implementation progress;
- resourcing and managing their Task Force secretariat.

3.3. Task Force Co-Chairs will support the Chair in the above functions, and by agreement between the Chair and Co-Chair, may take specific responsibility for particular tasks.

3.4. Membership of Task Forces will be kept to a manageable number to maximise efficiency.

Each Partner may nominate up to four members to each Task Force, comprising a mix of government and industry representatives as determined by the Partner. However the Committee recognises that additional members may be appropriate for some Task Forces and Partners. Alternates for each Task Force member may also be designated. Chair and secretariat positions may be additional to the four member positions.

3.5. Industry members shall be those engaging in activities related to the theme of the Task Force to which they are named by the relevant Partnership country. Industry members shall, to the extent possible, take into account the views of their industries in their countries. Partners are responsible for establishing appropriate consultation mechanisms, if deemed necessary, within their countries.

4. Meetings

4.1. Task Forces may determine their own meeting and communication arrangements, including face-to-face meetings, telephone and video conferences, and e-mail groups. The location and frequency of meetings will be determined by the Chair, in consultation with the Co-Chair and members. Task Force meetings may be held in conjunction with Committee meetings or other relevant international gatherings.

4.2. Members of the Committee may be observers on the Task Forces. Partner countries may send additional observers to attend Task Force meetings, subject to decision by the Task Force. Experts from Partner countries may be invited by the Task Force to participate in Task Force discussions. Task Forces may develop procedures, subject to approval by the Committee, for managing non-member participation, including designation of open and closed sessions.

4.3. Agendas and papers for Task Force meetings should be circulated to all members at least three weeks in advance of meetings, and records of decisions and actions should be circulated as soon after each meeting as practicable. Copies of agendas, papers, and records should be sent to the Partnership Administrative Support Group for distribution to members of the Committee and, as agreed, posting on the Partnership web site (***policy requirement***).

4.4. All Task Force decisions or recommendations are to be made by consensus. Matters that cannot be resolved within Task Forces may, at the request of the Task Force Chair, be referred to the Committee (***policy requirement***).

5. Action Plans and Reporting

5.1. Each Task Force should develop an Action Plan in accordance with the Partnership Action Plan Guidelines. Action Plans, and any significant variations, will be approved by the Committee (***policy requirement***).

5.2. Action Plans may include a range of projects and activities over several years, to be undertaken by Partners or other entities, as agreed Task Forces should, where possible, identify appropriate entities and means for implementing Action Plan projects and activities.

5.3. Task Forces should report periodically to the Committee on implementation of Action Plans (***policy requirement***).

6. Evaluation and Communication

6.1. The Committee will provide guidance to Task Forces on the preparation of evaluation plans for Task Force projects and activities.

6.2. The Committee will provide guidance to Task Forces on the public communication of Task Force activities and products.

7. Funding and Intellectual Property

7.1. The Partnership Charter established the following funding arrangements (***policy requirement***):

- all costs for Task Force participation should be met by the Partner that incurs them, unless otherwise agreed;
- Partners may, at their discretion, contribute funds, personnel and other resources to Partnership activities, subject to the laws, regulations, and policies of the Partner.

7.2. The direct cost for participation in Task Force meetings, projects and activities is the responsibility of each Partner, unless otherwise agreed by the Task Force (***policy requirement***).

7.3. The direct cost of hosting Task Force meetings will be borne by the host country, unless otherwise agreed by the Task Force (***policy requirement***).

7.4. The Partnership Charter established the following Intellectual Property arrangement (***policy requirement***):

- All matters relating to intellectual property and the treatment thereof arising from cooperative activities of the Partnership are to be addressed on a case-by-case basis within the specific context in which they appear, bearing in mind the purposes of the Partnership. Version 1.0 18 April 2006, http://www.dfat.gov.au/environment/climate/ap6/ — accessed 09/05/06.

CHAPTER 16

PROTECTING FORESTS TO MITIGATE GLOBAL CLIMATE CHANGE

Charlotte Streck*

Climate Focus, Minervahuis III, Rodezand 34, 3011 AN Rotterdam, NL, c.streck@climatefocus.com

Tropical deforestation is a major contributor to global climate change and is responsible for about 20% of global greenhouse gas emissions. Both the UN Framework Convention on Climate Change and the Kyoto Protocol list general obligations relating to land use, land use change and forestry, or "LULUCF". However, these rules are vague and have not triggered any significant reduction of LULUCF emissions. Forest and biodiversity conservation are intrinsically linked to climate change mitigation and adaptation. It is therefore necessary that a post-Kyoto regime includes a comprehensive carbon accounting mechanism that provides the necessary incentive framework for conserving not only temperate and boreal, but — and in particular — tropical rainforests. By assigning value to standing forests, carefully designed market mechanisms provide the most promising option to mobilize finances for conservation. Market mechanisms and emission trading schemes assigning value to tropical forests have to be complemented by measures that strengthen forest governance and law enforcement and are integrated in a robust international and

* This paper was initially presented at: *Forest governance and trade: exploring options.* 24 January 2007. Chatham House, UK. Updated in January 2008 for the Asia-Pacific Centre for Environmental Law, Faculty of Law, National University of Singapore.

national framework. Safeguards and independent control mechanisms must be put in place to avoid perverse incentives, "paper tons", and corrupt policies.

1. Introduction

Forests are our most important terrestrial storehouses of carbon and play an important role in controlling our climate. Yet, in many parts of the world, forests are degraded and destroyed to expand agricultural lands, gain timber or clear space for infrastructure or mining activities. Decreased harvesting and increased regeneration have made forests in temperate zones act as carbon sinks by sequestering more carbon from the atmosphere than they emit.[1] This effect is countered, however, by the rapid loss of tropical forests.[2] Tropical deforestation has severe consequences for loss of biodiversity, flooding, soil degradation and threats to the livelihoods and the cultural integrity of forest-dependent communities. It is also a major contributor to global climate change. On a worldwide scale, global change pressures (climate change, land-use practices and changes in atmospheric chemistry) are increasingly affecting the supply of goods and services from forests.

The international climate regime accounts for the role that forests play in influencing our climate. By storing carbon, forests

---

[1] Schulze, E.D., Lloyd, J., Kelliher, F.M., Wirth, C., Rebmann, C., Luhker, B., Mund, M., Knohl, A., Milyukova, I.M., Schulze, W., Ziegler, W., Varlagin, A.B., Sogachev, A.F., Valentini, R., Dore, S., Grigoriev, S., Kolle, O., Panfyorov, M.I., Tchebakova, N., and Vygodskaya, N.N. (1999), *Productivity of Forests in the Eurosiberian Boreal Region and their Potential to Act as a Carbon Sink — A Synthesis*, Global Change Biology, 5(6), 703–722.

[2] Houghton, R.A., Skole, D.L., Nobre, C.A., Hackler, J.L., Lawrence, K.T., and Chomentowski, W.H. (2000), *Annual Fluxes or Carbon From Deforestation and Regrowth in the Brazilian Amazon*, Nature, 403(6767), 301–304. It is important to note however that deforestation rates have been slowing in the last decade. However, Soares-Filho *et al.* describe a potential increment of deforestation in Amazonia for the next decades, B. Soares-Filho *et al.* (2006), *Modeling Conservation in the Amazon Basin,* Nature 440, pp. 520–523.

can help in the mitigation of climate change; by releasing carbon, they contribute to the increasing concentrations of greenhouse gases in the atmosphere. Both the UN Framework Convention on Climate Change ("UNFCCC") and the Kyoto Protocol list general obligations relating to land use, land use change and forestry, or "LULUCF" in the lingo of climate change negotiators. Parties to the treaties are mandated to promote sustainable forest management, forest conservation and the promotion of forests as sinks. However, these rules are neither binding nor sufficiently concrete to have triggered any significant forest conservation measures or policies. The significance of carbon storage and release in biological systems makes it therefore mandatory that a post-Kyoto climate regime includes incentives for forest conservation.

The present paper will look at the relationship between forest degradation, deforestation and climate change. It begins with the summary of the impact of climate change on forestry ecosystems on one hand, and tropical deforestation and its impact on climate change on the other. Moving from there to institutional and regulatory issues, the paper will then describe the current climate change regime and how it deals with the "LULUCF" issue, followed by a view into the future and how the incentive framework for conservation and afforestation should be set in a post-Kyoto regime. Finally, we will look at the connection between bioenergy and forest conservation. The paper concludes with an outlook of the possible impacts of an international climate regime on changes in land use patterns.

2. The Impact of Climate Change on Forests

Climate change can have a variety of impacts on forest ecosystems: Owing to the multiplicity of geographical, social, and economic conditions, change can bring positive effects to one area and community, while bringing negative effects to another. In addition, the impact of a changing environment on the individual plants needs to be distinguished from the impact on the surrounding ecosystem.

Forest ecosystems are affected by climate change in the following ways:[3]

- Hydrology and water resources: Change in climatic patterns result in modified rainfall patterns. In many regions, groundwater recharge will be affected. Deforestation and the degradation of forestry systems will increase the vulnerability of the water sector. On the other hand, reduced water supplies will affect forest ecosystems.
- Temperature increase may lead to increased photosynthetic activity of plants. It will also cause high levels of transpiration, leading to additional water loss.
- The impact of increased concentration of carbon dioxide (CO_2) in the atmosphere depends on the plant species and the surrounding ecosystem. Increased CO_2 concentration in the atmosphere has often had the fertilizing effect of stimulating photosynthetic activity. High CO_2 concentrations can also increase the water efficiency of plants. On the negative side, CO_2 concentrations may affect the plant tissue with potential detrimental effects on the metabolism of the affected plants.

Ecosystems are highly vulnerable to climate change. The combination of temperature and rainfall patterns characterizes an ecosystem in a certain area. A change in these variables will cause changes to that ecosystem as it adapts to the new conditions. The impact on the

[3] *See*, IPPC (2000), Land Use, Land-Use Change, and Forestry: IPPC Special Report (eds. Watson, R.T., Noble, I.R., Bolin, B., Ravindranath, N.H. Verardo, D.J., Dokken, D.J.), Cambridge University Press, Cambridge, UK; Schimmel, D. S. *et al.* (2001), Recent patterns and mechanism of carbon exchange by terrestrial ecosystems. Nature 414: 169–172; Robledo, C. and Forner, C. (2005), Adaptation of forest ecosystems and the forest sector to climate change, Forests and Climate Change Working Paper 2, FAO, Rome; Körner C. and Armone, J. (1992), Responses to elevated carbon dioxide in artificial tropical ecosystems, Science 257: 1672–1675; Kellomäki, S. and Wang, K.-Y. (2000), Modelling and measuring transpiration of Scots pine with increased temperature and carbon dioxide enrichment, Annals of Botany 85: 263–278.

competitiveness of the various species will lead to a change in the composition of species within a given ecosystem. Stress caused by a change in the conditions of the ecosystem may also increase its vulnerability to pests and fires.

3. Tropical Deforestation and Climate Change

Forestry and climate change have a multilayered, complex relationship. While changes in climatic patterns can increase or decrease the capacity of an ecosystem to store carbon, that same ecosystem can act either as emitter or as sink of greenhouse gases ("GHG"). The loss of biomass stored in forests counts among the factors that cause climate change. The sequestration of CO_2 in forests, on the other hand, addresses climate change by storing carbon.

Carbon sinks contribute to off-setting the effects of anthropogenic greenhouse gas emissions. Forest ecosystems contain the majority (approximately 60%) of the carbon stored in terrestrial ecosystems[4] and account for 90% of the annual carbon flux between the atmosphere and the Earth's land surface.[5] In many temperate ecosystems, forests act as net carbon sinks. Tropical deforestation however is responsible for up to 25% of the total human-caused greenhouse gas emissions each year.[6] Deforestation is also one of the main reasons for the rapid extinction of forest species.

---

[4] IPCC (2000), Land Use, Land-Use Change, and Forestry. A Special Report of the IPCC, (eds. Watson, R.T., Noble, I.R., Bolin, B., Ravindranath, N.H. Verardo, D.J., Dokken, D.J.), Cambridge University Press, Cambridge, United Kingdom and New York, NY, USA.

[5] Winjum, J.K., Dixon, R.K., and Schroeder, P.E. (1993), Forest management and carbon storage: An analysis of 12 key forest nations, Water, Air, & Soil Pollution, 70(1–4), 239–257.

[6] Using 1990 as the baseline and taking into account emissions of carbon dioxide, methane, nitrous oxide, and other chemically reactive gases that result from deforestation and subsequent uses of the land. See Houghton, R.A. (2005), Tropical Deforestation as a Source of Greenhouse Gas Emissions, Moutinho, P., and Schwartzman, S. (eds.), Tropical Deforestation and Climate Change, IPAM, Instituto de Pesquisa Ambiental da Amazonia, Environmental Defense, Belen, Brazil, Washington DC, USA.

Deforestation results from a variety of cultural and economic factors.[7] Demographic pressure, infrastructure and agricultural expansion are among the main drivers of conversion of forests to other land uses.[8] Another reason for the loss of forests is the unsustainable extraction of forestry products. This includes both legal and illegal logging activities as well as the harvesting of fuel wood and other forestry products. Where forests are poorly managed and timber and wood products are extracted without restoring the loss, forests degrade until they are lost.[9]

Deforestation is responsible for 90% of all greenhouse gas emissions related to LULUCF since 1850.[10] The highest deforestation rate can be observed in tropical Asia, followed by Africa and South America. Forests are cleared at the fastest rates in Brazil and Indonesia.[11] With current deforestation rates, Indonesia is expected to lose its primary forest by 2012. On the other hand, the forest coverage is growing in the Caribbean. Additional substantial GHG (methane) emissions result from the drainage of peatlands for palm oil and timber plantations.[12]

---

[7] Lovera, S. (2003), Going to the roots: addressing the underlying causes of deforestation and forest degradation, see: http://www.fao.org/DOCREP/ARTICLE/WFC/XII/MS12B-E.HTM; Lanly, J.-P. (2003); Deforestation and forest degradation factors, http://www.fao.org/DOCREP/ARTICLE/WFC/XII/MS12A-E.HTM; both accessed on 10 January 2007.

[8] Kaimowitz and Arild show that agricultural expansion (soy, beef) is one of the main causes of deforestation in Latin America. Kaimowitz, D. and Arild, A. (1998), Economic models of tropical deforestation: A review. Center for International Forestry Research, Bogor, Indonesia.

[9] According to Arild and Kaimowitz, cases where logging conduces to deforestation primarily occur in Asia, whereas unsustainable exploitation of forests for wood extraction primarily occurs in Africa. See footnote 5.

[10] Houghton, J., Ding, Y., Griggs, D., Noguer, M., Van der Linden, P., Dai, X., Maskell, K., and Johnson, C. (2001), Climate Change 2001: The scientific basis; summary for policy makers, Contribution of Working Group I to the Third Assessment Report of the Intergovernmental Panel on Climate Change (IPCC) Cambridge, CUP.

[11] Forest and Agricultural Organization, FAO (2001), Forest Resources Assessment 2000, Forestry Paper 140, FAO, Rome, Italy.

[12] Hooijer, A., Silvius, M., Wöste, H., Page, S., Peat-CO_2-Assesment of CO_2 emissions from drained peatlands in SE Asia, Delfts Hydraulic Report, Q3943 (2006), Delft, Netherlands.

Tropical forests account for slightly less than half of the world's forests. Yet, tropical forests store as much carbon in their soils and above-ground biomass as boreal and temperate forests combined. Trees in the tropics store, on average, about 50% more carbon per hectare than trees outside the tropics.[13] While temperate forests hold more carbon stored in the soil, tropical forest concentrate the carbon in the trees. Because only a fraction of soil carbon is lost when trees are felled, deforestation in tropical countries leads to a greater release of carbon than the equivalent deforestation in temperate zones.

Measures to manage and protect forests offer significant climate change mitigation potential:

(1) Biomass can store excess carbon. Afforestation, upgrading of ecosystems, regeneration, and sustainable forestry management leads to an absorption of CO_2 from the atmosphere. Immature forests act as carbon sinks even without human interference.
(2) Forests and terrestrial ecosystems need to be conserved and protected to reduce the GHG emissions from the release of CO_2 previously stored in terrestrial biomass (avoidance of further deforestation).
(3) The production of biomass can act as an energy source and biological products can substitute materials requiring energy-intensive production (e.g. aluminium and concrete). Sustainably harvested fuel wood is emissions neutral because the carbon released through burning the wood is compensated by an equivalent amount absorbed by forest growth.

However, accumulation of carbon should not be the only objective in forest management. Looking at forests exclusively from the carbon perspective does not do justice to the role of forests. Looking at them in this way is no different to looking at them from the

[13] Houghton, R.A. (2005), Tropical deforestation as a source of greenhouse gas emissions, Moutino, P., Schwartzman, S. (eds.), Tropical Deforestation and Climate Change, IPAM, Instituto de Pesquisa Ambiental da Amazonia, Environmental Defense, Belen, Brazil, Washington DC, USA.

perspective of the value of the timber. Managing forests and allowing the increase of forest area, forest age and tree size can have beneficial impact on the forest's biodiversity and ecosystem function. Additionally, natural forest fire regimes should not be altered. Even where activities are being undertaken to protect the carbon stored in a forest, these activities should meet a set of defined criteria to ensure that they are environmentally and socially robust for the long term.

4. LULUCF in the UNFCCC and the Kyoto Protocol

The UNFCCC and the Kyoto Protocol both acknowledge the vital role that forests play for the global climate. However, they fail to set an adequate incentive framework which would trigger significant afforestation on one hand, discourage deforestation and encourage conservation on the other.[14] During the negotiations leading up to and subsequent to the Kyoto Protocol, there was considerable concern that carbon off-sets may become void in cases where human action or natural events, such as wildfires, reversed the carbon benefits. This was called the permanence risk. The permanence risk is the main difference between emission removals generated by the LULUCF sector and emission reductions generated by the industrial sectors.

Parties decided that "direct human-induced", i.e. "net changes" in GHG and removals by sinks since 1990 may be used to meet part of the Parties emission commitments.[15] Countries have to (Article 3.3 of the Kyoto Protocol) or may (Article 3.4 of the Kyoto Protocol) account for the change in their carbon stock. Furthermore, Articles 6 and 12, which define the project-based mechanisms Joint Implementation ("JI") and the Clean Development Mechanism

[14] See for more detail: Streck, C. and Scholz, S. (2006), The role of forests in global climate change: whence we come and where we go, International Affairs 82(5): 861–879.

[15] Those 'net changes' must be "measured as verifiable changes in carbon stocks in each commitment period".

("CDM") refer directly (JI) or indirectly (CDM) to carbon sinks. With respect to JI, the Parties decided that the overall liability of a country to comply with its targeted amount of emissions would act as sufficient insurance against the potential loss of carbon due to 'non-permanence'.[16]

The issues are more complicated in the case of the CDM. It took almost six years of assessment and negotiation before a final agreement[17] on the definitions and modalities for the use of LULUCF projects in the CDM was reached at the 9th session of the Conference of the Parties to the UNFCCC in 2003 in Milan. LULUCF in the CDM was limited to afforestation and reforestation projects and the use of sinks projects in developing countries restricted to only one percent of a country's Kyoto obligations.[18] The protection of existing carbon pools (avoided deforestation) was finally declared to be non-eligible as a CDM project category.[19]

This means that while there is an incentive to restore and conserve temperate forests in industrialized countries, the most important source for emission from the LULUCF sector — tropical deforestation — is currently not covered under the Kyoto Protocol and there is no incentive for developing countries to protect their forest resources.

5. The Carbon Market and Payment for Environmental Services

Until today many of the benefits provided by forests are considered as global commons, freely available for everybody. Purification of air and water, stabilizing soil, creating the conditions for a rich biodiversity,

[16] These reductions in emissions and removals have to be both, 'additional' to what would otherwise have occurred without the project and 'supplemental' to domestic action aimed at meeting the Parties' respective reduction commitments.

[17] All rules governing A&R projects apply to only the first commitment period of the Kyoto Protocol and will have to be revisited before the second commitment period.

[18] *See*, Decision 17/CP.7 on 'Modalities and procedures for a clean development mechanism' FCCC/CP/2001/13/Add.2, paragraphs 7(a) and (b).

[19] The inclusion of this category in limited circumstances in small scale CDM projects is currently being debated amongst the Parties.

producing pharmaceutical substances and acting as carbon storehouses are considered to be unlimited and free services provided by forests. No ownership right and consequently no monetary value are being assigned to these services. The value of a forest is often captured alone in the value of those items that can be assigned and traded: the timber and the land on which the trees stand. This also means that those that control or have access to forests often have a greater incentive to clear the forest than to conserve it.

Applying the logic of the market, the whole range of forest services need to be priced in order for their true value to be acknowledged. Schemes that envisage the payment for "ecosystem services" try to address market failure by creating incentives to conserve, protect and restore forests. Assigning value to emission reductions or removals (carbon storage) by creating tradable carbon credits is the most developed and promising approach to assigning value to standing forests.

The carbon market relies on emission trading and the transfer of carbon credits. CDM and JI allow countries to invest in emissions-reducing or removing projects in countries where the abatement cost for emission reductions (or removals) are lower than in their own economies. In return for their payment, the investor or purchaser of carbon credits receives a right to the carbon credits generated by the project. These carbon credits can be used to meet compliance obligations under the international or national regulatory regimes. The carbon market created under the Kyoto Protocol and a number of regional and national emission trading schemes is worth billions of dollars each year.[20]

The blindness of the carbon market for any other forestry service beyond its function to store carbon has given reason for concern. Opponents of including carbon in the regime of the Kyoto Protocol and emission trading commonly refer to the ghostly vision of "carbon

[20] According to the World Bank, more than USD 30 billion have been traded in international carbon markets in 2005. Kapoor, C., Ambrosi, P., State and Trends of the Carbon Market 2007, IETA — The World Bank, Washington, DC. http://carbonfinance.org/docs/StateoftheCarbonMarket2007.pdf (Accessed 3 January 2008).

plantations"; monocultural lines of eucalyptus or pine trees that are biological and social deserts, bare of any biodiversity or community life. To counterbalance those fears, the negotiators of the Kyoto Protocol and various existing or emerging GHG trading schemes established safeguards that address these concerns by adopting a number of eligibility and approval criteria ranging from strict additionality testing, leakage management, environmental impact assessments, permanence strategies, and land eligibility criteria. While these safeguards indeed prevented any carbon plantations from being approved under the CDM or JI, they also made project development so costly that not only the bad but the good projects were also prevented from benefiting from carbon finance.

Because the Kyoto Protocol does not address forest conservation or the prevention of deforestation, tropical countries are restricted in their opportunities to benefit from the CDM. The Kyoto Protocol thus fails to set an incentive to protecting natural forests in developing countries. Most of the off-sets generated by LULUCF projects are therefore traded in the so-called voluntary or retail market. Companies invest in voluntary off-sets for marketing purposes, to meet certain Corporate Social Responsibility promises, or out of true environmental concern. More and more individuals seek to off-set their carbon "foot print".

6. Future Considerations of Forestry Under a Climate Regime

However, things have changed since 1997 when the Kyoto Protocol was negotiated. The issue of "Reducing Emissions from Deforestation and Degradation" ("REDD") has become a crucial point in the negotiations of a post-Kyoto climate regime and the 13th session of the Conference of the Parties to the UNFCCC has adopted a decision that leaves no doubt that REDD will be part of a comprehensive post-Kyoto deal. The decision is based on the acknowledgement of "the urgent need to take further meaningful action to reduce emissions from deforestation and degradation in developing countries" and encourages the contribution of capacity building funds and the development of demonstration activities. At the same time, it calls upon the subsidiary bodies under the Convention to evaluate positives incentives and approaches that reward any emission reduction by the avoidance of further deforestation. While action has to follow words and it is too early to celebrate success, the recent decision is a step into the right direction.

The long and often controversial history of the climate change negotiation shows that addressing the issue of deforestation has to take into consideration a number of important issues:

- The reluctance of developing countries to assume binding caps on emissions
- The right for a country to pursue its goals of sustainable development as it sees fit
- The consideration of forests as sovereign asset, and related problems to ownership and legal title of forests and tradable credits
- A science based approach that considers monitoring and methodological issues, permanence and liabilities
- Environmental integrity and effectiveness
- Avoidance of perverse incentives
- Economic efficiency.

Participation in emission trading should be an opportunity for developing countries rather than a constraint. Creating tradable carbon removal/reduction assets through voluntary participation in emission trading de-links the reward for achieving an environmental benefit from the obligation to achieve such benefit. By developing a scheme which relies on the creation of property rights associated with existing carbon stocks, developing countries can create additional income by protecting their forest resources.

In order to create an incentive framework which rewards activities in developing countries to increase biological carbon storage, an expansion of the CDM beyond afforestation and reforestation to include such activities as revegetation, ecological restoration, and improved forest and agricultural management is a necessary complement to including REDD into a future regime. Methodological insecurities that led to a limitation of the CDM to afforestation and reforestation in the earlier days have decreased. By only crediting narrowly-defined activities, the CDM is excluding many of the most important land use types, which have the potential to deliver significant GHG mitigation and ancillary benefits (including, conserving soil, restoring biodiversity habitat, protecting watersheds and improving water quality, creating sustainable livelihoods and increasing food production).

This is particularly important as sequestering carbon in afforestation and reforestation may result in a sub-optimal over-plantation of fast growing alien species with a potential negative impact on biodiversity: the realization of carbon plantations. Destruction of primary forests has historically been accompanied by an expansion of secondary and plantation forest.[21] In many parts of the tropics, extensive forest areas have been transformed into monoculture plantations. These plantations store carbon. However, they do not make up for the loss of biodiversity and forest ecosystem services.

The UNFCCC and the Kyoto Protocol do not advocate the use of an ecosystem approach which could be used to protect biodiversity needs. However, the Convention on Biological Diversity does provide relevant guidelines for the sustainable use of forests. To avoid the clearing of mature forests and their replacement them with younger fast-growing trees, the incentive to afforest needs to be complemented with a REDD incentive scheme.

While the design details are still under negotiation, it is important that any post-Kyoto LULUCF agreement takes into account the:

- Creation of a system that rewards:
  - — Decreasing deforestation
  - — Sustainable forest, land, and wetland management

---

[21] Smith, J., Sabogal, C., de Jong, W., Kaimowitz, D. (1998), Bosques Secundarios Como Recursos para el Desarrollo Rural y la Conservación Ambiental en los Trópicos de América Latina, Bogor, Indonesia, Center for International Forestry Research (CIFOR).

  — Restoring of forests
  — Sustainable use of biomass.[22]

- Establishing a reliable accounting system which includes the flux of biological carbon;[23]
- Promotion of sustainable development and an inclusive climate regime;
- Harmonization with the Convention on Biological Diversity and the Convention to Combat Desertification.

7. Managing Timber and Biomass

Forests are primarily managed for timber production. Unsustainably managed forests degrade and are eventually converted into other land uses. Legally or illegally, forests are primarily cleared for the extraction of timber and fuel wood. Logging does not necessarily trigger deforestation as logging companies are not seeking a change in the use of the land. However, wrong or weak regulatory frameworks, environmental neglect and economic considerations may facilitate not only the degradation of land but also land conversion as a consequence of logging.

Further improving institutional and governance structures and supporting developing countries in the enforcement of forest laws will be necessary to reduce illegal logging and establish robust and sustainable forest management frameworks. A carbon market mechanism could help to off-set the short time economic loss resulting from not cashing in the timber. Changing the harvesting patterns allowing the trees to grow longer and retaining older trees through successive harvests can significantly increase the carbon storage capacity of a

---

[22] Schlamadinger, B. *et al.* (2006), Options for including LULUCF activities in a post-2012 international climate agreement, A workshop report, pg. 10, http://www.joanneum.at/Carboinvent/post2012_/Bird/Schlamadinger_et_al_2005.pdf (Accessed 20 July 2006).

[23] Ward, M. (most recent version from February 2006), A Scenario For a Future Climate Change Regime For LULUCF. http://homepages.paradise.net.nz/murrayw3/documents/pdf/Future%20LULUCF%20scenario_May%202005.pdf (Accessed 16 July 2006).

forest.[24] The carbon market could set the right incentives to lengthen harvest cycles. It cannot however replace the strengthening of forestry governance and fighting the often rampant corruption that plagues the sector.

If wood products are used as a source of renewable energy and efficient materials, forests can also help to reduce energy related emissions. The simplest manner to cut GHG is to substitute wood for other energy sources. However, there is an inherent conflict between using the biomass and leaving carbon in a standing forest. While carbon stored in forests yields immediate climate benefits, the longer-term benefits may come with managed forests which allows for the production and extraction of biomass. Whereas forests with a particular biodiversity value should be fully protected, the most effective way to reduce GHG emissions may be found by exploring even more effective ways to use wood to cut fossil fuel emissions. If removed from sustainably managed forests, biomass generated energy can be carbon neutral as it does not release any additional carbon dioxide to the atmosphere. Wood based products can also be used to replace more energy intensive building and construction materials such as aluminium and concrete.

However, the large scale plantation of biofuel crops has lead to increased deforestation. Today, oil palm plantations cover millions of hectares across Malaysia, Indonesia, and Thailand. Palm oil and other biofuel plants are becoming an increasingly important agricultural product for tropical countries around the world, especially as crude oil prices top USD100 a barrel. According to a report by Friends of the Earth-Netherlands, Indonesia, currently the world's second largest producer of palm oil, had oil-palm plantations covering 5.3 million hectares of the country in 2004.[25] In recent years, vast areas of natural

[24] For the Pacific North West of the US see: Wayburn, L.A., Franklin, F.J., Gordon, J.C., Binkley, C.S., Mlandenoff, D.J., Christian Jr., N.L. (2000), Forest Carbon in the United States: Opportunities and Options for Private Lands, The Pacific Forest Trust, Inc., Santa Rosa, US.

[25] Milieudefensie, Friends of the Earth Netherlands and the Swedish Society for Nature Conservation (SSNC) (2006), Kalimantan Border Oil Palm Mega-Project. http://www.milieudefensie.nl/globalisering/publicaties/index.htm#palmolie-rapport (Accessed 8 January 2007).

forest have been cleared across tropical Asia for the rapid expansion of oil palm plantations. This conversion has reduced biodiversity, increased vulnerability to catastrophic fires, and affected local communities dependent on services and products provided by forest ecosystems.

Beyond the loss of forest ecosystems, the current production practices for palm oil often cause more damage to the environment. In 2001 Malaysia's production of 7 million tons of crude palm oil generated 9.9 million tons of solid oil wastes, palm fiber, and shells, and 10 million tons of palm oil mill effluent — a polluted mix of crushed shells, water, and fat residues that has been shown to have a negative impact on aquatic ecosystems. The extensive use of pesticides and herbicides adds additional pressure on the environment.[26]

To address the growing problems of sustainability in the palm oil industry, the "Roundtable on Sustainable Palm Oil" (RSPO) was established in Kuala Lumpur in 2003. A non-legally binding 'statement of intent' signed by over 40 companies and organisations was established to promote sustainable palm oil production through implementing better management practices. Its members comprise oil palm producers and growers, traders, 15 retailers including the Body Shop (a UK based ethical cosmetics chain), and environmental and social organisations, including the WWF. The RSPO develops training modules for plantation managers and small-holders, and monitors practices such as integrated pest management, land use planning and waste management.

8. Conclusion: The 2020 Scenario

Forest and biodiversity conservation are intrinsically linked to climate change mitigation and adaptation: Together with the forests, we lose our biggest terrestrial carbon storage and a system that regulates and influences the freshwater household and rainfall patterns. It is

[26] Butler, R.A., Why is oil palm replacing tropical rainforests? Why are biofuels fueling deforestation? mongabay.com. 25 April 2006, http://news.mongabay.com/2006/0425-oil_palm.html (Accessed 8 January 2007).

therefore necessary that a post-Kyoto regime includes a comprehensive carbon accounting mechanism that provides the necessary incentive framework for conserving not only temperate and boreal, but — and in particular — tropical rainforests.

Official development assistance ("ODA") and other public funding sources will not sufficiently provide the necessary financing to make a meaningful contribution to forest conservation. Neither have ODA-based funding mechanisms a history of leveraging significant amounts of private funds. It is therefore unlikely that such mechanisms will serve to create a robust incentive framework for longer term conservation measures.

By assigning value to the standing forests, carefully designed market mechanisms provide the most promising option to mobilize finances for conservation. Market mechanism and emission trading schemes assigning value to tropical forests have to be complemented by measures that strengthen forest governance and law enforcement and be integrated in a robust international and national framework. Safeguards and independent control mechanism must be put in place to avoid perverse incentives, "paper tons", and corrupt policies.

The LULUCF sector is often claimed to undermine the environment by promoting large-scale plantations, reducing biodiversity, or relying on genetically-modified or non-native species that replace rare native species. These kinds of projects should not benefit from environmental funding, and mechanisms have to be designed (or refined) to exclude them from being eligible for carbon finance. From a biodiversity perspective, carbon finance is one of the few funding sources of sufficient scale which is able to support the large-scale restoration and conservation of degraded habitat that is critical to the survival of threatened species.

Forests do not only store carbon, they also hold the potential to generate significant amounts of energy. Biomass systems are usually fueled by waste wood, from logging operations, forest thinning, low-grade wood, and/or sawmill residues. These systems assign commercial value for wood waste benefiting both the forest as well as the global climate.

Finally, forests and forest conservation will play a significant political role in the negotiations of a post-Kyoto agreement. Any effort to include the United States or major developing countries into the efforts to curb climate change will have to bring forestry and other LULUCF measures into the picture. The US has been a historic supporter of a full accounting of land-use related emissions under any climate regime. The participation of developing countries in reducing GHG emissions from REDD is also more advanced than in any other sector.

From a development perspective, LULUCF carbon finance projects represent one of the only means that many of the world's poorest people (including most of Africa) will be able to meaningfully participate in, and benefit from, the global carbon market. For the first time, they have the promise of being able to sustainably capture an ecosystem-service value associated with their land, instead of being forced to liquidate these natural resources just to survive.